14th Meeting on the Mathematics of Language

(MoL 2015)

Chicago, Illinois, USA
25 – 26 July 2015

ISBN: 978-1-5108-0815-7

Printed from e-media with permission by:

Curran Associates, Inc.
57 Morehouse Lane
Red Hook, NY 12571

CURRAN ASSOCIATES INC. proceedings.com

Some format issues inherent in the e-media version may also appear in this print version.

Printed by Curran Associates, Inc. (2015)

Association for Computational Linguistics
209 N. Eighth Street
Stroudsburg, Pennsylvania 18360

Phone: 1-570-476-8006
Fax: 1-570-476-0860

acl@aclweb.org

Additional copies of this publication are available from:

Curran Associates, Inc.
57 Morehouse Lane
Red Hook, NY 12571 USA
Phone: 845-758-0400
Fax: 845-758-2634
Email: curran@proceedings.com
Web: www.proceedings.com

MoL 2015

The 14th Meeting on the Mathematics of Language

Proceedings

July 25–26, 2015
Chicago, USA

Introduction

We are pleased to introduce the proceedings of the 14th Meeting on Mathematics of Language, MoL, to be held at the University of Chicago on July 25–26, 2015.

This volume contains eleven regular papers and two invited papers. The regular papers, which were selected by the Program Committee from a total of twenty-two submissions, feature a broad variety of work on mathematics of language, including phonology, formal language theory, natural language semantics, and language learning. The invited papers are presented by two distinguished researchers in the field: David McAllester, Professor and Chief Academic Officer at the Toyota Technological Institute at Chicago, and Ryo Yoshinaka, Assistant Professor at Kyoto University.

We would like to express our sincere gratitude to our colleagues on the Program Committee for the time and effort that they put into the reviewing of the papers, and to Min-Yen Kan for his help with the publishing of these proceedings in the ACL Anthology.

We wish you all a fruitful meeting.

Marco Kuhlmann, Makoto Kanazawa and Gregory M. Kobele (editors)

Program Chairs:

Marco Kuhlmann (Linköping University, Sweden)
Makoto Kanazawa (National Institute of Informatics, Japan)

Local Chair:

Gregory M. Kobele (University of Chicago, USA)

Program Committee:

Henrik Björklund (Umeå University, Sweden)
David Chiang (University of Notre Dame, USA)
Alexander Clark (King's College London, UK)
Shay Cohen (University of Edinburgh, UK)
Carlos Gómez-Rodríguez (University of A Coruña, Spain)
Jeffrey Heinz (University of Delaware, USA)
Gerhard Jäger (University of Tübingen, Germany)
Aravind Joshi (University of Pennsylvania, USA)
András Kornai (Hungarian Academy of Sciences, Hungary)
Giorgio Magri (CNRS, France)
Andreas Maletti (University of Stuttgart, Germany)
Jens Michaelis (Bielefeld University, Germany)
Gerald Penn (University of Toronto, Canada)
Carl Pollard (The Ohio State University, USA)
Jim Rogers (Earlham College, USA)
Mehrnoosh Sadrzadeh (Queen Mary University of London, UK)
Sylvain Salvati (INRIA, France)
Ed Stabler (University of California, Los Angeles, USA)
Mark Steedman (Edinburgh University, UK)
Anssi Yli-Jyrä (University of Helsinki, Finland)

Invited Speakers:

David McAllester (Toyota Technological Institute at Chicago, USA)
Ryo Yoshinaka (Kyoto University, Japan)

Table of Contents

Program

Saturday, July 25

09:30–10:15 *A Refined Notion of Memory Usage for Minimalist Parsing*
Thomas Graf, Brigitta Fodor, James Monette, Gianpaul Rachiele, Aunika Warren and Chong Zhang

10:15–11:00 *Abstract Categorial Parsing as Linear Logic Programming*
Philippe de Groote

11:00–11:15 Coffee Break

11:15–12:00 *Topology of Language Classes*
Sean A. Fulop and David Kephart

12:00–14:00 Lunch Break

14:00–14:20 S.-Y. Kuroda Prize Ceremony

14:20–15:05 *Individuation Criteria, Dot-types and Copredication: A View from Modern Type Theories*
Stergios Chatzikyriakidis and Zhaohui Luo

15:05–15:50 *Lexical Semantics and Model Theory: Together at Last?*
András Kornai and Marcus Kracht

15:50–16:35 *A Frobenius Model of Information Structure in Categorical Compositional Distributional Semantics*
Dimitri Kartsaklis and Mehrnoosh Sadrzadeh

16:35–16:50 Coffee Break

16:50–17:50 Invited Talk: *A Synopsis of Morphoid Type Theory*
David McAllester

Sunday, July 26

09:30–10:30 Invited Talk: *General Perspective on Distributionally Learnable Classes*
Ryo Yoshinaka

10:30–10:45 Coffee Break

10:45–11:30 *Canonical Context-Free Grammars and Strong Learning: Two Approaches*
Alexander Clark

11:30–12:15 *Output Strictly Local Functions*
Jane Chandlee, Rémi Eyraud and Jeffrey Heinz

12:15–13:00 *How to Choose Successful Losers in Error-Driven Phonotactic Learning*
Giorgio Magri and René Kager

13:00–15:00 Lunch Break

15:00–15:45 *A Concatenation Operation to Derive Autosegmental Graphs*
Adam Jardine and Jeffrey Heinz

15:45–16:30 *Syntactic Polygraphs. A Formalism Extending Both Constituency and Dependency*
Sylvain Kahane and Nicolas Mazziotta

16:30–17:30 Business Meeting

A Refined Notion of Memory Usage for Minimalist Parsing

Thomas Graf **Brigitta Fodor** **James Monette**
Gianpaul Rachiele **Aunika Warren** **Chong Zhang**

Stony Brook University, Department of Linguistics
mail@thomasgraf.net, {firstname.lastname}@stonybrook.edu

Abstract

Recently there has been a lot of interest in testing the processing predictions of a specific top-down parser for Minimalist grammars (Stabler, 2013). Most of this work relies on memory-based difficulty metrics that relate the shape of the parse tree to processing behavior. We show that none of the difficulty metrics proposed so far can explain why subject relative clauses are more easily processed than object relative clauses in Chinese, Korean, and Japanese. However, a minor tweak to how memory load is determined is sufficient to fully capture the data. This result thus lends further support to the hypothesis that very simple notions of resource usage are powerful enough to explain a variety of processing phenomena.

1 Introduction

One of the great advantages of mathematical linguistics is that its formal rigor allows for the exploration of ideas and questions that could not even be precisely formulated otherwise. A promising project along these lines is the investigation of syntactic processing from a computationally informed perspective (Joshi, 1990; Rambow and Joshi, 1995; Steedman, 2001; Hale, 2011; Yun et al., 2014). This requires I) an articulated theory of syntax that has sufficient empirical coverage to be applicable to a wide range of constructions, II) a sound and complete parser for the syntactic formalism, and III) a linking theory that derives psycholinguistic predictions from these two components. A successful model along these lines provides profound insights into the mechanisms of linguistic performance, and it can also rule out certain syntactic proposals as psycholinguistically inadequate. Unfortunately there are multiple choices for each one of the three components, which raises the question of which combinations are empirically adequate.

This paper explores this issue for Minimalist grammars (MGs), a formalization of the Chomskyan variety of generative grammar that informs a lot of psycholinguistic research nowadays. Taking as our vantage point Kobele *et al.* (2012; henceforth KGH) and their method for deriving structure-sensitive processing predictions from Stabler's (2013) MG top-down parser, we evaluate how well the parser captures the processing difficulty of relative clauses in Chinese, Japanese, and Korean — a phenomenon that escapes many processing models in the literature. By carefully modulating the set of syntactic assumptions as well as the linking hypotheses, we show that none of the memory-based proposals in the tradition of KGH yield the right predictions. The correct results are obtained, however, if the size of parse items also counts towards their memory usage. Our paper thus serves a dual purpose: it provides a positive result in the form of a more refined notion of memory usage that explains the observed processing behavior, and a negative one by eliminating many combinations of the three factors listed above.

Our discussion starts with two introductory sections that familiarize the reader with the research this paper follows up on. We first discuss MGs, the MG top-down parser, and how this parser has been used to model processing phenomena in re-

Proceedings of the 14th Meeting on the Mathematics of Language (MoL 14), pages 1–14,
Chicago, USA, July 25–26, 2015.

cent years. This is followed by a brief review of a long standing problem in syntactic processing: the preference for subject relative clauses (SRCs) over object relative clauses (ORCs) irrespective of cross-linguistic word order differences. We present two prominent relative clause analyses from the syntactic literature, and we discuss why the preference for SRCs over ORCs is surprising given current psycholinguistic models. In Sec. 4 we finally demonstrate that the MG parser cannot make the right predictions with any of the proposed metrics unless one refines their conception of memory load.

2 Minimalist Grammars for Processing

2.1 Minimalist Grammars

MGs (Stabler, 1997) are a formalization of the most recent iteration of transformational grammar, known as Minimalism. Since they formalize ideas that form the underpinning for the majority of contemporary research in theoretical syntax and syntactic processing, they act as a form of glue that makes these ideas amenable to more rigorous study. The main purpose of MGs in this paper is to provide a specific type of structure for the parser to operate on — derivation trees. Consequently the technical machinery is of interest only to the extent that it illuminates the connection between derivations and MG parsing, and we thus omit formal details where possible.

An MG is a finite set of lexical items (LIs), where every LI consists of a *phonetic exponent* and a finite, non-empty string of *features*. Each feature has a *positive* or *negative* polarity, and it is either a *Merge* feature (written in upper caps) or a *Move* feature (written in lower caps). MGs assemble LIs into trees via the structure-building operations *Merge* and *Move* according to the feature specifications of the LIs. Intuitively, Merge may combine two LIs if their respective first unchecked features are Merge features and differ only in their polarity. The LI with the positive polarity feature acts as the head of the assembled phrase. Move, on the other hand, removes a phrase from an already assembled tree and puts it in a different position; see Stabler (2011) for a formal definition. Figure 1 shows a simplified tree for *John, the girl likes*, with dashed lines indicating which positions certain phrases were displaced from.

The structure of a sentence is also fully encoded by its derivation tree, i.e. the record of how its phrase structure tree was assembled from the LIs via applications of Move and Merge. Every derivation tree corresponds to exactly one phrase structure tree, but the reverse does not necessarily hold. The main difference between the two types of tree is that moving phrases remain in their base position in the derivation tree — compare, for instance, the positions of *John* and *the girl* in the two trees in Fig. 1 (for the sake of clarity interior nodes have the same label as their counterpart in the phrase structure tree). As a result, derivation trees do not directly reflect the word order of a sentence, which must be derived by carrying out the movement steps.

In addition, an MG's set of well-formed derivation trees forms a regular tree language thanks to a specific restriction on Move that is known as the Shortest Move Constraint (Michaelis, 2001; Kobele et al., 2007; Salvati, 2011; Graf, 2012). The set of well-formed phrase structure trees, on the other hand, is supra-regular — a corollary of MGs' weak equivalence to MCFGs (Harkema, 2001; Michaelis, 2001). The fact that derivation trees do not need to directly encode linear order thus reduces their complexity significantly in comparison to phrase structure trees. Since derivation trees offer a complete regular description of the structure of a sentence, and because regular tree languages can be viewed as context-free grammars (CFGs) with an ancillary hidden alphabet (Thatcher, 1967), MGs turn out to be close relatives of CFGs with a more complex mapping from trees to strings. It is this close connection to CFGs that forms the foundation of Stabler (2013)'s top-down parser.

2.2 MG Parsing as Tree Traversal

Stabler (2013)'s parser for MGs builds on standard depth-first, top-down parsing strategies for CFGs but modifies them in three important respects: I) the parser is equipped with a search beam that discards the most unlikely analyses, thus avoiding the usual problems with left recursion, II) the parser constructs derivation trees rather than phrase structure trees, and III) since derivation trees do not directly reflect linear order, the parser moves through them in a particular fashion that would approximate a left-most, depth-first search in the corresponding phrase structure trees.

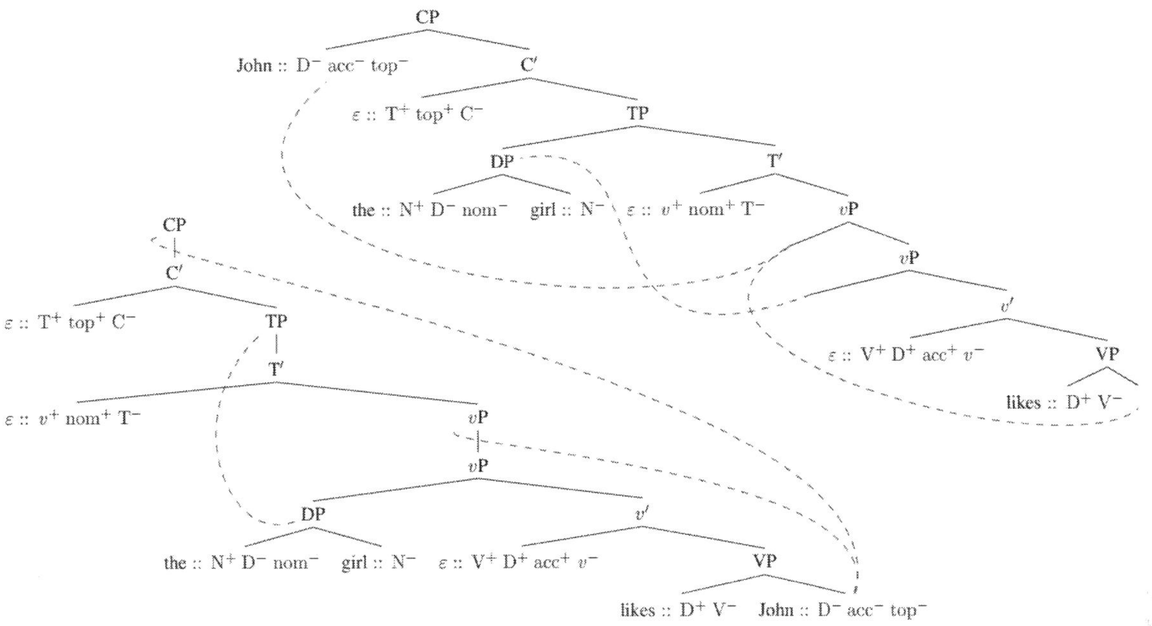

Figure 1: MG phrase structure tree and derivation tree for *John, the girl likes*; dashed branches indicate movement

We completely ignore the beam in this paper and instead adopt KGH's assumption that the parser is equipped with a perfect oracle so that it never makes any wrong guesses during the construction of the derivation tree. While psychologically implausible, this idealization is meant to stake out a specific research goal: processing effects must be explained purely in terms of the syntactic complexity of the involved structures, rather than the difficulty of finding these structures in a large space of alternatives. More pointedly, we assume that parsing difficulty *modulo* non-determinism is sufficient to account for the processing phenomena under discussion.

With non-determinism completely eliminated from the picture, the parse of some sentence s reduces to a specific traversal of the derivation tree of s. In general, the parser follows a left-most, depth-first strategy, where a node is left-most if it is a specifier or if it is a head with a complement. However, when a Move node is encountered, two things can happen, depending on whether the Move node is an *intermediary* landing site or a *final* one. Let p be a moving phrase and $m_1, \ldots, m_n$ the Move nodes that denote an instance of Move displacing p. Then m_i is a final landing site (or simply *final*) iff there is no m_j, $1 \leq j \leq n$, that properly dominates m_i in the derivation tree. A Move node is an intermediary landing site (or *intermediary*) iff there is no phrase in the derivation tree for which it is a final landing site. An intermediary Move node does not affect the parser's tree traversal strategy. A final Move node, on the other hand, causes the parser to take the shortest path to the phrase that will be displaced by this instance of Move. Once the root of that phrase has been reached, the parser traverses its subtree in the usual fashion and then returns to the point where it veered off the standard path.

The traversal is made fully explicit via a notation adopted from KGH where each node in the derivation tree has a superscripted *index* and a subscripted *outdex*. The index lists the point at which the parse item corresponding to the node is inserted into the parser's memory queue, whereas the outdex gives the point at which said parse item is removed from the queue. Both values can be computed in a purely tree-geometric fashion. Let *s[urface]-precedence* be the relation that holds between nodes m and n in a derivation tree iff their counterparts m' and n' in the corresponding phrase structure tree stand in the precedence relation (if m undergoes movement, its counterpart m' is the final landing site rather than its base position). Then indices and outdices can be inferred without knowledge of the parser by the following procedure (cf. Fig. 2 on page 7):

- The index of the root is 1. For every other node, its index is identical to the outdex of its mother.
- If nodes n and n' are distinct nodes with index i, and n reflexively dominates a node that is not s-preceded by any node reflexively dominated by n', then n has outdex $i + 1$.
- Otherwise, the outdex of node n with index i is $\max(i + 1, j + 1)$, where $j \geq 0$ is greatest among the outdices of all nodes that s-precede n but are not reflexively dominated by n.

2.3 Parsing Metrics

In order to allow for psycholinguistic predictions, the behavior of the parser must be related to processing difficulty via a parsing metric. There is no *a priori* limit on the complexity of metrics one may entertain, but the methodologically soundest position is to explore simple metrics before moving on to more complicated ones.

Extending KGH, Graf and Marcinek (2014; henceforth GM) evaluate a variety of memory-based metrics that measure I) how long a node is kept in memory (*tenure*), or II) how many nodes must be kept in memory (*payload*), or III) specific combinations of these two factors. Tenure and payload are easily defined using the node indexation scheme. A node's tenure is the difference between its index and outdex, and the payload of the derivation tree is equal to the number of nodes with a tenure strictly greater than 2 (in the derivation trees in Figs. 2–5, these nodes are boxed to highlight their contribution to the payload).

GM define three metrics, the first of which is adopted directly from KGH. Depending on the metric, the difficulty of a parse is given by

Max $\max(\{t \mid t \text{ is the tenure of some node } n\})$

Box $|\{n \mid n \text{ is a node with tenure } > 2\}|$

Sum $\sum_{n \text{ has tenure } >2} \text{tenure-of}(n)$

GM define an additional six variants by restricting the choice of nodes n to LIs and pronounced LIs, respectively. They then compare the predictions of these nine metrics with respect to right embedding VS center embedding, and nested dependencies VS crossing dependencies (both of which were originally analyzed in KGH), as well as two phenomena involving relative clauses: I) sentential complements containing a relative clause VS a relative clause containing a sentential complement, and II) the preference for subject relative clauses (SRCs) over object relative clauses (ORCs) in English. They conclude that the only metric that makes the right predictions in all four constructions is **Max** restricted to pronounced LIs.

Irrespective of the choice of metric, though, the psycholinguistic predictions of the MG parser vary with the choice of syntactic analysis. KGH use this fact for a persuasive demonstration of how processing data can be brought to bear on the distinction between so-called phrasal movement and head movement. It is unclear, however, whether this should be interpreted as support for a specific movement analysis or as evidence against the assumed difficulty metric. GM's comparison sheds little light on this because it presupposes a specific syntactic analysis for each phenomenon. A more elaborate comparison is required that varies both the parsing metric and the choice of syntactic analysis, ideally resulting in only a few empirically adequate combinations. The processing contrast between *prenominal* SRCs and ORCs is exactly such a case.

3 Surveying Relative Clauses

3.1 Syntax

The main idea of this paper is that the space of possible combinations of syntactic analyses and parsing metrics can be narrowed down quite significantly by looking at processing phenomena that have proven difficult to account for. As we will see next, the fact that SRCs are easier to parse than ORCs in Chinese, Korean, and Japanese constitutes such a problem. We first discuss how the two have been analyzed in the syntactic literature, while the next section explains why many well-known processing models have a hard time capturing the data.

Relative clauses (RCs) can be categorized according to two parameters. First, the *head noun*, i.e. the noun modified by the RC, may be the subject or the object of the RC, in which case we speak of an SRC and an ORC, respectively. Second, an RC is *postnominal* if it is linearly preceded by its head noun, and *prenominal* otherwise. Note that in prenominal languages the complementizer (if it is realized overtly) usually occurs at the right edge of the RC rather than the left edge. Whether RCs have such an overt complementizer is an ancillary parameter.

Most analyses of RCs were developed for languages like English, French, and German, where RCs are postnominal and have overt complementizers (which might be optional). The general template is [$_{DP}$ Det head-noun [$_{RC}$ complementizer subject verb object]], with either the subject or the object unrealized and the position of the verb depending on language-specific word order constraints.

(1) a. [$_{DP}$ The mayor [$_{RC}$ who _ invited the tycoon]] likes wine.

b. [$_{DP}$ The mayor [$_{RC}$ who the tycoon invited _]] likes wine.

The canonical account is the *wh-movement* analysis, according to which the complementizer fills the subject or object position, depending on the type of RC, and then moves into Spec,CP (Chomsky, 1965; Heim and Kratzer, 1998). Alternatively, the complementizer starts out as the C-head and instead a silent operator undergoes movement from the base position to Spec,CP. For the purposes of this paper the two variants of the wh-movement analysis are fully equivalent.

The *promotion* analysis is a well-known competing proposal (Vergnaud, 1974; Kayne, 1994). It combines the ideas above and posits that the complementizer starts out as the C-head, but instead of a silent operator it is the head noun that moves from the embedded subject/object position into Spec,CP. In contrast to the wh-movement analysis, the head noun is thus part of the RC. Crucially, though, all three proposals involve an element that fills the seemingly empty argument position of the verb and subsequently moves to Spec,CP.

Languages with prenominal RCs, such as Chinese, Japanese, and Korean, can be analyzed along these lines, but differences in word order lead to a significant increase in analytic complexity. Below is an example of the English sentence in (1) with Chinese word order.

(2) a. [$_{DP}$ [$_{RC}$ _ invited the tycoon who] the mayor] likes wine.

b. [$_{DP}$ [$_{RC}$ the tycoon invited _ who] the mayor] likes wine.

On a theoretical level, there are two major complications. First, while Chinese is an SVO language like English, Japanese and Korean are SOV languages, which requires movement of the object to Spec,vP, thereby adding at least one more movement step within each RC in these two languages. More importantly, the prenominal word order must be derived from the postnominal one via movement, which causes the wh-movement analysis and the promotion analysis to diverge more noticeably.

In the promotion analysis, the RC is no longer a CP, but rather a RelP that contains a CP (see also Yun et al., 2014). The head noun still moves from within the RC to Spec,CP, but this is followed by the TP moving to Spec,RelP so that one gets the desired word order with the complementizer between the rest of the RC in Spec,RelP and the head noun in Spec,CP. In the wh-movement analysis, the head noun is once again outside the RC, which is just a CP instead of a RelP. The complementizer starts out in subject or object position depending on the type of RC, and then moves into a right specifier of the CP. The CP subsequently moves to the specifier of the DP of the head noun, once again yielding the desired word order with the complementizer between the RC and the head noun.

In sum, the promotion analysis needs to posit a new phrase RelP but all movement is leftward and takes place within this phrase, whereas the wh-movement analysis sticks with a single CP but invokes one instance of rightward movement and moves the RC into Spec,DP, a higher position than Spec,RelP. Both accounts are fairly complicated due to the sheer number and intricate timing of movement steps — the reader is advised to carefully study the derivations in Figures 2 through 5.

Involved as they might be, both the promotion analysis and the wh-movement analysis are workable solutions for the kind of prenominal SRCs and ORCs found in Chinese, Korean, and Japanese. The latter two only add an additional movement step for each object to Spec,vP, and Japanese differs from Chinese and Korean in that the RC complementizer is never pronounced.

3.2 Psycholinguistics

SRCs and ORCs have been the subject of extensive psycholinguistic research, with overwhelming evidence pointing towards SRCs being easier to process than ORCs irrespective of whether RCs are prenominal or postnominal in a given language (Mecklinger et al., 1995; Gibson and Pearlmutter, 1998; Mak et al., 2002; Miyamoto and Nakamura, 2003; Gordon et al., 2006; Kwon et al., 2006; Mak et al., 2006; Ueno and Garnsey, 2008; Kwon et al., 2010; Miyamoto and Nakamura, 2013). The data is less clear-cut in Chinese (Lin and Bever, 2006), but it has recently been argued that this is only because of certain structural ambiguities (Gibson and Wu, 2013). Yun et al. (2014) even show how such an ambiguity-based account can be formalized via the MG parser. Recall, though, that we deliberately ignore ambiguities in this paper in an effort to find the simplest empirically adequate linking between derivations and processing behavior. For this reason, we assume that Chinese would also exhibit a uniform preference for SRCs over ORCs if it were not for the confound of structural ambiguity.

That language-specific differences in word order have no effect on the difficulty of SRCs relative to ORCs is unexpected under a variety of psycholinguistic models. Dependency Locality Theory (Gibson, 1998) and the Active-Filler strategy (Frazier, 1987), for example, contend that parsing difficulty increases with the distance between a filler and its gap due to a concomitant increase in memory load — an idea that is also implicit in KGH's **Max** metric. However, both models calculate distance over strings rather than trees. Since prenominal RCs put the object position (i.e. the gap) linearly closer to the head noun (the filler), while the subject is farther away, ORCs should be easier than SRCs.

The failure of string-based memory load models can be remedied in two ways. One is to abandon the notion that the SRC-ORC asymmetry derives from structural factors, replacing it by functional concepts such as Keenan and Comrie's (1977) accessibility hierarchy, which claims that objects are harder to manipulate than subjects irrespective of the construction involved. While certainly a valid hypothesis, a computationally informed perspective has little light to shed on it. We thus discard this option and focus instead on how a more elaborate concept of sentence structure may interact with memory-based concepts of parsing difficulty. More precisely: can the MG parser, when coupled with a suitable RC analysis and one of the metrics discussed in Sec. 2.3, explain why SRCs are easier to parse than ORCs?

4 Parser Predictions

4.1 Overview of Data

The annotated derivation trees for Chinese and Korean RCs are given in Figures 2 through 5. Japanese is omitted since it has exactly the same analysis as Korean except that the RC complementizer remains unpronounced. Interior nodes are labeled with projections instead of Merge and Move for the sake of increased readability, and a dashed branch spanning from node m to node n indicates movement of the whole subtree rooted in m to the specifier of n. For the wh-analysis, we use a dotted line instead of a dashed one if movement is to a right specifier rather than a left one. Since these notational devices make features redundant, they are omitted completely.

The tenure values for Chinese and Korean are summarized in Tables 1 and 2, respectively. The table subgroups nodes according to whether they are pronounced LIs, unpronounced LIs, or interior nodes. It also includes the summed tenure values for, respectively, pronounced LIs, all LIs, and all listed nodes. Once again we omit Japanese since it shows exactly the same behavior as Korean, except that the complementizer would be grouped under "lexical" and not "pronounced".

4.2 Evaluation of Metrics

All the metrics discussed in Sec. 2.3 fail insofar as they do not predict a consistent preference for SRC over ORC. On the other hand, some metrics fare worse than others because they predict the very opposite, ORC being easier than SRC. This is the case for **Sum**, which adds the tenure of all nodes that contribute to the derivation's payload. The problem is that ORCs have a smaller total tenure than SRCs in Korean and Japanese irrespective of the choice of analysis. Furthermore, if the tenure of phrasal nodes

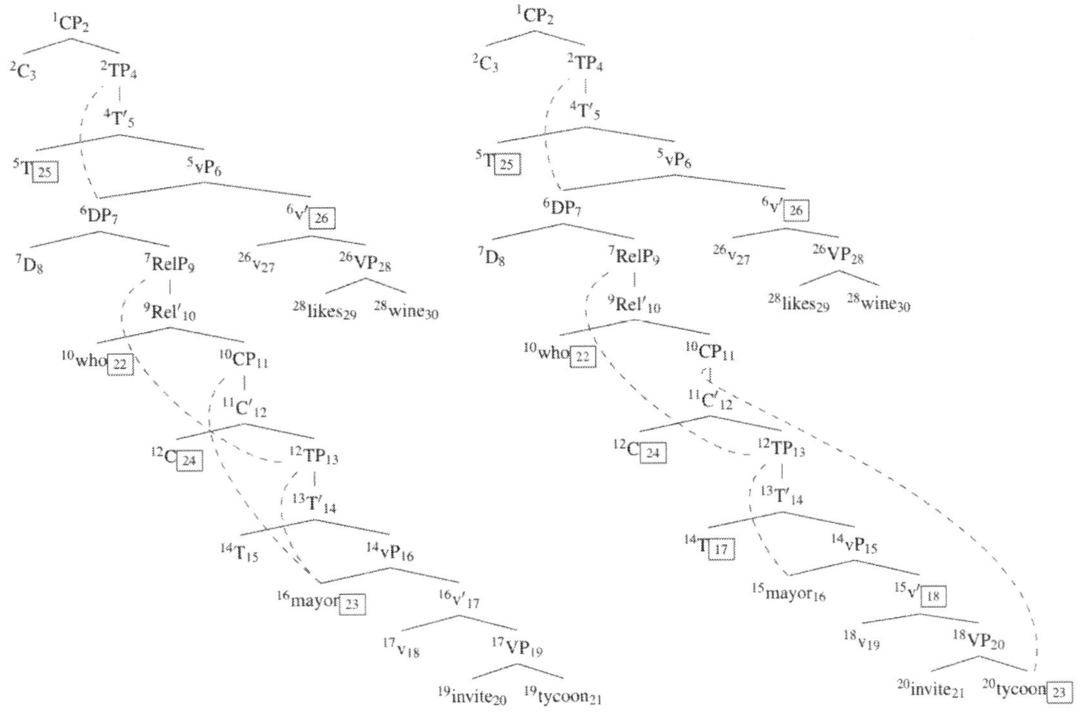

Figure 2: SRC and ORC in Chinese, promotion analysis

is ignored, then **Sum** also makes the wrong predictions for Chinese. This shows that all variants of **Sum** are completely unsuitable to account for the observed processing differences, corroborating previous findings by GM.

A more complicated picture emerges with pure payload, formalized as **Box**. Depending on the choice of analysis and which nodes count towards payload, **Box** predicts a preference for SRC, for ORC, or a tie. The unwanted preference for ORCs emerges I) with the wh-movement analysis in Korean if all nodes are taken into consideration, II) with both analyses in Korean if only LIs matter, and III) with both analyses in Korean and the wh-movement analysis in Chinese if only pronounced LIs are taken into account. The only defensible variant of **Box**, then, is the one that considers the full payload rather than its restriction to lexical or pronounced nodes. In combination with the promotion analysis, this predicts an SRC preference in Chinese and a tie in Korean.

Unfortunately, it has been shown by GM that **Box** fails to make a distinction in processing difficulty for crossing and nested dependencies, the latter of which are harder to parse despite their reduced computational complexity (Bach et al., 1986). Unless the relative ease of crossing dependencies can be explained by some other mechanism, an MG parser with **Box** cannot model all the phenomena that were already accounted for in KGH and GM.

Crossing dependencies were actually one of KGH's main arguments in support of **Max**— the maximum tenure among all nodes determines overall parsing difficulty — so if this metric fares just as well as **Box** for SRCs and ORCs, it is the preferable choice. Unfortunately, **Max** is ill-suited for the problem at hand. If one simply looks at the highest tenure value, **Max** predicts ties for SRCs and ORCs no matter which analysis or type of node is considered. If the metric is applied recursively such that derivation d is easier than d' iff they agree on the n highest tenure values and the $n+1$-th value of d is lower than the $n+1$-th value of d', then **Max** predicts ORC preferences under all combinations. So recursive application of **Max** leads from universal ties to a universal ORC preference.

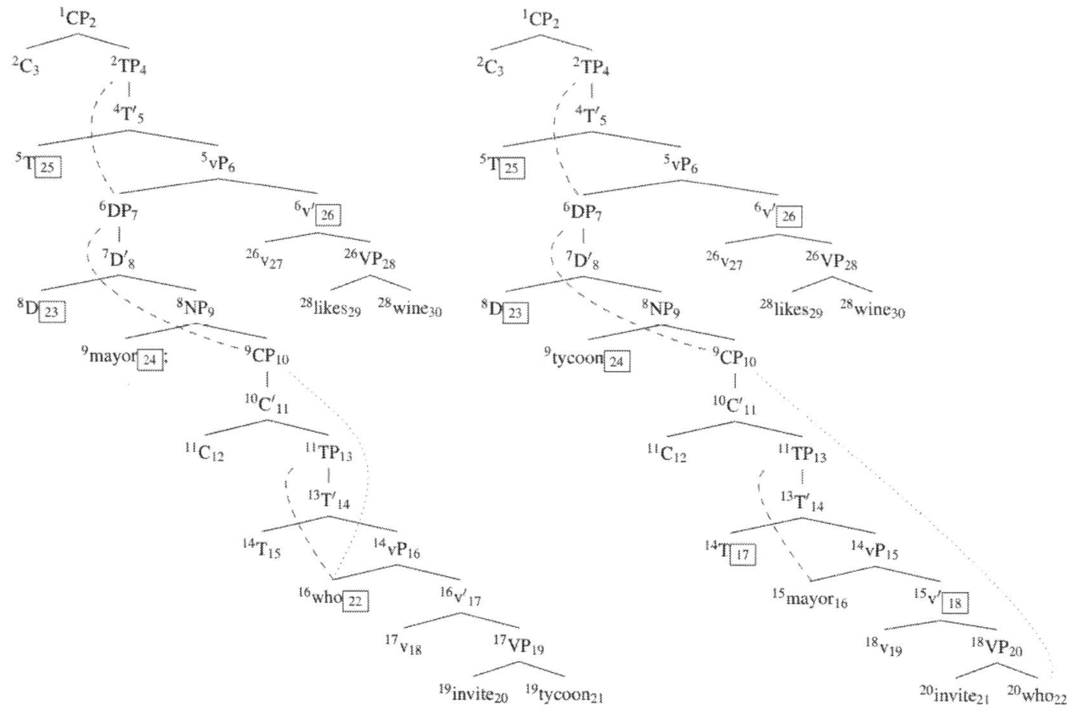

Figure 3: SRC and ORC in Chinese, wh-movement analysis

It seems, then, that we have a choice between an unrestricted version of **Box**, which works only with the promotion analysis and treats crossing and nested dependencies the same, and a non-recursive unrestricted version of **Max**, which treats prenominal SRCs and ORCs the same irrespective of the chosen analysis. Either metric needs to be supplemented by some additional principle to handle these cases. Recall, though, that **Box** predicts a tie for Korean under the promotion analysis. Furthermore, GM showed that the non-recursive version of **Max** is also unsuitable for postnominal RCs and fails to make a clear distinction between the easy case of a sentential complement containing an RC and the much harder case of an RC containing a sentential complement. So whatever additional principle one might propose, it must establish parsing preferences for a diverse range of phenomena.

4.3 A Refined Tenure Metric

On an intuitive level it is rather surprising that no metric grants a clear advantage to SRCs across the board. After all, SRC and ORC derivations differ only in the movement branch to the CP, which is much longer for ORCs than for SRCs as subjects occupy a higher structural position than objects. Since all the metrics home in on some aspect of memory load, one would expect at least one of them to pick up on this difference. That this does not happen is due to the very nature of tenure.

A node has high tenure if its corresponding parse item enters the parser's queue early but cannot be worked on for a long time. In the case of RCs, the complementizer (or alternatively the head noun in the wh-movement analysis) occupies a very high structural position, so that it is encountered early during the construction of the RC. At the same time, it cannot be removed from the queue until the full RC has been constructed, which means that the parser has to move all the way down to the verb and the object. But as long as the complementizer has not been removed from the queue, none of the nodes following it can be removed, either. The result is a "parsing bottle neck" that leads to high tenure on a large number of nodes. The difference between SRCs and ORCs has no effect because it does not change the need for the parser to build the entire RC

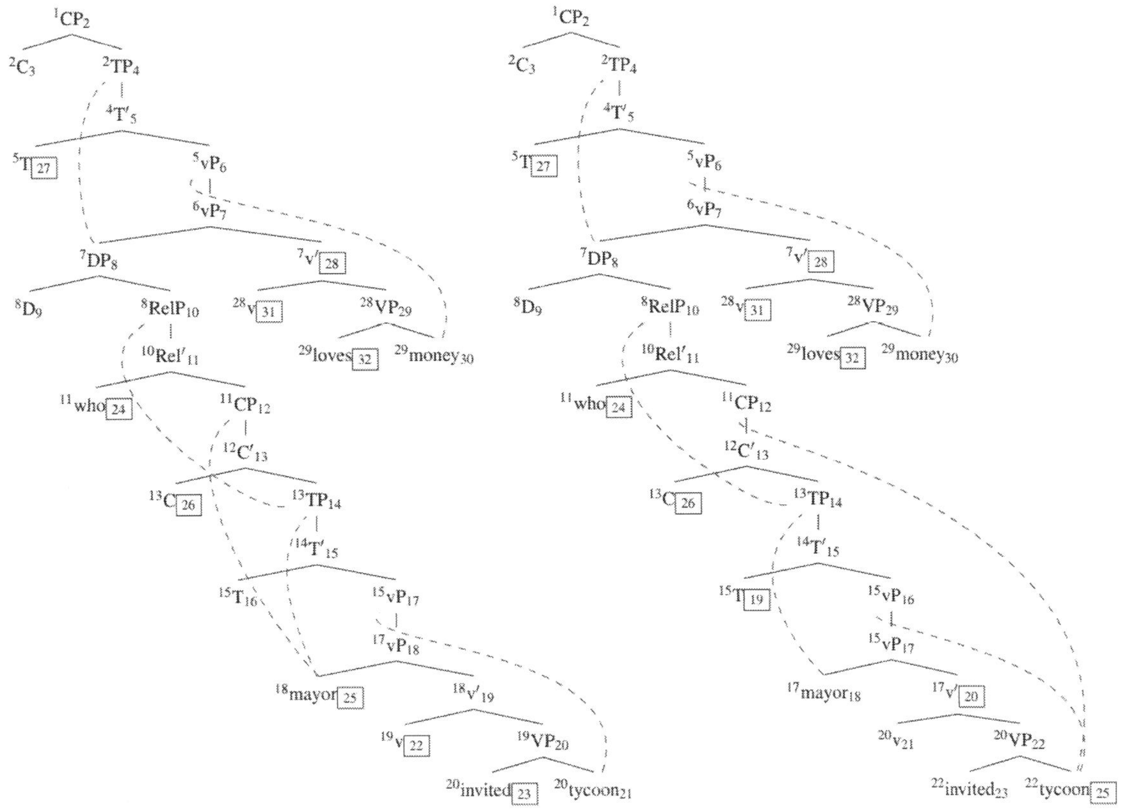

Figure 4: SRC and ORC in Korean, promotion analysis

before it can work on the complementizer, which is the actual cause for the bottle neck.

The central problem, then, is that the structural differences between SRC and ORC are too marginal to outweigh the effects of their shared structure on tenure. There are many conceivable ways around this, e.g. by combining payload and tenure so that each node's tenure from steps i to j is scaled relative to the overall payload from i to j. The most natural idea of multiplying tenure and payload leads to an ORC preference, but division seems to produce the correct results, even for the phenomena discussed in KGH and GM. However, such a step would take us away from the ideal of a simple metric. A less involved solution is to refine the granularity of tenure in a particular way.

Tenure measures how long a parse items remains in memory, but it does not take into account how much memory a given parse item consumes. Consider the parse item corresponding to the embedded CP of the SRC derivation in Fig. 2 on page 7. The step from CP to C′ corresponds to a specific inference rule in the parser that constructs the C′ parse item from the one for CP by adding a movement feature f^- to the list of movers that still need to be found. From here on out, f^- has to be passed around from parse item to parse item until it is finally instantiated on the object. All the parse items along this path would have been smaller if they did not have to carry along f^- in the list of movers. Therefore movement dependencies increase memory load to the extent that they increase the size of parse items (and thus the number of bits that are required for the encoding of said items).

From this perspective, the processing difference between SRCs and ORCs is due to the fact that the longer movement branch in ORCs means that some parse items are bigger in the ORC than their SRC

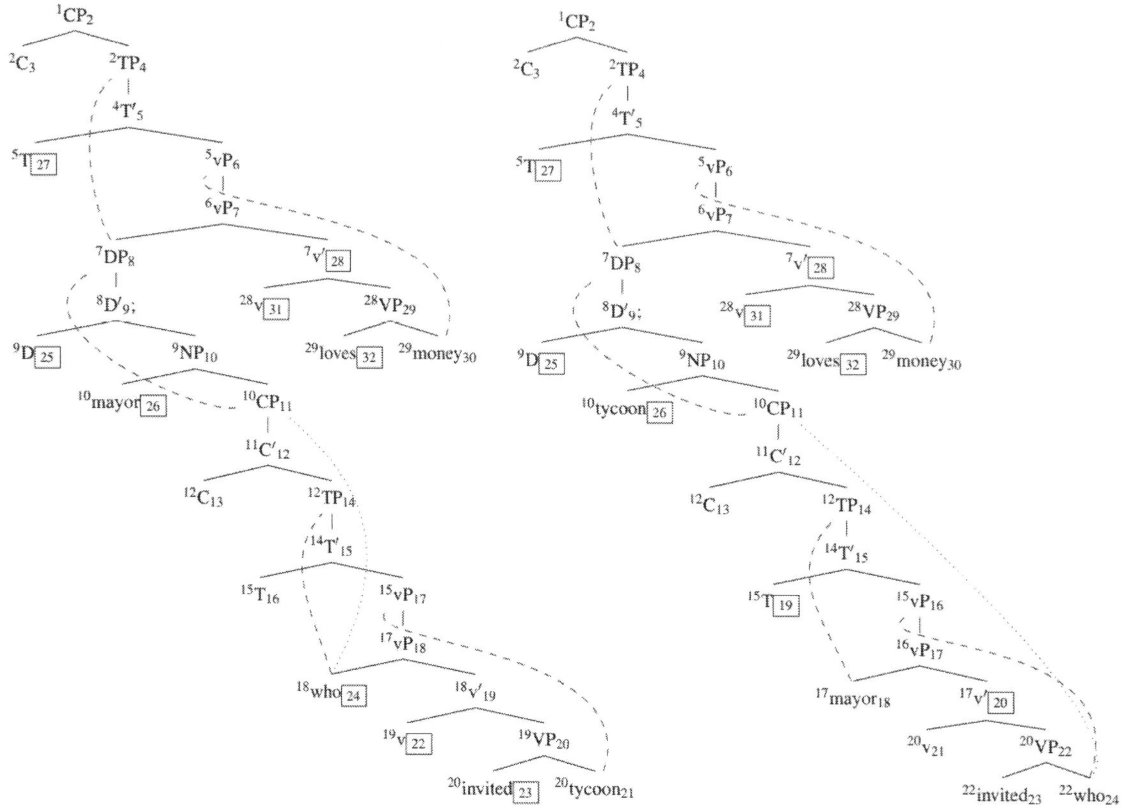

Figure 5: SRC and ORC in Korean, wh-movement analysis

counterparts. One must be careful, though, because only the features of final landing sites are passed along in this fashion — as defined in Stabler (2013), the parser handles the features of intermediary landing sites without increased memory usage. Once one controls for the fact that some final landing sites in the SRC are intermediate in the ORC, and the other way round, there still remains a small advantage for the SRC even in Korean. In both the SRC and the ORC in Figs. 4 and 5, all the interior nodes inside the embedded CP have to pass along at least one feature. More precisely, C$'$, TP, v' and VP pass along exactly one feature, while both vPs carry exactly two features. Only T$'$ shows a difference: in the SRC it hosts only the negative feature that triggers movement of the subject, whereas in the ORC it must also pass along the feature for the object.

This comparison is rather involved, but it can be approximated via the index-based metric **Gap** (inspired by filler-gap dependencies), where i_p is the index of moving phrase p and f_p the index of the final landing site:

Gap $\sum_{p \text{ a moving phrase}} f_p - i_p$

Both **Box** and non-recursive **Max** as discussed above now make the right predictions in conjunction with **Gap** as a secondary metric to resolve ties (this includes also the constructions investigated in KGH and GM). Such a system will grant an advantage to SRCs as long as subjects occur in a higher position than objects. Consequently, it argues against proposals where subjects start out lower than objects (Sigurðsson, 2006). **Box** furthermore favors the promotion analysis over wh-movement, while **Max** remains agnostic.

Conclusion

We showed that the MG parser does not make the right predictions for prenominal SRCs and ORCs

Language	Analysis	RC Type	Node Type	Node	Index	Outdex	Tenure
Chinese	Promotion	SRC	pronounced	who	10	22	12
				mayor	16	23	7
			lexical	matrix T	5	25	*20*
				C	12	24	12
			interior	matrix v'	6	26	*20*
				Summed tenure:	19	51	71
		ORC	pronounced	who	10	22	12
				mayor	20	23	3
			lexical	matrix T	5	25	*20*
				C	12	24	12
				embedded T	14	17	3
			interior	matrix v'	6	26	*20*
				embedded v'	15	18	3
				Summed tenure:	15	50	73
	Wh	SRC	pronounced	mayor	9	24	15
				who	16	22	6
			lexical	matrix T	5	25	*20*
				D	8	23	15
			interior	matrix v'	5	25	*20*
				Summed tenure:	21	56	76
		ORC	pronounced	mayor	9	24	15
			lexical	matrix T	5	25	*20*
				D	8	23	15
				embedded T	14	17	3
			interior	matrix v'	6	26	*20*
				embedded v'	15	18	3
				Summed tenure:	15	53	76

Table 1: Tenure of nodes for Chinese, grouped by analysis; maximum tenure values are in *italics*

under any of the tree-geometric metrics that have been proposed in the literature so far. However, the observed processing effects can be explained if one also take the memory requirements of movement dependencies into account, formalized via the metric **Gap**. The next step will be to test this hypothesis against recent data from Basque (Carreiras et al., 2010), where a uniform preference for ORCs has been observed. Basque is an ergative language, for which it has been argued that subject and object might occur in different positions. If so, the observed behavior may fall out naturally from slightly different movement patterns and their effect on the size of parse items.

A more pressing concern, though, is the mathematical investigation of the parser — a sentiment that is also expressed by KGH. The current method of testing various metrics against numerous constructions is essential for mapping out the space of empirically pertinent alternatives, but it is needlessly labor intensive due to the usual pitfalls of combinatorial explosion. Nor does it enjoy the elegance and generality of a proof-based approach. We believe that true progress in this area hinges on a sophisticated understanding of the tree traversal algorithm instantiated by the parser and how exactly this tree traversal interacts with specific metrics to prefer particular tree shapes over others. Our insistence on simple metrics, free from complicating aspects like probabilities, stems from this desire to keep the parser as open to future mathematical inquiry as possible.

Acknowledgments

We are greatly indebted to John Drury, Jiwon Yun, and the three anonymous reviewers for their com-

Language	Analysis	RC Type	Node Type	Node	Index	Outdex	Tenure
Korean	Promotion	SRC	pronounced	who	11	24	13
				tycoon	18	25	7
				invited	20	23	3
				loves	29	32	3
			lexical	matrix T	5	27	*22*
				C	13	26	13
				embedded v	19	22	3
				matrix v	28	31	3
			interior	matrix v'	7	28	21
				Summed tenure:	26	67	88
		ORC	pronounced	who	11	24	13
				mayor	22	25	3
				loves	29	32	3
			lexical	matrix T	5	27	*22*
				C	13	26	13
				embedded T	15	19	4
				matrix v	28	31	3
			interior	embedded v'	17	20	3
				matrix v'	7	28	21
				Summed tenure:	19	61	85
	Wh	SRC	pronounced	tycoon	10	26	16
				who	18	24	6
				loves	28	31	3
				invited	20	23	3
			lexical	matrix T	5	27	*22*
				D	9	25	16
				embedded v	19	22	3
				matrix v	27	30	3
			interior	matrix v'	7	28	21
				Summed tenure:	28	72	93
		ORC	pronounced	tycoon	10	26	16
				loves	29	32	3
			lexical	matrix T	5	27	*22*
				D	9	25	16
				embedded T	15	19	4
				matrix v	28	31	3
			interior	embedded v'	17	20	3
				matrix v'	7	28	21
				Summed tenure:	19	64	88

Table 2: Tenure of nodes for Korean, grouped by analysis; maximum tenure values are in *italics*

ments and remarks that allowed us to streamline essential parts of this work and improve the presentation of the material.

References

Emmon Bach, Colin Brown, and William Marslen-Wilson. 1986. Crossed and nested dependencies in German and Dutch: A psycholinguistic study. *Language and Cognitive Processes*, 1:249–262.

Manuel Carreiras, Jon Andoni Duñabeitia, Marta Vergara, Irene de la Cruz-Pavía, and Itziar Laka. 2010. Subject relative clauses are not universally easier to process: Evidence from Basque. *Cognition*, 115:79–92.

Noam Chomsky. 1965. *Aspects of the Theory of Syntax*. MIT Press, Cambridge, Mass.

Lyn Frazier. 1987. Sentence processing: A tutorial review.

Edward Gibson and Neal J. Pearlmutter. 1998. Constraints on sentence comprehension. *Trends in Cognitive Sciences*, 2(7):262–268.

Edward Gibson and H.-H. Iris Wu. 2013. Processing Chinese relative clauses in context. *Language and Cognitive Processes*, 28(1-2):125–155.

Edward Gibson. 1998. Linguistic complexity: Locality of syntactic dependencies. *Cognition*, 68:1–76.

Peter C. Gordon, Randall Hendrick, Marcus Johnson, and Yoonhyoung Lee. 2006. Similarity-based interference during language comprehension: Evidence from eye tracking during reading. *Journal of Experimental Psychology: Learning, Memory, and Cognition*, 32(6):1304.

Thomas Graf and Bradley Marcinek. 2014. Evaluating evaluation metrics for minimalist parsing. In *Proceedings of the 2014 ACL Workshop on Cognitive Modeling and Computational Linguistics*, pages 28–36.

Thomas Graf. 2012. Locality and the complexity of minimalist derivation tree languages. In Philippe de Groot and Mark-Jan Nederhof, editors, *Formal Grammar 2010/2011*, volume 7395 of *Lecture Notes in Computer Science*, pages 208–227, Heidelberg. Springer.

John Hale. 2011. What a rational parser would do. *Cognitive Science*, 35:399–443.

Henk Harkema. 2001. A characterization of minimalist languages. In Philippe de Groote, Glyn Morrill, and Christian Retoré, editors, *Logical Aspects of Computational Linguistics (LACL'01)*, volume 2099 of *Lecture Notes in Artificial Intelligence*, pages 193–211. Springer, Berlin.

Irene Heim and Angelika Kratzer. 1998. *Semantics in Generative Grammar*. Blackwell, Oxford.

Aravind Joshi. 1990. Processing crossed and nested dependencies: An automaton perspective on the psycholinguistic results. *Language and Cognitive Processes*, 5:1–27.

Richard S. Kayne. 1994. *The Antisymmetry of Syntax*. MIT Press, Cambridge, Mass.

Edward L. Keenan and Bernard Comrie. 1977. Noun phrase accessiblity and universal grammar. *Linguistic Inquiry*, 8:63–99.

Gregory M. Kobele, Christian Retoré, and Sylvain Salvati. 2007. An automata-theoretic approach to minimalism. In James Rogers and Stephan Kepser, editors, *Model Theoretic Syntax at 10*, pages 71–80.

Gregory M. Kobele, Sabrina Gerth, and John T. Hale. 2012. Memory resource allocation in top-down minimalist parsing. In *Proceedings of Formal Grammar 2012*.

Nayoung Kwon, Maria Polinsky, and Robert Kluender. 2006. Subject preference in Korean. In *Proceedings of the 25th West Coast Conference on Formal Linguistics*, pages 1–14. Cascadilla Proceedings Project Somerville, MA.

Nayoung Kwon, Peter C. Gordon, Yoonhyoung Lee, Robert Kluender, and Maria Polinsky. 2010. Cognitive and linguistic factors affecting subject/object asymmetry: An eye-tracking study of prenominal relative clauses in Korean. *Language*, 86(3):546–582.

Chien-Jer Charles Lin and Thomas G. Bever. 2006. Subject preference in the processing of relative clauses in Chinese. In *Proceedings of the 25th West Coast Conference on Formal Linguistics*, pages 254–260. Cascadilla Proceedings Project Somerville, MA.

Willem M. Mak, Wietske Vonk, and Herbert Schriefers. 2002. The influence of animacy on relative clause processing. *Journal of Memory and Language*, 47(1):50–68.

Willem M. Mak, Wietske Vonk, and Herbert Schriefers. 2006. Animacy in processing relative clauses: The hikers that rocks crush. *Journal of Memory and Language*, 54(4):466–490.

Axel Mecklinger, Herbert Schriefers, Karsten Steinhauer, and Angela D. Friederici. 1995. Processing relative clauses varying on syntactic and semantic dimensions: An analysis with event-related potentials. *Memory & Cognition*, 23(4):477–494.

Jens Michaelis. 2001. Transforming linear context-free rewriting systems into minimalist grammars. *Lecture Notes in Artificial Intelligence*, 2099:228–244.

Edson T. Miyamoto and Michiko Nakamura. 2003. Subject/object asymmetries in the processing of relative clauses in Japanese. In *Proceedings of WCCFL*, volume 22, pages 342–355.

Edson T. Miyamoto and Michiko Nakamura. 2013. Unmet expectations in the comprehension of relative clauses in Japanese. In *Proceedings of the 35th Annual Meeting of the Cognitive Science Society*.

Owen Rambow and Aravind Joshi. 1995. A processing model for free word order languages. Technical Report IRCS-95-13, University of Pennsylvania.

Sylvain Salvati. 2011. Minimalist grammars in the light of logic. In Sylvain Pogodalla, Myriam Quatrini, and Christian Retoré, editors, *Logic and Grammar — Essays Dedicated to Alain Lecomte on the Occasion of His 60th Birthday*, number 6700 in Lecture Notes in Computer Science, pages 81–117. Springer, Berlin.

Halldór Ármann Sigurðsson. 2006. The nominative puzzle and the low nominative hypothesis. *Linguistic Inquiry*, 37:289–308.

Edward P. Stabler. 1997. Derivational minimalism. In Christian Retoré, editor, *Logical Aspects of Computational Linguistics*, volume 1328 of *Lecture Notes in Computer Science*, pages 68–95. Springer, Berlin.

Edward P. Stabler. 2011. Computational perspectives on minimalism. In Cedric Boeckx, editor, *Oxford Handbook of Linguistic Minimalism*, pages 617–643. Oxford University Press, Oxford.

Edward P. Stabler. 2013. Two models of minimalist, incremental syntactic analysis. *Topics in Cognitive Science*, 5:611–633.

Mark Steedman. 2001. *The Syntactic Process*. MIT Press, Cambridge, Mass.

James W. Thatcher. 1967. Characterizing derivation trees for context-free grammars through a generalization of finite automata theory. *Journal of Computer and System Sciences*, 1:317–322.

Mieko Ueno and Susan M Garnsey. 2008. An ERP study of the processing of subject and object relative clauses in Japanese. *Language and Cognitive Processes*, 23(5):646–688.

Jean-Roger Vergnaud. 1974. *French Relative Clauses*. Ph.D. thesis, MIT.

Jiwon Yun, Zhong Chen, Tim Hunter, John Whitman, and John Hale. 2014. Uncertainty in processing relative clauses across East Asian languages. *Journal of East Asian Linguistics*, pages 1–36.

Abstract Categorial Parsing as Linear Logic Programming

Philippe de Groote
Inria Nancy - Grand Est
France
Philippe.deGroote@inria.fr

Abstract

This paper shows how the parsing problem for general Abstract Categorial Grammars can be reduced to the provability problem for Multiplicative Exponential Linear Logic. It follows essentially a similar reduction by Kanazawa, who has shown how the parsing problem for second-order Abstract Categorial Grammars reduces to datalog queries.

1 Introduction

Kanazawa (2007; 2011) has shown how parsing and generation may be reduced to datalog queries for a class of grammars that encompasses mildly context-sensitive formalisms. These grammars, which he calls *context-free λ-term grammars*, correspond to second-order abstract categorial grammars (de Groote, 2001).

In this paper, we show how Kanazawa's reduction may be carried out in the case of abstract categorial grammars of a degree higher than two. The price to pay is that we do not end up with a datalog query, but with a provability problem in multiplicative exponential linear logic (Girard, 1987). This is of course a serious difference. In particular, it is not known whether the multiplicative exponential fragment of linear logic is decidable.

The paper is organized as follows. Section 2 presents some mathematical preliminaries concerning the linear λ-calculus. We then introduce, in Section 3, the notion of abstract categorial grammar. Section 4 is the core of the paper, where we explain Kanazawa's reduction. To this end, we proceed by stepwise refinement. We first introduce an obviously correct but inefficient parsing algorithm. We then improve it by successive correctness-preserving transformations. Finally, we conclude in Section 5.

2 Linear λ-calculus

We assume from the reader some acquaintance with the basic concepts of the (simply typed) λ-calculus. Nevertheless, in order to fix the terminology and the notations, we briefly reminds the main definitions and properties that will be needed in the sequel. In particular, we review the notions *linear implicative types*, *higher-order linear signature*, and *linear λ-terms* built upon a higher-order linear signature.

Let A be a set of atomic types. The set $\mathscr{T}(A)$ of *linear implicative types* built upon A is inductively defined as follows:

1. if $a \in A$, then $a \in \mathscr{T}(A)$;
2. if $\alpha, \beta \in \mathscr{T}(A)$, then $(\alpha \multimap \beta) \in \mathscr{T}(A)$.

Given two sets of atomic types, A and B, a mapping $h : \mathscr{T}(A) \to \mathscr{T}(B)$ is called a *type homomorphism* (or a *type substitution*) if it satisfies the following condition:

$$h(\alpha \multimap \beta) = h(\alpha) \multimap h(\beta)$$

A type substitution that maps atomic types to atomic types is called a *relabeling*.

In order to save parentheses, we use the usual convention of right association, i.e., we write $\alpha_1 \multimap \alpha_2 \multimap \cdots \alpha_n \multimap \alpha$ for $(\alpha_1 \multimap (\alpha_2 \multimap \cdots (\alpha_n \multimap \alpha) \cdots))$.

A *higher-order linear signature* consists of a triple $\Sigma = \langle A, C, \tau \rangle$, where:

Proceedings of the 14th Meeting on the Mathematics of Language (MoL 14), pages 15–25,
Chicago, USA, July 25–26, 2015. ©2015 Association for Computational Linguistics

1. A is a finite set of atomic types;

2. C is a finite set of constants;

3. $\tau : C \to \mathscr{T}(A)$ is a function that assigns to each constant in C a linear implicative type in $\mathscr{T}(A)$.

Given, a higher-order linear signature Σ, we write A_Σ, C_Σ, and τ_Σ, for its respective components.

The above notion of linear implicative type is isomorphic to the usual notion of simple type. Consequently, there is no technical difference between a higher-order linear signature and a higher-order signature. The only reason for using the word *linear* is to emphasize that we will be concerned with the typing of the *linear λ-terms*, i.e., the λ-terms whose typing system corresponds to the implicative fragment of multiplicative linear logic (Girard, 1987).

Let X be a infinite countable set of λ-variables. The set $\Lambda(\Sigma)$ of *linear λ-terms* built upon a higher-order linear signature Σ is inductively defined as follows:

1. if $c \in C_\Sigma$, then $c \in \Lambda(\Sigma)$;

2. if $x \in X$, then $x \in \Lambda(\Sigma)$;

3. if $x \in X$, $t \in \Lambda(\Sigma)$, and x occurs free in t exactly once, then $(\lambda x.\, t) \in \Lambda(\Sigma)$;

4. if $t, u \in \Lambda(\Sigma)$, and the sets of free variables of t and u are disjoint, then $(t\, u) \in \Lambda(\Sigma)$.

$\Lambda(\Sigma)$ is provided with the usual notions of capture-avoiding substitution, α-conversion, β-reduction, and η-reduction (Barendregt, 1984). Let t and u be linear λ-terms. We write $t \twoheadrightarrow_\beta u$ and $t =_\beta u$ for the relations of β-reduction and β-conversion, respectively, We use similar notations for the relations of reduction and conversion induced by η and $\beta\eta$.

Let Σ_1 and Σ_2 be two signatures. We say that a mapping $h : \Lambda(\Sigma_1) \to \Lambda(\Sigma_2)$ is a λ-term homomorphism if it satisfies the following conditions:

$$\begin{aligned} h(x) &= x \\ h(\lambda x.\, t) &= \lambda x.\, h(t) \\ h(t\, u) &= h(t)\,(h(u)) \end{aligned}$$

Given a higher-order linear signature Σ, each linear λ-term in $\Lambda(\Sigma)$ may possibly be assigned a linear implicative type in $\mathscr{T}(A_\Sigma)$. This type assignment obeys the following typing rules:

$$\vdash_\Sigma c : \tau_\Sigma(c) \quad \text{(CONS)}$$

$$x : \alpha \vdash_\Sigma x : \alpha \quad \text{(VAR)}$$

$$\frac{\Gamma, x : \alpha \vdash_\Sigma t : \beta}{\Gamma \vdash_\Sigma (\lambda x.\, t) : (\alpha \multimap \beta)} \text{(ABS)}$$

$$\frac{\Gamma \vdash_\Sigma t : (\alpha \multimap \beta) \qquad \Delta \vdash_\Sigma u : \alpha}{\Gamma, \Delta \vdash_\Sigma (t\, u) : \beta} \text{(APP)}$$

where $\mathrm{dom}(\Gamma) \cap \mathrm{dom}(\Delta) = \varnothing$.

We end this section by reviewing some properties that will turn out to be useful in the sequel.

The set of linear λ-terms being a subset of the set of simply typed λ-terms, it inherits the universal properties of the latter (e.g., strong normalization, or existence of a principal type scheme). It also satisfies the usual subject-reduction property.

Proposition 1 *Let Σ, t, u, Γ, and α be such that $\Gamma \vdash_\Sigma t : \alpha$ and $t \twoheadrightarrow_\beta u$. Then $\Gamma \vdash_\Sigma u : \alpha$.* □

The set of simply typed λ-terms, which is not closed in general under β-expansion, is known to be closed under *linear* β-expansion. Consequently, the set of linear λ-terms satisfies the subject-expansion property.

Proposition 2 *Let Σ, t, u, Γ, and α be such that $\Gamma \vdash_\Sigma u : \alpha$ and $t \twoheadrightarrow_\beta u$. Then $\Gamma \vdash_\Sigma t : \alpha$.* □

The subject-reduction property also holds for the relation of $\beta\eta$-reduction. This is not the case, however, for the subject-expansion property. This possible difficulty may be circumvented by using the notion of η-long form.

A linear λ-term is said to be in η-long form when every of its sub-terms of functional type is either a λ-abstraction or the operator of an application. The set of linear λ-terms in η-long forms is closed under both β-reduction and β-expansion. Consequently, the following proposition holds.

Proposition 3 *Let t and u be λ-terms in η-long forms. Then, $t =_{\beta\eta} u$ if and only if $t =_\beta u$.* □

In the sequel, we will often assume that the linear λ-terms under consideration are in η-long forms. This will allow us to only consider β-reduction and β-expansion, while using the relation of $\beta\eta$-conversion as the notion of equality between linear λ-terms.

Finally, it is known from a categorical coherence theorem that every *balanced* simple type is inhabited by at most one λ-term up to $\beta\eta$-conversion (see (Babaev and Solov'ev, 1982; Mints, 1981)). It is also known that the principal type of a pure linear λ-term is balanced (Hirokawa, 1991). Consequently, the following property holds.

Proposition 4 *Let t be a pure linear λ-term (i.e., a linear λ-term that does not contain any constant), and let $\Gamma \vdash t : \alpha$ be its principal typing. If u is a pure linear λ-term such that $\Gamma \vdash u : \alpha$, then $t =_{\beta\eta} u$.* □

3 Abstract Categorial Grammar

This section gives the definition of an abstract categorial grammar (ACG, for short) (de Groote, 2001).

We first define a *lexicon* to be a morphism between higher-order linear signatures. Let $\Sigma_1 = \langle A_1, C_1, \tau_1 \rangle$ and $\Sigma_2 = \langle A_2, C_2, \tau_2 \rangle$ be two higher-order signatures. A lexicon $\mathscr{L} : \Sigma_1 \to \Sigma_2$ is a realization of Σ_1 into Σ_2, i.e., an interpretation of the atomic types of Σ_1 as types built upon A_2, together with an interpretation of the constants of Σ_1 as linear λ-terms built upon Σ_2. These two interpretations must be such that their homomorphic extensions commute with the typing relations. More formally, a *lexicon* $\mathscr{L}$ from Σ_1 to Σ_2 is defined to be a pair $\mathscr{L} = \langle F, G \rangle$ such that:

1. $F : A_1 \to \mathscr{T}(A_2)$ is a function that interprets the atomic types of Σ_1 as linear implicative types built upon A_2;

2. $G : C_1 \to \Lambda(\Sigma_2)$ is a function that interprets the constants of Σ_1 as linear λ-terms built upon Σ_2;

3. the interpretation functions are compatible with the typing relation, i.e., for any $c \in C_1$, the following typing judgement is derivable:

$$\vdash_{\Sigma_2} G(c) : \hat{F}(\tau_1(c)) \qquad (1)$$

where $\hat{F}$ is the unique homomorphic extension of F.

Remark that Condition (1) compels $G(c)$ to be typable with respect to the empty typing environment. This means that G interprets each constant c as a closed linear λ-term. Now, defining $\hat{G}$ to be the unique homomorphic extension of G, Condition (1) ensures that the following commutation property holds for every $t \in \Lambda(\Sigma_1)$:

$$\text{if} \quad \vdash_{\Sigma_1} t : \alpha \quad \text{then} \quad \vdash_{\Sigma_2} \hat{G}(t) : \hat{F}(\alpha)$$

In the sequel, given such a lexicon $\mathscr{L} = \langle F, G \rangle$, $\mathscr{L}(a)$ will stand for either $\hat{F}(a)$ or $\hat{G}(a)$, according to the context.

We now define an *abstract categorial grammar* as quadruple, $\mathscr{G} = \langle \Sigma_1, \Sigma_2, \mathscr{L}, S \rangle$, where:

1. Σ_1 and Σ_2 are two higher-order linear signatures; they are called the *abstract vocabulary* and the *object vocabulary*, respectively;

2. $\mathscr{L} : \Sigma_1 \to \Sigma_2$ is a lexicon from the abstract vocabulary to the object vocabulary;

3. S is an atomic type of the abstract vocabulary; it is called the *distinguished type* of the grammar.

Every ACG $\mathscr{G}$ generates two languages: an *abstract language*, $\mathcal{A}(\mathscr{G})$, and an object language $\mathcal{O}(\mathscr{G})$.

The abstract language, which may be seen as a set of abstract parse structures, is the set of closed linear λ-terms built upon the abstract vocabulary and whose type is the distinguished type of the grammar. It is formally defined as follows:

$$\mathcal{A}(\mathscr{G}) = \{t \in \Lambda(\Sigma_1) : \vdash_{\Sigma_1} t : S \text{ is derivable}\}$$

The *object language*, which may be seen as the set of surface forms generated by the grammar, is defined to be the image of the abstract language by the term homomorphism induced by the lexicon.

$$\mathcal{O}(\mathscr{G}) = \{t \in \Lambda(\Sigma_2) : \exists u \in \mathcal{A}(\mathscr{G}).\ t =_{\beta\eta} \mathscr{L}(u)\}$$

Both the abstract language and the object language generated by an ACG are sets of linear λ-terms. This allows more specific data structures such as strings, trees, or first-order terms to be represented. A string of symbols, for instance, can be encoded as a composition of functions. Consider an

$$\begin{aligned}
\text{MAN} &: N \\
\text{WOMAN} &: N \\
\text{WISE} &: N \multimap N \\
\text{A}_s &: N \multimap (NP_s \multimap \overline{S}) \multimap S \\
\text{A}_o &: N \multimap (NP_o \multimap \overline{S}) \multimap \overline{S} \\
\text{SEEK} &: ((NP_o \multimap S) \multimap \overline{S}) \multimap NP_s \multimap S \\
\text{INJ} &: S \multimap \overline{S}
\end{aligned}$$

Figure 1: The abstract vocabulary Σ_1

$$\begin{aligned}
\text{MAN} &:= \boldsymbol{man} \; : \; \sigma \\
\text{WOMAN} &:= \boldsymbol{woman} \; : \; \sigma \\
\text{WISE} &:= \lambda x.\, \boldsymbol{wise} + x \; : \; \sigma \multimap \sigma \\
\text{A}_s &:= \lambda xp.\, p\,(\boldsymbol{a} + x) \; : \; \sigma \multimap (\sigma \multimap \sigma) \multimap \sigma \\
\text{A}_o &:= \lambda xp.\, p\,(\boldsymbol{a} + x) \; : \; \sigma \multimap (\sigma \multimap \sigma) \multimap \sigma \\
\text{SEEK} &:= \lambda px.\, p\,(\lambda y.\, x + \boldsymbol{seeks} + y) \; : \; ((\sigma \multimap \sigma) \multimap \sigma) \multimap \sigma \multimap \sigma \\
\text{INJ} &:= \lambda x.\, x \; : \; \sigma \multimap \sigma
\end{aligned}$$

Figure 2: The lexicon $\mathscr{L} : \Sigma_1 \rightarrow \Sigma_2$

arbitrary atomic type s, and define $\sigma \triangleq s \multimap s$ to be the type of strings. Then, a string such as '$abbac$' may be represented by the linear λ-term:

$$\lambda x.\, a\,(b\,(b\,(a\,(c\,x)))),$$

where the atomic strings 'a', 'b', and 'c' are declared to be constants of type σ. In this setting, the empty word is represented by the identity function:

$$\epsilon \triangleq \lambda x.\, x$$

and concatenation is defined to be functional composition:

$$\_ + \_ \triangleq \lambda\alpha.\, \lambda\beta.\, \lambda x.\, \alpha\,(\beta\, x),$$

which is indeed an associative operator that admits the identity function as a unit.

We end this section by giving a fragment of a categorial grammar that will serve as a running example throughout the rest of this paper.[1]

The abstract vocabulary, which specifies the abstract parse structures, is given in Fig. 1. In this signature, the atomic types (N, NP_s, NP_o, $\overline{S}$, S) must be thought of as atomic syntactic categories. The lexicon, which is given in Fig. 2, allows the abstract structures to be transformed in surface forms. These surface forms are strings that are built upon an object vocabulary, Σ_2, which includes the following atomic strings as constants of type σ: $\boldsymbol{man}, \boldsymbol{woman}, \boldsymbol{wise}, \boldsymbol{a}, \boldsymbol{seeks}$.

For such a grammar, the parsing problem consists in deciding whether a possible surface form (i.e.,

[1]This grammar, which follows the categorial type-logical tradition (Moortgat, 1997), has been devised in order to present the main difficulties encountered in ACG parsing: it is higher order (it assigns third-order types to the quantified noun phrases, and a fourth-order type to an intensional transitive verb such as *seek*); it is lexically ambiguous (it assigns two different lexical entries to the indefinite determiner); and it includes a non-lexicalized entry (the coercion operator INJ).

term $t \in \Lambda(\Sigma_2)$) belongs to the object vocabulary of the grammar. Spelling it out, is there an abstract parse structure (i.e., a term $u \in \Lambda(\Sigma_1)$ of type S) whose image through the lexicon is the given surface form (i.e., $\mathscr{L}(u) = t$).

Consider, for instance, the following string:

$$\boldsymbol{a} + \boldsymbol{wise} + \boldsymbol{woman} + \boldsymbol{seeks} + \boldsymbol{a} + \boldsymbol{wise} + \boldsymbol{man} \quad (2)$$

One can show that it belongs to the object language of the grammar. Indeed, when applying the lexicon to the following abstract term:

$$\begin{array}{l} \text{A}_s\,(\text{WISE WOMAN}) \\ \quad (\lambda x.\, \text{INJ}\,(\text{SEEK}\,(\lambda p.\, \text{A}_o\,(\text{WISE MAN}) \\ \qquad\qquad\qquad (\lambda y.\, \text{INJ}\,(p\,y))) \\ \qquad\qquad x)) \end{array} \quad (3)$$

one obtains a λ-term that is $\beta\eta$-convertible to (2). In fact, it is even the case that (2) is ambiguous in the sense that there is another abstract term, essentially different from (3), whose image through the lexicon yields (2).[2] This abstract term is the following:

$$\begin{array}{l} \text{A}_s\,(\text{WISE WOMAN}) \\ \quad (\lambda x.\, \text{A}_o\,(\text{WISE MAN}) \\ \qquad (\lambda y.\, \text{INJ}\,(\text{SEEK}\,(\lambda p.\, \text{INJ}\,(p\,y)) \\ \qquad\qquad x))) \end{array} \quad (4)$$

4 Development of the parsing algorithm

In this section, we develop a parsing algorithm based on proof-search in the implicative fragment of linear logic. We start with a simple non-deterministic algorithm, which is rather inefficient but whose correctness and semi-completeness are obvious. Then, we proceed by stepwise refinement, preserving the correctness and semi-completeness of the algorithm.

By correctness, we mean that if the parsing algorithm answers positively then it is indeed the case that the input term belongs to the object language of the grammar. By semi-completeness, we mean that if the input term belongs to the object language of the grammar, then the parsing algorithm will eventually give a positive answer.

In the present state of knowledge, semi-completeness is the best we may expect. Indeed, the ACG membership problem is known to be equivalent to provability in multiplicative exponential logic (de Groote et al., 2004; Yoshinaka and Kanazawa, 2005), the decidability of which is still open.

4.1 Generate and test

Our starting point is a simple *generate and test* algorithm:

1. *derive S using the rules of implicative linear logic with the types of the abstract constants* (Fig. 3) *as proper axioms;*
2. *interpret the obtained derivation as a linear λ-term (through the Curry-Howard isomorphism);*
3. *apply the lexicon to the resulting λ-term, and check whether it yields a term $\beta\eta$-convertible to the input term.*

$$\begin{array}{l} N \quad (\text{MAN}) \\ N \quad (\text{WOMAN}) \\ N \multimap N \\ N \multimap (NP_s \multimap \overline{S}) \multimap S \\ N \multimap (NP_o \multimap \overline{S}) \multimap \overline{S} \\ ((NP_o \multimap S) \multimap \overline{S}) \multimap NP_s \multimap S \\ S \multimap \overline{S} \end{array}$$

Figure 3: The type of the abstract constants as proper axioms

The above algorithm is obviously correct. It is also semi-complete because it enumerates all the terms of the abstract language. Now, if the input term belongs to the object language of the grammar then its abstract parse structure(s) will eventually appear in the enumeration.

4.2 Type-driven search

The generate and test algorithm proceeds by trial and error without taking into account the form of the input term. In order to improve our algorithm, we must focus on the construction of an abstract term

[2] If the grammar was provided with a Montague semantics, the abstract parse structures (3) and (4) would correspond to the *de dicto* and *de re* readings, respectively.

whose image by the lexicon would be the input term. To this end, we take advantage of Proposition 4.

In general, the input term is not a pure λ-term. Consequently, in order to apply Proposition 4, we must consider each occurrence of a constant in the input term as a fresh free variable. Applying this idea to our example, we obtain the following principal typing that characterizes uniquely the input string (in η-long β-normal form):

$$\begin{array}{l} \boldsymbol{a}_1 : s_1 \multimap s_0,\ \boldsymbol{wise}_1 : s_2 \multimap s_1, \\ \boldsymbol{woman} : s_3 \multimap s_2, \boldsymbol{seeks} : s_4 \multimap s_3, \\ \boldsymbol{a}_2 : s_5 \multimap s_4,\ \boldsymbol{wise}_2 : s_6 \multimap s_5,\ \boldsymbol{man} : s_7 \multimap s_6 \\ \vdash \lambda z.\, \boldsymbol{a}_1\, (\boldsymbol{wise}_1\, (\boldsymbol{woman} \\ \qquad (\boldsymbol{seeks}\, (\boldsymbol{a}_2\, (\boldsymbol{wise}_2\, (\boldsymbol{man}\, z)))))) : s_7 \multimap s_0 \end{array}$$

The types assigned to the constant occurrences of the input term induce a new specialized object vocabulary, which we will call Σ_2^S. We take for granted the definition of the forgetful homomorphism

$$|\cdot| : \Sigma_2^S \to \Sigma_2$$

that allows to project Σ_2^S on Σ_2. Roughly speaking, this forgetful homomorphism consists simply in identifying the several occurences of a same object constant. Remark that at the level of the types, this forgetful homomorphism is a relabeling because the input string has been given in η-long form. In our case, this relabeling is the following one:

$$|s_i| = s \quad (0 \le i \le 7)$$

The next step is to adapt the abstract vocabulary and the lexicon to this specialized object vocabulary. We start with the abstract atomic types. Let $a \in A_{\Sigma_1}$, and define the set $\xi(a)$ as follows:

$$\xi(a) = \{\alpha \in \mathscr{T}(A_{\Sigma_2^S}) : |\alpha| = \mathcal{L}(\tau_{\Sigma_1}(a))\}$$

For instance, we have:

$$\xi(N) = \{s_i \multimap s_j : 0 \le i \le 7 \,\&\, 0 \le j \le 7\}$$

Then we define the set of atomic types of the specialized abstract signature as follows:

$$A_{\Sigma_1^S} = \{a_\alpha : a \in A_{\Sigma_1} \,\&\, \alpha \in \xi(a)\}$$

and we let

$$\mathcal{L}^S(a_\alpha) = \alpha$$

Back to our example, it means that the specialised abstract signature contains 64 copies of N:

$$\begin{array}{cccc} N_{s_0 \multimap s_0}, & N_{s_0 \multimap s_1}, & \ldots & N_{s_0 \multimap s_7}, \\ \vdots & \vdots & \ddots & \vdots \\ N_{s_7 \multimap s_0}, & N_{s_7 \multimap s_1}, & \ldots & N_{s_7 \multimap s_7}. \end{array}$$

In order to accommodate the abstract constants, we look at the lexicon. Consider the first two lexical entries. Their typing, according to the specialized object vocabulary, is as follows:

$$\begin{array}{rcl} \text{MAN} & := & \lambda z.\, \boldsymbol{man}\, z \ : \ s_7 \multimap s_6 \\ \text{WOMAN} & := & \lambda z.\, \boldsymbol{woman}\, z \ : \ s_3 \multimap s_2 \end{array}$$

Accordingly, we let the specialized abstract vocabulary contain the following two constants:

$$\begin{array}{rl} \text{MAN} : & N_{s_7 \multimap s_6} \\ \text{WOMAN} : & N_{s_3 \multimap s_2} \end{array}$$

Consider now the third entry:

$$\text{WISE} := \lambda xz.\, \boldsymbol{wise}\, (x\, z) \ : \ (s \multimap s) \multimap s \multimap s$$

There are two ways of specializing it. On the one hand, the object constant ***wise*** may be replaced by its first occurrence ($\boldsymbol{wise}_1$) or by its second one ($\boldsymbol{wise}_2$). On the other hand, each occurrence of the atomic type s may be instantiated by one of s_0, s_1, ..., s_7. This give rise to 8,192 a priori possibilities. These possibilities, however, do not all correspond to actual typing judgements. Filtering out the ill-typed ones (which is effective since typing is decidable), we are left with 16 new lexical entries which obey the following schemes:

$$\begin{array}{rcl} \text{WISE}_{1i} & := & \lambda xz.\, \boldsymbol{wise}_1\, (x\, z) \ : \\ & & (s_i \multimap s_2) \multimap s_i \multimap s_1 \quad (0 \le i \le 7) \\ \text{WISE}_{2i} & := & \lambda xz.\, \boldsymbol{wise}_2\, (x\, z) \ : \\ & & (s_i \multimap s_6) \multimap s_i \multimap s_5 \quad (0 \le i \le 7) \end{array}$$

and we add the following 16 constants to the specialized abstract vocabulary:

$$\begin{array}{rll} \text{WISE}_{1i} : & N_{s_i \multimap s_2} \multimap N_{s_i \multimap s_1} & (0 \le i \le 7) \\ \text{WISE}_{2i} : & N_{s_i \multimap s_6} \multimap N_{s_i \multimap s_5} & (0 \le i \le 7) \end{array}$$

By proceeding in the same way with the other lexical entries, we obtain a new specialized abstract signature Σ_1^S together with a new specialized lexicon:

$$\mathscr{L}^S : \Sigma_1^S \to \Sigma_2^S$$

Clearly, there exists a forgetful homomorphism between Σ_1^S and Σ_1, and the specialized abstract signature and specialized lexicon are such that the following diagram commutes:

$$\begin{array}{ccc} \Sigma_1^S & \xrightarrow{\mathscr{L}^S} & \Sigma_2^S \\ \big\downarrow |\cdot| & & \big\downarrow |\cdot| \\ \Sigma_1 & \xrightarrow{\mathscr{L}} & \Sigma_2 \end{array}$$

We may now use the specialized grammar to drive the proof-search on which the generate and test algorithm is based. Remember that the specialized object type assigned to the input string is $s_7 \multimap s_0$. Our parsing problem is then reduced to the following proof-search problem:

> *derive $S_{s_7 \multimap s_0}$ using the rules of implicative linear logic with the types of the specialized abstract constants as proper axioms.*

Now, suppose that we derive $S_{s_7 \multimap s_0}$, and that $t \in \Lambda(\Sigma_1^S)$ is the specialized abstract linear λ-term corresponding to this derivation. By construction of the specialized grammar, we have that:

$$\vdash_{\Sigma_2^S} \mathscr{L}^s(t) : s_7 \multimap s_0 \tag{5}$$

Then, by Proposition 4, we have that

$$\mathscr{L}^s(t) =_{\beta\eta} \lambda z.\, \boldsymbol{a}_1\, (\boldsymbol{wise}_1\, (\boldsymbol{woman}\, (\boldsymbol{seeks}\, (\boldsymbol{a}_2\, (\boldsymbol{wise}_2\, (\boldsymbol{man}\, z)))))) \tag{6}$$

because

$$\vdash_{\Sigma_2^S} \lambda z.\, \boldsymbol{a}_1\, (\boldsymbol{wise}_1\, (\boldsymbol{woman}\, (\boldsymbol{seeks}\, (\boldsymbol{a}_2\, (\boldsymbol{wise}_2\, (\boldsymbol{man}\, z)))))) : s_7 \multimap s_0 \tag{7}$$

amounts to a principal typing. Finally, by taking $t' = |t|$, we obtain a term $t' \in \Lambda(\Sigma_1)$ such that:

$$\mathscr{L}(t') =_{\beta\eta} \lambda z.\, \boldsymbol{a}\, (\boldsymbol{wise}\, (\boldsymbol{woman}\, (\boldsymbol{seeks}\, (\boldsymbol{a}\, (\boldsymbol{wise}\, (\boldsymbol{man}\, z)))))) \tag{8}$$

This shows the correctness of the algorithm.

To establish its semi-completeness, suppose that there exists an abstract linear λ-term $t' \in \Lambda(\Sigma_1)$ such that (8). From this, one can easily construct a term $t \in \Lambda(\Sigma_1^S)$ of type S such that $|t| = t'$ and Equation (6) holds. Since the lexical entries are given in η-long forms, so is $\mathscr{L}^s(t)$. Then, because the specialized input term is in η-long β-normal form, by Proposition 3, we have that:

$$\mathscr{L}^s(t) \twoheadrightarrow_\beta \lambda z.\, \boldsymbol{a}_1\, (\boldsymbol{wise}_1\, (\boldsymbol{woman}\, (\boldsymbol{seeks}\, (\boldsymbol{a}_2\, (\boldsymbol{wise}_2\, (\boldsymbol{man}\, z)))))) \tag{9}$$

Then, (5) follows from (7) and (9) by Proposition 2. From this, it is not too difficult to establish that t is of type $S_{s_7 \multimap s_0}$.

4.3 Proof-search in the implicative fragment of linear logic

The type-driven algorithm that we have sketched presents two serious defects. On the one hand, the construction of the specialized grammar is both time and space consuming. For our simple running example, for instance, we would obtain 6,226 specialized lexical entries. On the other hand, the reduction depends upon the input string.

In order to circumvent these difficulties, consider again the specialized lexical entries corresponding to the third lexical entry of the original grammar:

$$\begin{array}{rcl}
\mathrm{WISE}_{10} & := & \lambda xz.\, \boldsymbol{wise}_1\, (x\, z) : \\
 & & (s_0 \multimap s_2) \multimap s_0 \multimap s_1 \\
\mathrm{WISE}_{11} & := & \lambda xz.\, \boldsymbol{wise}_1\, (x\, z) : \\
 & & (s_1 \multimap s_2) \multimap s_1 \multimap s_1 \\
 & & \vdots \\
\mathrm{WISE}_{17} & := & \lambda xz.\, \boldsymbol{wise}_1\, (x\, z) : \\
 & & (s_7 \multimap s_2) \multimap s_7 \multimap s_1 \\
\mathrm{WISE}_{20} & := & \lambda xz.\, \boldsymbol{wise}_2\, (x\, z) : \\
 & & (s_0 \multimap s_6) \multimap s_0 \multimap s_5 \\
\mathrm{WISE}_{21} & := & \lambda xz.\, \boldsymbol{wise}_2\, (x\, z) : \\
 & & (s_1 \multimap s_6) \multimap s_1 \multimap s_5 \\
 & & \vdots \\
\mathrm{WISE}_{27} & := & \lambda xz.\, \boldsymbol{wise}_2\, (x\, z) : \\
 & & (s_7 \multimap s_6) \multimap s_7 \multimap s_5
\end{array}$$

In fact, all the specialized object types assigned to these lexical entries are instances of the principal typing of the corresponding lexical entry of the original lexicon:

$$\boldsymbol{wise} : j \multimap i \ \vdash\ \lambda xz.\, \boldsymbol{wise}\, (x\, z) : (k \multimap j) \multimap k \multimap i$$

This means that if the specialized object vocabulary assigns the constant ***wise*** with the following type:

$$\boldsymbol{wise} : j \multimap i \tag{10}$$

then the specialized abstract vocabulary should contain abstract constants obeying the following type scheme:

$$N_{k \multimap j} \multimap N_{k \multimap i} \tag{11}$$

$$\boldsymbol{man}[i,j] \vdash N[i,j]$$
$$\boldsymbol{woman}[i,j] \vdash N[i,j]$$
$$\boldsymbol{wise}[i,j] \vdash N[j,k] \multimap N[i,k]$$
$$\boldsymbol{a}[i,j] \vdash N[j,k] \multimap (NP_s[i,k] \multimap \overline{S}[l,m]) \multimap S[l,m]$$
$$\boldsymbol{a}[i,j] \vdash N[j,k] \multimap (NP_o[i,k] \multimap \overline{S}[l,m]) \multimap \overline{S}[l,m]$$
$$\boldsymbol{seeks}[i,j] \vdash ((NP_o[j,k] \multimap S[l,k]) \multimap \overline{S}[m,n]) \multimap NP_s[l,i] \multimap S[m,n]$$
$$\vdash S[i,j] \multimap \overline{S}[i,j]$$

Figure 4: The lexicon as a linear logic program

$$\frac{\Gamma \vdash \boldsymbol{man}[i,j]}{\Gamma \vdash N[i,j]}\ (\mathrm{M}) \qquad \frac{\Gamma \vdash \boldsymbol{woman}[i,j]}{\Gamma \vdash N[i,j]}\ (\mathrm{W}) \qquad \frac{\Gamma \vdash \boldsymbol{wise}[i,j] \quad \Delta \vdash N[j,k]}{\Gamma, \Delta \vdash N[i,k]}\ (\mathrm{WI})$$

$$\frac{\Gamma \vdash \boldsymbol{a}[i,j] \qquad \Delta \vdash N[j,k] \qquad \Theta, NP_s[i,k] \vdash \overline{S}[l,m]}{\Gamma, \Delta, \Theta \vdash S[l,m]}\ (\mathrm{A}_s)$$

$$\frac{\Gamma \vdash \boldsymbol{a}[i,j] \qquad \Delta \vdash N[j,k] \qquad \Theta, NP_o[i,k] \vdash \overline{S}[l,m]}{\Gamma, \Delta, \Theta \vdash \overline{S}[l,m]}\ (\mathrm{A}_o)$$

$$\frac{\Gamma \vdash \boldsymbol{seeks}[i,j] \qquad \Delta, NP_o[j,k] \multimap S[l,k] \vdash \overline{S}[m,n] \qquad \Theta \vdash NP_s[l,i]}{\Gamma, \Delta, \Theta \vdash S[m,n]}\ (\mathrm{S})$$

$$\frac{\Gamma \vdash S[i,j]}{\Gamma \vdash \overline{S}[i,j]}\ (\mathrm{I})$$

Figure 5: The lexicon as a set of inference rules

Writing $N[j,k]$ for $N_{k \multimap j}$ and representing (10) by the predicate $\boldsymbol{wise}[i,j]$, we may represent the dependence between 10 and 11 by the following linear logic sequent:[3]

$$\boldsymbol{wise}[i,j] \vdash N[j,k] \multimap N[i,k]$$

Applying the same process to the other lexical entries, we end up with the set of sequents given in Fig. 4. Our parsing problem amounts then to a proof-search problem in linear logic:

derive

$$\boldsymbol{a}[0,1], \boldsymbol{wise}[1,2], \boldsymbol{woman}[2,3], \boldsymbol{seeks}[3,4], \boldsymbol{a}[4,5], \boldsymbol{wise}[5,6], \boldsymbol{man}[6,7] \vdash S[0,7]$$

using the rules of implicative linear logic with the set of sequents of Fig. 4 as proper axioms.

We give in Fig. 6 and Fig. 7 (in the annex) the derivations corresponding to the *de dicto* parsing (3) and to the *de re* parsing (4). These two derivations use the inference rules given in Fig. 5, which are equivalent to the sequents of Fig. 4.

Acknowledgments

I am grateful to Makoto Kanazawa for fruitful discussions about his work on parsing as datalog

[3]Following (Kanazawa, 2011) and (Kanazawa, 2007), when writing $N[j,k]$ for $N_{k \multimap j}$, we write the variables in the reverse order.

queries. A preliminary version of the results reported in this paper has been presented in a talk given in June 2007 at the Colloquium in Honor of Gérard Huet on the occasion of his 60th birthday. This work has been supported by the French agency *Agence Nationale de la Recherche* (ANR-12-CORD-0004).

References

A.A. Babaev and S.V. Solov'ev. 1982. A coherence theorem for canonical morphisms in cartesian closed categories. *Journal of Soviet Mathematics*, 20(4):2263–2279. *Original in Russian: Zapiski Nauchnykh Seminarov Leningradskogo Otdeleniya Matematicheskogo Instituta imeni V. A. Steklova Akademii Nauk SSSR (LOMI), 88:3–29, 1979.*

H.P. Barendregt. 1984. *The lambda calculus, its syntax and semantics*. North-Holland, revised edition.

Ph. de Groote, B. Guillaume, and S. Salvati. 2004. Vector addition tree automata. In *Proceedings of the 19th annual IEEE symposium on logic in computer science*, pages 64–73.

Ph. de Groote. 2001. Towards abstract categorial grammars. In *Association for Computational Linguistics, 39th Annual Meeting and 10th Conference of the European Chapter, Proceedings of the Conference*, pages 148–155.

J.-Y. Girard. 1987. Linear logic. *Theoretical Computer Science*, 50:1–102.

J.R. Hindley. 1969. The principal type-scheme of an object in combinatory logic. *Transaction of the American Mathematical Society*, 146:29–60.

S. Hirokawa. 1991. Principal type-schemes of bci-lambda-terms. In T. Ito and A.R. Meyer, editors, *Theoretical Aspects of Computer Software, TACS'91*, volume 526 of *Lecture Notes in Computer Science*, pages 633–650. Springer-Verlag.

M. Kanazawa. 2007. Parsing and generation as datalog queries. In *Proceedings of the 45th Annual Meeting of the Association for Computational Linguistics*, pages 176–183. Association for Computational Linguistics.

M. Kanazawa. 2011. Parsing and generation as datalog query evaluation. Last revised August 26, 2011. 74 pages. (Under review).

G.E. Mints. 1981. Closed categories and the theory of proofs. *Journal of Soviet Mathematics*, 15(1):45–62. *Original in Russian: Zapiski Nauchnykh Seminarov Leningradskogo Otdeleniya Matematicheskogo Instituta imeni V. A. Steklova Akademii Nauk SSSR (LOMI), 68:83–114, 1977.*

M. Moortgat. 1997. Categorial type logics. In J. van Benthem and A. ter Meulen, editors, *Handbook of Logic and Language*, chapter 2. Elsevier.

R. Yoshinaka and M. Kanazawa. 2005. The complexity and generative capacity of lexicalized abstract categorial grammars. In Philippe Blache, E. Stabler, J. Busquets, and R. Moot, editors, *Logical Aspects of Computational Linguistics, LACL 2005*, volume 3492 of *Lecture Notes in Computer Science*, pages 330–346. Springer-Verlag.

$$\dfrac{\textbf{wise}[5,6] \vdash \textbf{wise}[5,6] \qquad \dfrac{\textbf{man}[6,7] \vdash \textbf{man}[6,7]}{\textbf{man}[6,7] \vdash N[6,7]}\ (\text{M})}{\textbf{wise}[5,6], \textbf{man}[6,7] \vdash N[5,7]}\ (\text{WI}) \quad (1)$$

$$\dfrac{\dfrac{NP_o[4,7] \multimap S[0,7] \vdash NP_o[4,7] \multimap S[0,7] \qquad NP_o[4,7] \vdash NP_o[4,7]}{NP_o[4,7] \multimap S[0,7], NP_o[4,7] \vdash S[0,7]}\ (\text{APP})}{NP_o[4,7] \multimap S[0,7], NP_o[4,7] \vdash \overline{S}[0,7]}\ (\text{I}) \quad (2)$$

$$\dfrac{\textbf{seeks}[3,4] \vdash \textbf{seeks}[3,4] \qquad \dfrac{\textbf{a}[4,5] \vdash \textbf{a}[4,5] \qquad \begin{array}{c}\vdots\ (1)\\ \textbf{wise}[5,6], \textbf{man}[6,7] \vdash N[5,7]\end{array} \qquad \begin{array}{c}\vdots\ (2)\\ NP_o[4,7] \multimap S[0,7], NP_o[4,7] \vdash \overline{S}[0,7]\end{array}}{\textbf{a}[4,5], \textbf{wise}[5,6], \textbf{man}[6,7], NP_o[4,7] \multimap S[0,7] \vdash \overline{S}[0,7]}\ (\text{A}_o) \qquad NP_s[0,3] \vdash NP_s[0,3]}{\textbf{seeks}[3,4], \textbf{a}[4,5], \textbf{wise}[5,6], \textbf{man}[6,7], NP_s[0,3] \vdash S[0,7]}\ (\text{S}) \quad (3)$$

$$\dfrac{\textbf{a}[0,1] \vdash \textbf{a}[0,1] \qquad \dfrac{\textbf{wise}[1,2] \vdash \textbf{wise}[1,2] \qquad \dfrac{\textbf{woman}[2,3] \vdash \textbf{woman}[2,3]}{\textbf{woman}[2,3] \vdash N[2,3]}\ (\text{W})}{\textbf{wise}[1,2], \textbf{woman}[2,3] \vdash N[1,3]}\ (\text{WI}) \qquad \dfrac{\begin{array}{c}\vdots\ (3)\\ \textbf{seeks}[3,4], \textbf{a}[4,5], \textbf{wise}[5,6], \textbf{man}[6,7], NP_s[0,3] \vdash S[0,7]\end{array}}{\textbf{seeks}[3,4], \textbf{a}[4,5], \textbf{wise}[5,6], \textbf{man}[6,7], NP_s[0,3] \vdash \overline{S}[0,7]}\ (\text{I})}{\textbf{a}[0,1], \textbf{wise}[1,2], \textbf{woman}[2,3], \textbf{seeks}[3,4], \textbf{a}[4,5], \textbf{wise}[5,6], \textbf{man}[6,7] \vdash S[0,7]}\ (\text{A}_s)$$

Figure 6: The derivation corresponding to $\text{A}_s\,(\text{WISE WOMAN})\,(\lambda x.\, \text{INJ}\,(\text{SEEK}\,(\lambda p.\, \text{A}_o\,(\text{WISE MAN})\,(\lambda y.\, \text{INJ}\,(p\,y)))\,x))$

$$
\dfrac{\boldsymbol{seeks}[3,4] \vdash \boldsymbol{seeks}[3,4] \qquad \dfrac{\dfrac{NP_o[4,7] \multimap S[0,7] \vdash NP_o[4,7] \multimap S[0,7] \qquad NP_o[4,7] \vdash NP_o[4,7]}{NP_o[4,7], NP_o[4,7] \multimap S[0,7] \vdash S[0,7]}\;(\mathrm{APP})}{NP_o[4,7], NP_o[4,7] \multimap S[0,7] \vdash \overline{S}[0,7]}\;(\mathrm{I}) \qquad NP_s[0,3] \vdash NP_s[0,3]}{\boldsymbol{seeks}[3,4], NP_s[0,3], NP_o[4,7] \vdash S[0,7]}\;(\mathrm{S}) \quad (1)
$$

$$
\dfrac{\boldsymbol{a}[4,5] \vdash \boldsymbol{a}[4,5] \qquad \dfrac{\boldsymbol{wise}[5,6] \vdash \boldsymbol{wise}[5,6] \qquad \dfrac{\boldsymbol{man}[6,7] \vdash \boldsymbol{man}[6,7]}{\boldsymbol{man}[6,7] \vdash N[6,7]}\;(\mathrm{M})}{\boldsymbol{wise}[5,6], \boldsymbol{man}[6,7] \vdash N[5,7]}\;(\mathrm{WI}) \qquad \dfrac{\begin{matrix} \vdots\;(1) \\ \boldsymbol{seeks}[3,4], NP_s[0,3], NP_o[4,7] \vdash S[0,7] \end{matrix}}{\boldsymbol{seeks}[3,4], NP_s[0,3], NP_o[4,7] \vdash \overline{S}[0,7]}\;(\mathrm{I})}{\boldsymbol{seeks}[3,4], \boldsymbol{a}[4,5], \boldsymbol{wise}[5,6], \boldsymbol{man}[6,7], NP_s[0,3] \vdash \overline{S}[0,7]}\;(\mathrm{A}_o) \quad (2)
$$

$$
\dfrac{\boldsymbol{a}[0,1] \vdash \boldsymbol{a}[0,1] \qquad \dfrac{\boldsymbol{wise}[1,2] \vdash \boldsymbol{wise}[1,2] \qquad \dfrac{\boldsymbol{woman}[2,3] \vdash \boldsymbol{woman}[2,3]}{\boldsymbol{woman}[2,3] \vdash N[2,3]}\;(\mathrm{W})}{\boldsymbol{wise}[1,2], \boldsymbol{woman}[2,3] \vdash N[1,3]}\;(\mathrm{WI}) \qquad \begin{matrix} \vdots\;(2) \\ \boldsymbol{seeks}[3,4], \boldsymbol{a}[4,5], \boldsymbol{wise}[5,6], \boldsymbol{man}[6,7], NP_s[0,3] \vdash \overline{S}[0,7] \end{matrix}}{\boldsymbol{a}[0,1], \boldsymbol{wise}[1,2], \boldsymbol{woman}[2,3], \boldsymbol{seeks}[3,4], \boldsymbol{a}[4,5], \boldsymbol{wise}[5,6], \boldsymbol{man}[6,7] \vdash S[0,7]}\;(\mathrm{A}_s)
$$

Figure 7: The derivation corresponding to $\mathrm{A}_s\,(\mathrm{WISE\ WOMAN})\,(\lambda x.\,\mathrm{A}_o\,(\mathrm{WISE\ MAN})\,(\lambda y.\,\mathrm{INJ}\,(\mathrm{SEEK}\,(\lambda p.\,\mathrm{INJ}\,(p\,y))\,x)))$

Topology of Language Classes

Sean A. Fulop
Dept. of Linguistics
California State University Fresno
sfulop@csufresno.edu

David Kephart
Link-Systems International
dkephart@link-systems.com

Abstract

The implications of a specific pseudometric on the collection of languages over a finite alphabet are explored. In distinction from an approach in (Calude et al., 2009) that relates to collections of infinite or bi-infinite sequences, the present work is based on an adaptation of the "Besicovitch" pseudometric introduced by Besicovitch (1932) and elaborated in (Cattaneo et al., 1997) in the context of cellular automata. Using this pseudometric to form a metric quotient space, we study its properties and draw conclusions about the location of certain well-understood families of languages in the language space. We find that topologies, both on the space of formal languages itself and upon quotient spaces derived from pseudometrics on the language space, may offer insights into the relationships, and in particular the distance, between languages over a common alphabet.

1 Introduction

The question of distance between languages, and comparison of possible definitions, has relatively less consideration in the literature than other language issues, with notable exceptions being (Berstel, 1973) and (Salomaa and Soittola, 1978). This may seem surprising, considering that the current digital climate necessitates the measurement of likeness between texts and languages, for instance in search engine entries and results. *Ad hoc* measures of differences exist based upon rooted tree distances, but these are more like attempts to incorporate the intuitive notion of differences between words than overall differences between languages. In Linguistics, as well, there is as yet no accepted way of measuring the distance between two dialects of a language, with each employing the same vocabulary.

This paper borrows a pseudometric from cellular automata theory to use language density and form a topology on the set of languages (consisting of words of finite length over a fixed alphabet). A similar pseudometric is discussed in (Cattaneo et al., 1997). Our goal is to continue a systematic review and categorization of language distances, with a view to determining what gives rise to apparent weaknesses and strengths of each. As seen in (Salomaa and Soittola, 1978) and (Yu, 1997) language *density* is understood as the number of words in a language, conceived of as a function $\varrho(n)$ of word length n. This is shown to convey information about the nature of a language. Analysis of language density over finite words may be confined to the treatment of regular languages (Yu, 1997; Kozik, 2005), or seen as a probability density of distance between infinite sequences (Kozik, 2006).

Herein we continue the approach of (Kozik, 2006) of capturing distances between arbitrary languages, specifically by looking at features of the topology generated by each. We consider that languages—natural or formal—are most beneficially understood as potential or actually infinite objects. As such, language patterns may or may not be adequately defined syntactically. We continue the work of Kozik in grasping language differences as word density of distinctions at the limit. Since such a limit may not exist, we look at a pseudometric inspired by Besicovitch (1932) that captures, in fact, the upper density limit of language differences.

Next, we consider "where," in the resulting "Besicovitch" topology of the language space, individual

Proceedings of the 14th Meeting on the Mathematics of Language (MoL 14), pages 26–38,
Chicago, USA, July 25–26, 2015.

languages lie. We also look at how this relates to the Chomsky hierarchy of languages. We find that the pseudometric space of languages is not complete, and look at the lifting of the pseudometric to a quotient metric space. The hope is that this consideration may contribute to a list of relative advantages and disadvantages associated with various candidate language topologies. Our contribution is thus conceived as a part of a broader exploration in search of the most useful topology of language spaces, with eventual application to linguistic problems like measuring the distance between dialects over a common vocabulary. We have tried to study the Besicovitch topology and its quotients in some detail, but some proofs have been condensed to outlines due to space limitations.

1.1 Early approaches

Nelson (Nelson, 1980) elaborated work by Walter (Walter, 1975) which constructed a topological space from a space of rewriting grammars by means of successive divisors of grammatical derivations. The resulting topologies of both languages and grammatical derivations are equivalent to quasi-ordered sets, and have the property that each point has a smallest open neighborhood. If such a topology is T_1 then it is discrete.

An equivalence relation between languages was suggested by Marcus (Marcus, 1966; Marcus, 1967) based on equivalence of word contexts. Improved and elaborated by Dincă (Dincă, 1976), this approach treats the space of languages as a semigroup over the alphabet, and a distance in the quotient space (dividing by context equivalence) measures the distance between context classes of strings with respect to some chosen language.

The above described approaches, while not without interest where linguistic applications are in view, do not yield a "sufficiently smooth" topology of a language space. The first approach similar in spirit to our main thread was published by Vianu (1977), who applied the metric proposed earlier by Bodnarchŭk (1965). This approach has a number of variants, but we will point out the most important conclusions to be drawn from them as well as possible limitations of this approach.

1.2 Current literature on language topologies and distances

Language spaces allowing infinitary words, on the other hand, can be more easily endowed with adequate topologies arising out of the word topology (Calude et al., 2009), but this will not be a topic of discussion here because there seems to be no application of infinitary word languages to the study of natural human languages.

2 Preliminaries

In this section we review some basic definitions from formal language theory and review the best-known approach to language distance, namely, what we will call the Cantor metric.

2.1 Notation and Definitions

For the most part, we adopt notation common to formal language theory. There are a few modifications in the interest of brevity and, hopefully, clarity of expression. We consider a language as a set of *words* which are concatenated from symbols in an *alphabet* Σ with finite cardinality α. We will deal only with words of finite length (as opposed to the words discussed, for instance, in (Calude et al., 2009)). By a *language space* we mean the collection of all possible languages, namely, 2^{Σ^*}.

Sets. We frequently employ *the symmetric set difference of sets A and B*, denoted $A \triangle B$.

Words. The length of word w will be denoted $|w|$ and will always be non-negative. The empty word, which is the unique word of length zero, will, as usual, be denoted λ. When we need to refer to the ith symbol of the word w, we will denote this by $w_{[i]}$, preserving ordinary subscripts for the enumeration of words. The fundamental operation on symbols is (non-commutative) concatenation, which is represented multiplicatively. We use the Kleene-$*$ (-star) and -$+$ (-plus) operations in the usual way. Moreover, Σ^n denotes the set of all words of length n, and $\Sigma^{<n}$ denotes the set of all words of lengths up to $n-1$.

Languages. The empty language is simply the null set, $\emptyset$. Concatenation extends from words to languages. That is, if L and M are languages, then $LM = \{uv : u \in L, v \in M\}$. Suppose L is a language over Σ and $n \in \mathbb{N} \cup \{0\}$. Then we denote by L^n (respectively $L^{<n}$, for $n > 0$) the set $L \cap \Sigma^n$ (respectively, $\bigcup_{i=0}^{n-1} L^i$). For example, L^0 is either $\emptyset$ or $\{\lambda\}$. The *density of language* L is the sequence $\{\varrho_n\}_{n\in\mathbb{N}}$ such that $\varrho_i = |L^i|$. Then $|L^{<n}|$ is the nth partial sum of the series $\sum \varrho$. Finally, given languages L and M, we denote by $L \triangle^n M$ (respectively, $L \triangle^{<n} M$) the symmetric set difference between words of length n (respectively, less than n) in the two languages.

Remark 1. *Note that*

$$|\Sigma^{<n}| = \frac{\alpha^n - 1}{\alpha - 1}. \quad (1)$$

2.2 Language norms, metrics and the Cantor space

In setting out to find ways to adequately express the "distance" between two languages, we consider how to adapt the notions of size and separation into the realm of formal symbols. We already observe that the first defined language distance, i.e., the distance between two languages, in the literature, derives from the density of their symmetric set difference. The metric mentioned by (Vianu, 1977) is based on the shortest word in $L \triangle M$. Indeed, this leads to a full metric, and a metric topology on 2^{Σ^*}. By analogy to the norm in a normed space representing distance from a zero point, and hence magnitude, a "language norm" can be elaborated from a pseudometric.

The reader will recall that a pseudometric d on space X is a function that maps $X \times X$ to $\mathbb{R}^{\geq 0}$, such that $d(x,x) = 0$, $d(x,y) = d(y,x)$ and $d(x,y) + d(y,z) \geq d(x,z)$. We call a pseudometric a *language distance* just in case it additionally is such that, if $L \cap N = \emptyset$ and $M \subseteq N$, then $d(L,M) \leq d(L,N)$.

Then, to every language distance we may associate a function $\|\cdot\|_d : 2^{\Sigma^*} \to \mathbb{R}^{\geq 0}$ by defining $\|L\|_d = d(L,\emptyset)$. Note that $\|\cdot\|_d$ has the following properties:

$$\|\emptyset\|_d = 0 \quad (2)$$

$$\|L \cup M\|_d \leq \|L\|_d + \|M\|_d \quad (3)$$

$$L \subseteq M \implies \|L\| \leq \|M\| \quad (4)$$

We define a *language norm* as any such function on languages.

Lemma 2. *To each language norm* $\|\cdot\|$ *there corresponds a unique language distance* d *such that* $d(L,M) = \|L \triangle M\|$.

The contrapositive of Lemma 2 also holds. That is, for any language distance d on 2^{Σ^*}, the function $\|\cdot\| : 2^{\Sigma^*} \to \mathbb{R}^{\geq 0}$ such that $\|L\| = d(L,\emptyset)$ defines a unique language norm.

2.3 The Cantor language metric and topology on 2^{Σ^*}

A Cantor language space

Two languages can be compared by beginning with the shortest word in each language and proceeding to longer words. A first notion of distance is obtained using the word-length of the first distinction between languages so observed. To this end, let the language space then be normed by assigning the norm 0 to $\emptyset$ and by associating each non-empty language to a power of $1/2$, as follows.

Definition 3. *The language norm* $\|\cdot\|_1 : 2^{\Sigma^*} \to \mathbb{R}$ *is as follows: for* $L \in 2^{\Sigma^*}$,

$$\|L\|_1 = \begin{cases} 0 & \text{if } L = \emptyset, \\ 2^{-\min\{|w| : w \in L\}} & \text{otherwise.} \end{cases} \quad (5)$$

Observe that $-\log_2 \|L\|_1 \in \mathbb{N}$ for all non-empty L.

To this language norm corresponds the following language metric.

Definition 4. *The function* $d_1 : 2^{\Sigma^*} \times 2^{\Sigma^*} \to \mathbb{R}$ *is a metric, where, for* L *and* M *in* 2^{Σ^*}, $d_1(L,M) = \|L \triangle M\|_1$

To see that d_1 is in fact not only a pseudometric but a metric, consider that $d_1(L,M) = 0$ iff $L \triangle M = \emptyset$, i.e., iff $L = M$. Let τ_1 be the metric topology induced on 2^{Σ^*} by d_1. For reasons to be made clear below, we call $\|\cdot\|_1$, d_1, and τ_1 the *Cantor norm, distance,* and *topology,* respectively, on a language space.

The open neighborhoods of radius $\epsilon > 0$ around some language $L \in 2^{\Sigma^*}$, denoted $\mathcal{B}_\epsilon(L) = \{M \in 2^{\Sigma^*} : d_1(L,M) < \epsilon\}$, form the standard basis for τ_1. Since distances between distinct languages are powers of $1/2$, it follows that elements of the standard metric basis for $(2^{\Sigma^*}, \tau_1)$ form the collection

$$\mathcal{C} = \{\mathcal{B}_{2^{-n}}(L) : n \in \mathbb{N}, L \in 2^{\Sigma^*}\}. \quad (6)$$

Definition 5 ((Vianu, 1977; Genova and Jonoska, 2006)). *The* language cylinder set $C_{L,k}$ *of length* $k \in \mathbb{N}$ *around language* $L \in 2^{\Sigma^*}$ *is:*

$$C_{L,k} \stackrel{\text{def}}{=} \{M \in 2^{\Sigma^*} : L \cap \Sigma^{\leq k} = M \cap \Sigma^{\leq k}\}. \quad (7)$$

Now let $\mathcal{C}_k \stackrel{\text{def}}{=} \{C_{L,k} : L \in 2^{\Sigma^*}\}$ be the collection of all language cylinder sets of length k. From (6) and (7) it follows that the collection $\mathcal{C} \stackrel{\text{def}}{=} \bigcup_{k \in \mathbb{N}} \mathcal{C}_k$, comprising all language cylinder sets, is the standard basis for $(2^{\Sigma^*}, \tau_1)$.

Cantor topology on a language space

As it turns out, τ_1 is equivalent to the topology of the Bodnarchŭk metric space discussed in (Vianu, 1977). We quickly recap the properties of this topology on a language space, as proven in (Vianu, 1977) and (Genova and Jonoska, 2006).

Lemma 6. *In* $(2^{\Sigma^*}, \tau_1)$*, every cylinder set is both closed and open.*

Lemma 7. *A sequence* $\{L_i\}_{i \in \mathbb{N}} \subset 2^{\Sigma^*}$ *converges to language* L *in* $(2^{\Sigma^*}, \tau_1)$ *iff for all* $m \in \mathbb{N}$, $|(L_i \triangle^m L)| = 0$ *for all but finitely many* i. *In this case we write* $L_i \to L$.

From this and the fact $L \cap \Sigma^{\leq i} \to L$ we also have:

Corollary 8. *The finite languages are dense in a space of languages under the* τ_1 *topology.*

Lemma 9. *The topological space* $(2^{\Sigma^*}, \tau_1)$ *is homeomorphic to the Cantor space.*

Thus the terminology "Cantor language space, topology," etc.[1]

Corollary 10. $(2^{\Sigma^*}, \tau_1)$ *is compact, perfect, and totally disconnected.*

3 Besicovitch pseudometric, language norm and topology

We now consider a language distance that is in many respects more satisfactory than the Cantor distance d_1, by exploiting the general philosophy of comparing languages by comparing finite sections of languages. We then show several results, including that neither finite nor locally testable languages are dense in the topology induced. We call this alternative pseudometric the Besicovitch distance, denoted by d_ς. Under the topology induced, a language space is not compact. Rather, it has a geometry which becomes apparent from the vantage point of a metric quotient space.

The original Besicovitch pseudometric expressed the distance between two almost-periodic real-valued functions (Besicovitch, 1932) $\phi, \psi \in \ell^1$ as

$$d_{B^p}(\phi, \psi) \stackrel{\text{def}}{=} \limsup_{n \to \infty} \frac{1}{2n+1} \sum_{-n}^{n} |\phi(x) - \psi(x)|.$$

Because this pseudometric depends on the evaluation of the two functions only at discrete intervals, it is naturally adaptable to expressing distances between objects with a bound proportion of differences, as with the distance between cellular automata (Cattaneo et al., 1997); our adaptation to languages is in some sense a generalization thereof.

3.1 A Besicovitch pseudometric on language spaces

We begin by defining a Besicovitch-style language norm. Rather than halting at a particular term of the density of the symmetric set difference between two languages, this norm considers the derived infinite series $|L \triangle^{<n} M|$ in ratio to the total possible words over Σ^* (given by (1)) as n goes to infinity.

Definition 11. *Let* $\|\cdot\|_\varsigma$ *for fixed alphabet* Σ *be the function defined:*

$$\|L\|_\varsigma \stackrel{\text{def}}{=} \limsup_{k \to \infty} \frac{|(L \cap \Sigma^{<k})|}{|\Sigma^{<k}|} \quad (8)$$

We call $\|\cdot\|_\varsigma$ *the* Besicovitch language norm.

Then let d_ς *be the function mapping* (L, M) *to* $\|L \triangle M\|_\varsigma$*, and call it the* Besicovitch language distance.

By Lemma 2, distance d_ς is a language pseudometric.

Remark 12. *The Besicovitch distance* d_ς *between languages*

[1] As pointed out by an anonymous reviewer, the Cantor topology can be understood as the profinite completion of an algebra of recognizable languages (Gehrke, 2009). While this does not modify the topological characteristics of the space, it does raise the interesting point that the Besicovitch topology, the main subject of this paper, likely cannot be so conceived: the Besicovitch metric (quotient) space is not complete, by Corollary 31.

1. *can be described as the upper density of their set-difference;*
2. *turns out to constitute (like the Besicovitch language norm) a continuous, surjective mapping of 2^{Σ^*} into the unit interval $[0,1]$;*
3. *for a language and its complement is 1, since $(L \cap \Sigma^k) \triangle (\neg L \cap \Sigma^k) = \Sigma^k$ for every $k \in \mathbb{N}$;*
4. *constitutes a strict pseudo-metric if $|\Sigma| > 1$, since, for instance, $\|M\|_\zeta = 0$ where $M = \{a\}$, so $d_\zeta(M, \emptyset) = 0$ even though $M \neq \emptyset$;*
5. *given languages $L, M \in 2^{\Sigma^*}$, can be written as follows:*

$$d_\zeta(L, M) = \limsup_{k\to\infty} \left| L \triangle^{<k} M \right| \left(\frac{\alpha - 1}{\alpha^k - 1} \right)$$

We present the following without proof.

Lemma 13. *If L and M are disjoint languages in 2^{Σ^*}, then $\|L\|_\zeta + \|M\|_\zeta = \|L \cup M\|_\zeta$.*

Corollary 14. *For $L, M \in 2^{\Sigma^*}$, $\|L\|_\zeta + \|M\|_\zeta = \|L \cup M\|_\zeta$ if and only if $\|L \cap M\|_\zeta = 0$.*

Corollary 15. *For all $L, M \in 2^{\Sigma^*}$, $\|L\|_\zeta + \|M\|_\zeta \geq \|L \cup M\|_\zeta$.*

The conclusion here is that $\|\cdot\|_\zeta$ is truly "norm-like." For the remainder of this section, we drop most subscripts ζ.

To establish surjectivity, we first need a way to construct a language with a specified arbitrary norm.

Definition 16. *Given $0 \leq r \leq 1$, consider the sequence $\check{r}_{\langle\alpha\rangle} \stackrel{\text{def}}{=} \{\lfloor r\alpha^k \rfloor\}_{k\in\mathbb{N}}$. Then we call $\mathbf{L}_r$ the set of r-*simple languages *in 2^{Σ^*}, defined as follows:*

$$\mathbf{L}_r \stackrel{\text{def}}{=} \left\{ L \in 2^{\Sigma^*} : \{|(L \cap \Sigma^i)|\}_{i\in\mathbb{N}} = \check{r}_{\langle\alpha\rangle} \right\}.$$

Lemma 17. *For $r \in [0,1]$ there is at least one r-simple language; moreover for each particular value r, every r-simple language has norm r.*

Proof. By construction, for each $r \in [0,1]$ the sequence $\check{r}_{\langle\alpha\rangle}$ exists. We can select $\check{r}_k$ words in Σ^k for all k. This amounts to the construction of a language L in $\mathbf{L}_r$ for each $r \in [0,1]$. But then $\|L\| = r$, which establishes the claim. □

Now we have established our hoped-for result.

Corollary 18. *The Besicovitch language norm is a surjective mapping from 2^{Σ^*} onto $[0,1]$.*

In addition, it is relatively easy to see that diagonalization yields that there are uncountably many r-simple languages for each $r \in [0,1]$.

3.2 Besicovitch distance quotient space

The Besicovitch distance equivalence induces a quotient space on 2^{Σ^*}

We next form collections of languages at distance zero from each other and map each such collection to a point in a quotient space, which can then be metrized. So, given $L, M \in 2^{\Sigma^*}$, let $L \sim_\zeta M$ if $d_\zeta(L, M) = 0$.

Proposition 19. *The relation $\sim_\zeta$ is an equivalence on 2^{Σ^*}.*

Proof. Reflexivity and symmetry are apparent and, if $L, M, N \in 2^{\Sigma^*}$ such that $L \sim M$ and $M \sim N$, then $0 = d(L, M) + d(M, N) \geq d(L, N)$. From Remark 12(1), $L \sim N$. □

The collection of $\sim$ equivalence classes will be called the *Besicovitch quotient space* over 2^{Σ^*}, denoted $\mathcal{Q}_\zeta^\Sigma$. Here we will drop the ζ subscript for notational clarity and assume the language space 2^{Σ^*} unless otherwise noted. Elements of the quotient space (points in $\mathcal{Q}$) will be denoted with sans-serif letters $\mathsf{L}, \mathsf{M}, \mathsf{N}, \ldots$, while collections of such points will be denoted with corresponding bold letters $\mathbf{\mathsf{L}}, \mathbf{\mathsf{M}}, \mathbf{\mathsf{N}}, \ldots$. Let $\eta \colon 2^{\Sigma^*} \to \mathcal{Q}_\zeta^\Sigma$ denote the quotient mapping which takes a language to its $\sim$ equivalence class.

Remark 20. *As a partition of 2^{Σ^*}, the mapping η is well-defined and surjective, but not injective since it is a quotient mapping. The set operations of union, intersection and complementation are preserved by mappings from collections of points in $\mathcal{Q}$ to the sets of languages of which they are equivalence classes. In particular, every topology on $\mathcal{Q}_\zeta^\Sigma$ is the quotient of a topology on 2^{Σ^*}.*

When language L is a member of language family $\mathbf{L}$, and every member of $\mathbf{L}$ is contained in an equivalence class in the collection of points $\mathbf{\mathsf{L}} \subseteq \mathcal{Q}_\zeta$, we will write $L \in \mathsf{L}$ and $\mathbf{L} \subseteq \mathbf{\mathsf{L}}$ instead of the more tedious $\eta(L) = \mathsf{L}$ and $\eta(\mathbf{L}) \subseteq \mathbf{\mathsf{L}}$.

Lemma 21. *For languages $L, M \in 2^{\Sigma^*}$, $L \nsim M$ iff there exists a positive integer m such that, for infinitely many word-lengths n, $|(L \triangle^{<n} M)| \geq |\Sigma^{<n-m}|$.*

Proof. ($\Rightarrow$) Suppose there is no such m. That would mean that, for each $m \in \mathbb{N}$, there is a word length n_m such that $k > n_m$ implies $|(L \triangle^{<k} M)| < |\Sigma^{k-m}|$. We can then construct an increasing sequence $\{k_i\}_{i \in \mathbb{N}}$ where $k_0 = 0$ and k_i $(i > 0)$ is the least integer greater than k_{i-1} such that $|(L \triangle^{<k'} M| > |\Sigma^{<k'-i}| = \sum_{j=0}^{k'-i-1} |\Sigma^j| = \frac{\alpha^{k'-i}-1}{\alpha-1}$ if $k' > k_i$. But, if this were true, then the Besicovitch distance between the two languages would be 0, since a straightforward calculation shows that, for each $m \in \mathbb{N}$, $d_\zeta(L, M)$ is bounded above by α^{-m}.

($\Leftarrow$) Assume that, for some $m \in \mathbb{N}$ and for n sufficiently large, $|(L \triangle^{<n} M)| \geq |\Sigma^{n-m}|$. Then a similarly straightforward calculation shows that $\|L \triangle M\| = \limsup_{k \to \infty} \frac{|L \triangle^{<k} M|}{|\Sigma^{<k}|}$ is bounded below by, for instance, $\alpha^{-m}/2$. Thus, $L \nsim_\zeta M$.

Note that when two languages are similar, the sequence used in the first part of the proof is *finite*. □

Definition 22. *Given languages $L, M \in 2^{\Sigma^*}$, we will denote by $K_\zeta(L, M)$ the (possibly finite) increasing integer sequence $\{k_i\}_{i \in \mathbb{N}}$ in accord with the above lemma. Indeed, $K_\zeta(L, M)$ is infinite precisely when $L \sim M$.*

We note that if $K_\zeta(L, M)$ has at least i terms, then $k_i > i$ and, by considering the words in $L \triangle M$ of length greater than m_i, we have a first estimate of the distance between two languages, namely, α^{-i}.

By Lemma 21, the unique sequence $K_\zeta(L, M)$ expresses the relative location of languages in the quotient space $\mathcal{Q}_\zeta$.

The quotient space has a natural metric quotient topology

We define the metric $\mathbf{d}_\zeta$ on the Besicovitch quotient space as the lifting of Besicovitch distance d.

Definition 23. *Let the distance $\mathbf{d}_\zeta$ between points L and M in $\mathcal{Q}_\zeta$ be set equal to* $\inf\{d(L, M) \colon L \in \eta^{-1}(\mathsf{L}), M \in \eta^{-1}(\mathsf{M})\}$.

Lemma 24. *For $\mathsf{L}, \mathsf{M} \in \mathcal{Q}_\zeta$, $\mathbf{d}_\zeta(\mathsf{L}, \mathsf{M}) = 0$ iff $\mathsf{L} = \mathsf{M}$.*

Proof. Only the implication left to right requires a proof. Suppose that $\mathbf{d}(\mathsf{L}, \mathsf{M}) = 0$. Now suppose, contrary to the claim, that there is some language $L \in \mathsf{L}$ but not M. We conclude from the preceding definition 23 that there exists $M \in \mathsf{M}$ such that $d(L, M) = \epsilon > 0$. But then, for arbitrary languages $L' \in \mathsf{L}$ and $M' \in \mathsf{M}$, the triangle inequality provides us that

$$\begin{aligned} \epsilon \leq d(L, M) &\leq d(L, L') + d(L', M') + d(M', M) \\ &\leq d(L', M'). \end{aligned} \tag{9}$$

Thus $\mathbf{d}(\mathsf{L}, \mathsf{M}) \geq \epsilon/2 > 0$ by the preceding definition, Q.E.A. □

Corollary 25. *For $L, M \in 2^{\Sigma^*}$ the diagram below commutes, showing the isometry between Besicovitch language space and quotient space.*

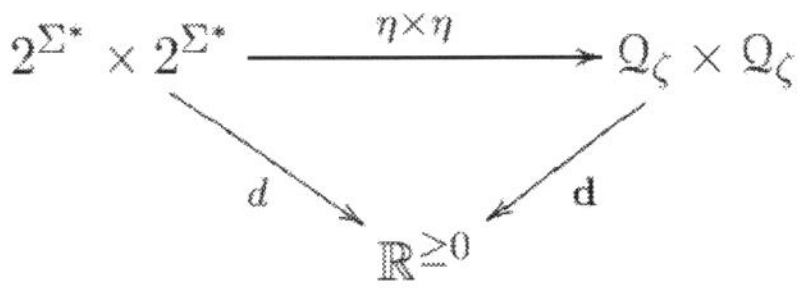

It is by now evident that the Besicovitch quotient space is a metric space under distance $\mathbf{d}$. Moreover we have:

Corollary 26. *If languages L, M are in point $\mathsf{L} \in \mathcal{Q}_\zeta$, then $\|L\| = \|M\|$.*

This just means that the $\mathbf{d}$ metric topology on $\mathcal{Q}_\zeta$ is the quotient of the pseudo-metric topology induced by d on 2^{Σ^*}. Let $\tilde{\tau}_\zeta$ denote the collection of open sets in $\mathcal{Q}_\zeta$ under the $\mathbf{d}$ metric topology, and let τ_ζ denote the collection of language sets in 2^{Σ^*} such that $\eta(\tau_\zeta) = \tilde{\tau}_\zeta$. We will call τ_ζ the *Besicovitch language topology.*

Convergence has a novel interpretation in the quotient space

From Remark 12, the Besicovitch language topology is not T_1, and so convergence to a language is not well-defined in $(2^{\Sigma^*}, \tau_\zeta)$. But there is no such difficulty with the quotient space.

Lemma 27. *A sequence $\{\mathsf{L}_i\}_{i \in \mathbb{N}}$ in $\mathcal{Q}_\zeta$ converges to the point $\mathsf{L} \in \mathcal{Q}_\zeta$ iff the following: $\forall m \in \mathbb{N}\ \exists k_m \in \mathbb{N}$ such that $i > k_m$ means that, if language $L_i \in \mathsf{L}_i$ and $L \in \mathsf{L}$ then there exists integer N_i for which $k > N_i$ implies $|(L \triangle^k L_i)| < \alpha^{k-m}$.*

Note that, unlike the case of the Cantor space, in the Besicovitch language (quotient) topology, a convergent sequence of points converges to $\sim$ equivalence with the (languages in the) limit point.

The quotient space is perfect but not compact

We can next address the compactness question for the Besicovitch quotient space, since it is a metric space, by determining whether every infinite sequence of points has a convergent subsequence. We ultimately show here that neither $\mathcal{Q}_\zeta$ nor 2^{Σ^*} is compact, although $\mathcal{Q}_\zeta$ is a perfect set.

We first establish the latter property using the following fact.

Lemma 28. *Every point in* $(\mathcal{Q}_\zeta, \tilde{\tau})$ *is a condensation point.*

Proof. Every open set M in $(\mathcal{Q}_\zeta, \tilde{\tau})$ includes the image of an open interval in the subset topology of $[0, 1]$ and, by a diagonalization, is uncountable. Each number in M has a distinct open inverse image in $(\mathcal{Q}_\zeta, \tilde{\tau})$. □

It then follows immediately that $\mathcal{Q}_\zeta$ is perfect. To make progress on the compactness question, we construct a family of two-sided *word ideals* in Σ^* which, when split into non-disjoint right ideals, yields an infinite sequence in the quotient space with no convergent subsequence. We will call I a right, left, or two-sided *word ideal* of the monoid Σ^* just in case there is a word $w \in \Sigma^*$ such that $I = w\Sigma^*$, $I = \Sigma^* w$, or $I = \Sigma^* w \Sigma^*$ respectively. Note this is just like the definition of ideal in a monoid (Howie, 1995) except we are restricting the reference to a singleton set containing particular word w. Now let J_w denote the two-sided word ideal $\Sigma^* w \Sigma^*$. Then for $k \in \mathbb{N}$, the kth section of J_w is $\Sigma^k w \Sigma^* \subsetneq J_w$, which is denoted $J_{w,k}$.

Lemma 29. *For* $i, j \in \mathbb{N}$ *and where* $|w| = l$,

$$d(J_{w,i}, J_{w,j}) = 2\alpha^{-l}. \tag{10}$$

We can also compute the norm of $J_{w,i}$ when $|w| = l$:

$$\begin{aligned}
\|J_{w,i}\| &= \limsup_{k\to\infty} \frac{|\Sigma^i w \Sigma^{<k-i-l}|}{|\Sigma^{<k}|} \\
&= \limsup_{k\to\infty} \frac{\sum_{s=0}^{k-i-l-1} \alpha^{i+s}}{|\Sigma^{<k}|} \\
&= \limsup_{k\to\infty} \frac{\frac{\alpha^{k-l}-\alpha^i}{\alpha-1}}{\frac{\alpha^k-1}{\alpha-1}} \\
&= \limsup_{k\to\infty} \alpha^{-l}\left[\frac{\alpha^k-1}{\alpha^k-1} - \frac{\alpha^{i+l}-1}{\alpha^k-1}\right] \\
&= \alpha^{-l}.
\end{aligned} \tag{11}$$

From Lemma 29 and the above calculation, taking $\mathsf{J}_{w,i} \stackrel{\text{def}}{=} \eta(J_{w,i})$, the sequence $\{\mathsf{J}_{w,i}\}_{i\in\mathbb{N}}$ is such that no subsequence can converge, yet every language in each point of the sequence has the same norm.

Lemma 30. *The Besicovitch quotient space* $\mathcal{Q}_\zeta$ *is not compact.*

Proof. It is sufficient to display an infinite sequence of languages belonging to distinct $\sim$ equivalence classes separated from each other by a distance greater than some fixed $\epsilon > 0$. Then the η-images of these languages will form an infinite sequence in $\mathcal{Q}_\zeta$ which has no convergent subsequence.

To this end, consider the language sequence $\mathbf{J}_a \stackrel{\text{def}}{=} \{J_{a,i}\}_{i\in\mathbb{N}}$ where $a \in \Sigma$. Two distinct terms $J_{a,i}$ and $J_{a,j}$ are at distance $2\alpha^{-1}$, from the previous lemma, so consider the sequence $\mathbf{L} = \{\mathsf{L}_i\}_{i\in\mathbb{N}}$, where $J_{a,i} \in \mathsf{L}_i$ for all $i \in \mathbb{N}$. By Corollary 25, there is no convergent subsequence of $\mathbf{L}$, since $\mathbf{d}(\mathsf{L}_i, \mathsf{L}_j) > \alpha^{-1}$ if $i \neq j$. □

Since sequential compactness is not defined in the pseudo-metric language space, we exhibit the following result to clear up any remaining doubts about compactness there.

Corollary 31. *The metric* $\mathbf{d}$ *is not complete.*

Proof. (Outline) It suffices to exhibit a sequence of points which are Cauchy convergent in $\mathcal{Q}_\zeta$, but which do not converge to any point in $\mathcal{Q}_\zeta$. We then produce a sequence, Cauchy in $\mathcal{Q}$, but containing the non-convergent sequence $\mathbf{L}$ from the proof of Lemma 30 as a subsequence.

□

Corollary 32. *A language space is not compact under the Besicovitch topology.*

Proof. Let $\mathcal{O}$ be an open cover of 2^{Σ^*} defined by

$$\mathcal{O} = \{\{M : d(L, M) < \alpha^{-1}\} : L \in 2^{\Sigma^*}\}.$$

We then have by Lemma 30 that any finite subset of $\mathcal{O}$ contains at most finitely many languages in $\mathbf{J}_a$. Therefore $\mathcal{O}$ has no finite subcover. □

While establishing noncompactness has been important, it will also be useful to establish a relation to a known compact space. This is the subject of the next subsection.

3.3 A second lifting of the quotient space

To obtain a compact space for exploring the most general features of the Besicovitch topology on language spaces, we define the language norm $\|\cdot\|_\zeta$ as a quotient map from $\mathcal{Q}_\zeta$ into $[0,1]$. This will result in a total of three spaces: the non-T_1 language space under the topology induced by Besicovitch distance, the quotient space topologized by the metric quotient topology, and a compact upper quotient space with a well-known topology. We proceed as with the definition of $\mathcal{Q}_\zeta$ by defining an equivalence relation, the equivalence classes, and the quotient map which takes points in $\mathcal{Q}_\zeta$ to their equivalence classes. We call the collection of equivalence classes the *upper Besicovitch quotient space,* denoted $\mathcal{N}_\zeta$. We ultimately show that the topological space $\mathcal{N}_\zeta$ under the quotient topology is homeomorphic to the unit interval.

Take points $\mathsf{L},\mathsf{M} \in \mathcal{Q}_\zeta$. Let $\mathsf{L} \equiv_\zeta \mathsf{M}$ if $\|L\|_\zeta = \|M\|_\zeta$ for all $L \in \mathsf{L}$ and $M \in \mathsf{M}$; let $\langle \mathsf{L} \rangle_\zeta = \{\mathsf{M} \in \mathcal{Q}_\zeta : \mathsf{M} \equiv_\zeta \mathsf{L}\}$, and denote by $\mathcal{N}_\zeta$ the collection $\{\langle \mathsf{L} \rangle_\zeta : \mathsf{L} \in \mathcal{Q}_\zeta\}$, write elements of $\mathcal{N}_\zeta$ in calligraphy font $\mathcal{L},\mathcal{M},\mathcal{N},\ldots$, and denote collections of such elements in corresponding bold letters $\boldsymbol{\mathcal{L}},\boldsymbol{\mathcal{M}},\boldsymbol{\mathcal{N}},\ldots$; let κ be the map from $\mathcal{Q}_\zeta$ to $\mathcal{N}_\zeta$ which takes L to its equivalence class $\langle \mathsf{L} \rangle_\zeta$. Finally, for $r \in [0,1]$, let $\mathbf{r}_\zeta$ denote $\{\mathsf{L} \in \mathcal{Q}_\zeta : \|L\|_\zeta = r \; \forall L \in \mathsf{L}\}$.

Remark 33. *It is obvious that $\equiv$ is an equivalence relation. Moreover, the quotient map κ is well-defined, by Corollary 26. Since $\mathbf{r}_\zeta = \langle \mathsf{M} \rangle_\zeta$ for each $\mathsf{M} \in \mathbf{r}_\zeta$, this implies by Remark 12 that $\mathbf{r}_\zeta = \mathcal{M}$ for precisely one $\mathcal{M} \in \mathcal{N}_\zeta$.*

We next equip the upper quotient space with a metric. Let the distance function $\rho : \mathcal{N}_\zeta \times \mathcal{N}_\zeta \to [0,1]$ be defined such that, if $\mathcal{L} = \mathbf{r}_\zeta$ and $\mathcal{M} = \mathbf{s}_\zeta$ for some $r,s \in [0,1]$, then $\rho(\mathcal{L},\mathcal{M}) = |r-s|$ as a metric on $\mathcal{N}_\zeta$. The collection $\mathbf{U}$ of basis sets under the induced topology equals:

$$\{\{\boldsymbol{\mathcal{L}} \subset \mathcal{N}_\zeta : \mathbf{r}_\zeta \in \boldsymbol{\mathcal{L}} \text{ if } |r-s| < \epsilon > 0\} : s \in [0,1]\}. \tag{12}$$

Remark 34. *The set $\mathbf{U}$ is apparently equivalent to the subset topology on the unit interval. To wit, there is a homeomorphism between $\mathcal{N}_\zeta$ and $[0,1]$ if the function ρ induces the quotient topology on $\mathcal{N}_\zeta$.*

We continue to abuse the notation as was done with languages and the quotient space, and write $L \in \mathbf{r}_\zeta$ or equivalently $L \in \boldsymbol{\mathcal{L}}$ to mean language L is found in points of the equivalence class $\mathbf{r}_\zeta$. We write $\mathbf{L} \subseteq \mathbf{r}_\zeta$ to mean that each language in the class $\mathbf{L}$ is in a point (not necessarily all in the same point) in the equivalence class $\mathbf{r}_\zeta$. We write $\mathbf{L} \subseteq \boldsymbol{\mathcal{L}}$ to mean that the image $\kappa(\eta(\mathbf{L}))$ is a subset of the collection of elements $\boldsymbol{\mathcal{L}} \subseteq \mathcal{N}_\zeta$. We will next show that, with exactly two exceptions, $\mathbf{r}_\zeta$ is always an uncountable subset of $\mathcal{Q}_\zeta$.

Lemma 35. *The $\equiv$ equivalence classes $\mathbf{0}_\zeta$ and $\mathbf{1}_\zeta$ are singletons in $\mathcal{Q}_\zeta$.*

Proof. The $\equiv$ class $\mathbf{0}_\zeta$ contains only the $\sim$ class $\eta(\emptyset)$, since $\|L\| = d(L,\emptyset)$. Thus, $L \in \mathbf{0}_\zeta$ implies $d(L,\emptyset) = 0$, which implies $L \sim \emptyset$.

On the other hand, suppose languages L and M and points L and M are such that $L \in \mathsf{L}$ and $M \in \mathsf{M}$ and $\mathsf{L},\mathsf{M} \in \mathbf{1}_\zeta$. By Remark 12, $\|\neg L\| = \|\neg M\| = 0$, which we have just seen means $\neg L \sim \neg M$. But since $L\backslash M = \neg M \backslash \neg L$, it is true that $L \triangle M = \neg L \triangle \neg M$. Therefore $d(L,M) = \|L \triangle M\| = \|\neg L \triangle \neg M\| = d(\neg L, \neg M) = 0$. Hence, $L \sim M$. Since L, M were arbitrary, it follows that $\mathsf{L} = \mathsf{M}$ and that $\mathbf{1}_\zeta$ contains just a single point, viz. the equivalence class $\eta(\Sigma^*)$. □

Since $\mathbf{1}$ is a singleton, given a point L there is exactly one point in $\mathcal{Q}_\zeta$ at distance 1. If $\mathsf{L},\mathsf{M} \in \mathcal{Q}_\zeta$ and $\mathbf{d}(\mathsf{L},\mathsf{M}) = 1$, then points L,M will be called *antipodes,* which we denote as $\mathsf{L} = \neg\mathsf{M}$.

Lemma 36. *Every point $\mathsf{L} \in \mathcal{Q}_\zeta$ has a unique antipode in the Besicovitch quotient space.*

Proof. From Corollary 25 this is the same as claiming that, if two languages are at distance 1 from the same language $L \in \mathsf{L}$, then they are $\sim$-equivalent. But this is a consequence of the identity

$$(L \triangle M_1) \triangle (L \triangle M_2) = M_1 \triangle M_2. \tag{13}$$

We can show this because if $d(L, M_1) = 1$ and $d(L, M_2) = 1$, it follows that $L \triangle M_1$ and $L \triangle M_2$ are in $\mathbf{1}$ (from Def. 11), implying by Lemma 35 that $d(L \triangle M_1, L \triangle M_2) = 0$, requiring $M_1 \sim M_2$. □

Corollary 37. *For $L \in 2^{\Sigma^*}$, $\|\neg L\| = 1 - \|L\|$.*

In addition we note that $L \in \mathbf{0}$ iff $\neg L \in \mathbf{1}$, and also that $\langle \neg \mathsf{L} \rangle = \langle \mathsf{L} \rangle$ if and only if $\|L\| = \frac{1}{2}$ for any language $L \in \mathsf{L}$.

For each point $\mathsf{L} \in \mathcal{Q}_\zeta$, the L-rotation of point $\mathsf{M} \in \mathcal{Q}_\zeta$, denoted $\eta_{\mathsf{L}}(\mathsf{M})$, is defined as the point $\eta(L \triangle M)$ for some language $L \in \mathsf{L}$. The L-rotation of the Besicovitch quotient space, denoted $\mathcal{Q}_\zeta^{\Sigma,\mathsf{L}}$, is then the collection $\{\eta_{\mathsf{L}}(\mathsf{M}) : \mathsf{M} \in \mathcal{Q}_\zeta\}$. The L-rotation of the $\equiv$-equivalence class $\mathbf{r}$, denoted $\mathbf{r}_{\mathsf{L}}$, is defined as the set $\{M \in \mathcal{Q}_\zeta : \mathbf{d}(\mathsf{M}, \mathsf{L}) = r\}$. The L-rotation of the upper Besicovitch quotient space, meaning the collection $\{\mathbf{r}_{\mathsf{L}} : r \in [0, 1]\}$, will be denoted $\mathcal{N}_{\zeta,\mathsf{L}}$.

Lemma 38. *$\mathcal{Q}_\zeta^{\Sigma,\mathsf{L}}$ is equivalent as a set to $\mathcal{Q}_\zeta$, and L-rotation is a bijection of the quotient space onto itself. Moreover, $\mathcal{N}_{\zeta,\mathsf{L}}$ is a bijection with $\mathcal{N}_\zeta$.*

There are uncountably many $\equiv$ equivalence classes, because the norm $\|\cdot\|$ is surjective onto the unit interval. In addition, we now proceed to show that no open set in $\mathcal{Q}_\zeta$ is contained in a single $\equiv$ equivalence class. This is the essential condition for the proof that ρ is the quotient of $\mathbf{d}$. We begin with a straightforward proposition.

Proposition 39. *For $L \in 2^{\Sigma^*}$ where $\|L\| = r$ and $0 \leq s \leq r (\leq 1)$, there exists language $M \subset L$ such that $\|M\| = s$.*

Proof. For $s = 0$, let $M = \emptyset$, Q.E.D. For $s = r$, let $M = L$, Q.E.D. Now assume $s \in (0, r)$. Note that $s/r > 0$; form language sequence $\mathbf{L} = \{L \cap \Sigma^i\}_{i \in \mathbb{N}}$, and using this define the integer sequence $\{m_i\}_{i \in \mathbb{N}}$ such that

$$m_i = \lfloor (s/r) |(L \cap \Sigma^i)| \rfloor. \qquad (14)$$

There exists a language sequence $\{M_i\}_{i \in \mathbb{N}}$ such that $M_i \subseteq L \cap \Sigma^i$ and $|M_i| = m_i$. Then we can calculate that $0 \leq (s/r)|(L \cap \Sigma^{<k})| = |(M \cap \Sigma^{<k})| < k$, so $\|M\| = (s/r)\|L\| = s$, and $M \subseteq L$; Q.E.D. □

Remark 40. *The above result can be reversed, in that if $0 \leq r \leq s \leq 1$, then for any language $L \in \mathbf{r}$ there exists language $M \supseteq L$ such that $M \in \mathbf{s}$. The target language is L in case $0 = r = s$, Σ^* in case $s = 1$, and in case $s \in (0, 1)$ may be constructed as in Proposition 39 by inverting the fractions in (14) et seq.*

Lemma 41. *No open set in $\mathcal{Q}_\zeta$ is a subset of a $\equiv$ equivalence class.*

Proof. Since $\mathcal{Q}_\zeta$ is perfect, this follows for the classes $\mathbf{0}_\zeta$ and $\mathbf{1}_\zeta$ directly from Lemmas 35 and 28. Otherwise, suppose $L \in 2^{\Sigma^*}$ and $\mathsf{L} \in \mathcal{Q}_\zeta$ such that $L \in \mathsf{L} \in \mathbf{r}$. For any open set $\mathbf{L} \subset \mathcal{Q}_\zeta$ containing L, there is a number $\epsilon' > 0$ such that $\mathbf{d}(\mathsf{L}, \mathsf{M}) < \epsilon'$ implies $\mathsf{M} \in \mathbf{L}$.

It is sufficient to exhibit a language $M \in \mathsf{M}$ such that $\|M\| \neq \|L\|$ and $d(L, M) < \epsilon'$. Let $\epsilon = \min\{r/2, \epsilon'/2\}$. Note that $\epsilon' > \epsilon > 0$. Our selection of ϵ guarantees the following: $0 < \epsilon < r \leq 1$, which implies that

$$0 < r - \epsilon < r \qquad (15)$$

Then by Proposition 39 there is a language $M \subset L$ such that $\|M\| = r - \epsilon$. But since $r - \epsilon < r$, $\|M\| \neq \|L\|$. It also follows that $d(L, M) = \|L \triangle M\| = \|L \backslash M\|$. However, $\|L\| = \|M\| + \|L \backslash M\|$ from Corollary 14. Thus $d(L, M) = \|L\| - \|M\| = r - (r - \epsilon) = \epsilon$. □

Corollary 42. *If $\mathbf{L} \in \tau_\zeta$ is an open set in the Besicovitch topological space and language $L \in \mathbf{L}$, then there exists $\epsilon > 0$ such that for every real number*

$$r \in (\|L\| - \epsilon, \|L\| + \epsilon) \cap [0, 1]$$

there exists language $M \in \mathbf{L}$ such that $M \in \mathbf{r}$.

This corollary states that, under the Besicovitch topology, representatives of some continuous interval of norm values can be found in every open set in the language space. This means that, as was claimed in Remark 12(1), the language norm $\|\cdot\|_\zeta$ is a continuous map from 2^{Σ^*} onto $[0, 1]$.

Theorem 43. *The upper quotient space $\mathcal{N}_\zeta$ is homeomorphic to (and so essentially is) the unit interval $[0, 1]$.*

Ideals simplify exploration of the elements of the upper quotient space

Earlier we defined the (word) ideals of Σ^*. To elaborate on this, recall the earlier discussion of

r-simple languages (v. Def. 16), and consider the monoid ideals of Σ^*.

Lemma 44. *If real number $r \in [0,1]$, there exists a right ideal of Σ^* in $\mathbf{L}_r$.*

Proof. If $r = 1$ then $w = \lambda$ trivially satisfies the lemma. So we assume $r \in (0,1)$. Since by Def. 16 $0 \leq \check{r}_1 < \alpha$, there is a subset I_1 of Σ (actually, at least α subsets) such that $|I_1| = \check{r}_1$. Note from the definition of $\mathbf{L}_r$ that $\check{r}_k \leq r\alpha^k < \check{r}_k + 1$ for all $k \in \mathbb{N}$. Multiplying through by α gives the inequality

$$\check{r}_k\alpha \leq r\alpha^{k+1} < \check{r}_k\alpha + \alpha. \tag{16}$$

But for $k+1$ we have

$$\check{r}_{k+1} = \lfloor r\alpha^{k+1} \rfloor \leq r\alpha^{k+1} < \check{r}_{k+1} + 1. \tag{17}$$

Since all values are non-negative integers we can combine the preceding two equations to yield

$$\check{r}_k\alpha \leq \check{r}_{k+1} < \check{r}_k\alpha + \alpha. \tag{18}$$

It follows that $\check{r}_{k+1} = \check{r}_k\alpha + t_k$ for some $t_k \in \mathbb{N}$ such that $0 \leq t_k < \alpha$. Therefore for all $k \in \mathbb{N}$, $\check{r}_k\alpha \leq \alpha^{k+1} - \alpha$.

Thus there exists language $T_1 \subseteq \Sigma^2 \backslash I_1\Sigma$ such that $|T_1| = t_1$, so that $|I_1\Sigma \cup T_1| = \check{r}_2$. Set $I_2 = I_1\Sigma \cup T_1$. Continuing in this fashion, let T_k for each $k \in \mathbb{N}$ be a language such that $T_k \subseteq \Sigma^{k+1}\backslash I_k\Sigma$ and $|T_k| = t_k$. Finally, for $k \in \mathbb{N}$ define language $I \in 2^{\Sigma^*}$ such that $I \cap \Sigma^k = I_k$, which is to say let $I = \bigcup_{i\in\mathbb{N}} I_i$. Then by construction, $I \in \mathbf{L}_r$, and $w\Sigma^j \subseteq I$ for all $w \in I$ and every $j \in \mathbb{N}$. Thus $I\Sigma^* \subseteq I$. □

The preceding result provides further evidence that right ideals are ubiquitous in the Besicovitch topological space. We now develop our understanding of the ideals to comprehend the elements of the upper quotient space. We begin by extending the notion of "sections of a word ideal."

Definition 45. *An n-word ideal in the monoid Σ^* is a language J_F such that*

$$J_F = \Sigma^* w_1 \Sigma^* w_2 \ldots \Sigma^* w_n \Sigma^*$$

for some finite language $F = \{w_1, w_2, \ldots, w_n\}$ over Σ^*. Then $f_F = \sum_{i=1}^{n}|w_i|$ is the length of F. If $\mathbf{v} = (v_1, \ldots, v_n)$ is a vector over $\mathbb{N}^{1\times n}$, then the $\mathbf{v}$-section of J_F is denoted $J_{F,\mathbf{v}}$ and is the right ideal defined as:

$$J_{F,\mathbf{v}} = \Sigma^{v_1} w_1 \Sigma^{v_2} w_2 \Sigma^{v_3} \cdots \Sigma^{v_n} w_n \Sigma^*.$$

Lemma 46. *For every vector $\mathbf{v}$ over $\mathbb{N}^{1\times n}$ and every language F such that $|F| = n$, $\|J_{F,\mathbf{v}}\| = \alpha^{-f_F}$.*

Proof. Let $v_1 + v_2 + \ldots + v_n = S$. Then when $k \geq f_F + S$, $|J_{F,\mathbf{v}} \cap \Sigma^k| = \alpha^{k-f_F}$. Therefore,

$$\begin{aligned}
&\lim_{k\to\infty} \frac{\sum_{i=0}^{k-1} \alpha^{i-f_F} - |J_{F,\mathbf{v}} \cap \Sigma^{<k}|}{\sum_{i=0}^{k-1} \alpha^i} \\
&= \lim_{k\to\infty} \frac{\sum_{i=0}^{f_F+S-1} \alpha^{i-f_F}}{\sum_{i=0}^{k-1} \alpha^i} \\
&= \lim_{k\to\infty} \frac{\frac{\alpha^S-1}{\alpha-1}}{\frac{\alpha^k-1}{\alpha-1}} = 0,
\end{aligned}$$

which implies that $\|J_{F,\mathbf{v}}\| = \alpha^{-f_F}$. □

In addition to the above result, it is possible to extend the proofs of Lemmas 44 and 30 to the n-word ideals by induction. Taken together, these results tell us that points in the upper quotient space contain languages that "closely resemble" unions of sections of ideals of Σ^*, in the following sense: cardinality of sections of these languages (as word length increases) must approximate the cardinality of the unions of (sections of) ideals.

We conclude this section by showing that all $\equiv$ classes except $\mathbf{0}_\zeta$ and $\mathbf{1}_\zeta$ are uncountable.

Lemma 47. *For any real number $r \in (0,1)$, the element $\mathbf{r} \in \mathcal{N}_\zeta$ is uncountable.*

Proof. From Lemma 44, there is an r-simple language L. In fact, there exist at least two r-simple languages, since for each $r \in (0,1)$,

$$\begin{aligned}
0 \leq |L \cap \Sigma^k| &< \left[r + \frac{1-r}{2}\right]\alpha^k \\
&= \left(\frac{r+1}{2}\right)\alpha^k < r\alpha^k.
\end{aligned}$$

This means that for $k \in \mathbb{N}$ there exists a subset of $\Sigma^k \backslash L = \neg(L \cap \Sigma^k)$ consisting of the lesser of either $\lfloor r\alpha^k \rfloor$ or $\lfloor \left(\frac{1-r}{2}\right)\alpha^k \rfloor$ words, and there exists a subset of $L \cap \Sigma^k$ consisting of the same number of words. This means there exists an r-simple language at distance $s = \min\{2r, 1-r\}$ from L. We now construct this language in the following way: let $t_k = \min\left\{\lfloor r\alpha^k \rfloor, \left\lfloor \left(\frac{1-r}{2}\right)\alpha^k \right\rfloor\right\}$; let T_k be a language such that $|T_k| = t_k$ and $T_k \subseteq \neg(L \cap \Sigma^k)$, which is possible since $|\neg(L \cap \Sigma^k)| \geq 2t_k$; and let $F_k \subseteq L$ be such that $|F_k| = t_k$, which is possible since $t_k \leq |L \cap \Sigma^k|$. Let $T = \bigcup_{i\in\mathbb{N}} T_i$, $F = \bigcup_{i\in\mathbb{N}} F_i$, and let $N = L \backslash F$. Then language $L' \stackrel{\text{def}}{=} N \cup T$ is the language formed by exchanging t_k words in L for t_k words in $\neg L$. Thus the number of words in $L \triangle^k L'$ is $2t_k = s\alpha^k$ for each $k \in \mathbb{N}$. Hence, $d(L, L') = s$ and, since L and L' contain the same number of words of each length, they have the same norm. Since L is r-simple, so is L'.

For all $t \in \mathbb{R}$ such that $0 \leq t \leq s$, since $s \leq r$ there exists language $F' \subseteq F \subseteq L$ such that $\|F'\| = t/2$, and there exists language $T' \subseteq T = L' \cap \neg L$ such that $\|T'\| = t/2$ (by Proposition 39). Then it can be shown that if language $L_t \stackrel{\text{def}}{=} (L\backslash F') \cup T'$ is such that

$$L_t = (L\backslash F) \cup (F\backslash F') \cup T' = N \cup (F\backslash F') \cup T',$$

so $L_t \triangle L' = (T\backslash T') \cup (F\backslash F')$. Thus $d(L_t, L') = s - t$. □

3.4 The Chomsky hierarchy

In this final section we show a few results which relate our Besicovitch topologies to the classical language classes.

The finite and locally testable languages are not dense

A major inadequacy of the Cantor topology was the density of the finite languages. By contrast, these are confined to a single $\sim$-equivalence class in the Besicovitch topology.

Lemma 48. *The finite languages are all in* $\mathbf{0}_\varsigma$.

Proof. If language L is finite, there exists $N \in \mathbb{N}$ such that $n > N$ implies $L \cap \Sigma^n = \emptyset$, and hence also that $|L \triangle^n \emptyset| = 0$. □

This naturally leads to the question, addressed presently, what happens if the description of an infinite language is entirely finitary?

We first remind the reader that a language L is *locally testable* just in case there is a fixed integer k (called a window length) and a proper subset $F \subsetneq \Sigma^k$ such that, if every factor of word w of length k is in F then $w \in L$. The important thing about the locally testable family is that the membership question "Is $w \in L$?" is decidable by inspecting subsequent k-length factors of w. We next define a larger class of "generally testable" languages with the property that every locally testable language is a subset of some generally testable language.

Definition 49. *A language L is* generally testable *if there exists a window length $n \in \mathbb{N}$ and a set of permitted factors $S \subseteq \Sigma^n$, where $L = S^*\Sigma^{<n}$.*

From this definition we see that word $w \in L$ if and only if $w \in \Sigma^{<n}$ or w can be written $u_1u_2\cdots u_tv$, where $u_i \in S$ for all $i \in \mathbb{N}_t$ and $v \in \Sigma^{<n}$. It is interesting that the size of a generally testable language is not really limited, but yet we have the following result.

Lemma 50. *Every generally testable language in* 2^{Σ^*} *is in* $\mathbf{0}_\varsigma$ *with the exception of* Σ^*, *which is in* $\mathbf{1}_\varsigma$.

Proof. (Outline) Let the permitted factors of a word in L be $S \subseteq \Sigma^n$. If $|S| = s$, suppose $s = \alpha^n$. But then $S = \Sigma^n$, $L = \Sigma^*$, and therefore $L \in \mathbf{1}_\varsigma$.

On the other hand, if $s < \alpha^n$, and word $w \in L$, there exist unique non-negative integers q and r, such that $|w| = nq + r$ and $0 \leq r < n$, and words $u_1, u_2, \ldots, u_q$ in S, and word v in Σ^r such that $w = u_1u_2\ldots u_qv$We deduce that $|L \cap \Sigma^{|w|}| = s^q\alpha^r$.

We can therefore easily see that the proportion of the number of words $L^{<i}$ to those in $\Sigma^{<k}$ is maximized at word lengths where $q = n-1$, i.e., where $i = nk + n - 1$. We conclude the following:

$$\|L\| \leq \limsup_{k\to\infty} \frac{\sum i = 0^k s^i |(|\Sigma^{<n}}{|\Sigma^{<kn+n}|}. \tag{19}$$

By our assumption, $s \leq \alpha^n - 1$. Straightforward calculation shows the right side of the above equation tends to zero, because it is bound above by

$$\limsup_{k\to\infty} \frac{1}{2}\frac{(\alpha^n-1)^k}{(\alpha^n)^k}.$$

Thus, $L \in \mathbf{0}_\varsigma$. □

Corollary 51. *Every locally testable language belongs to $\mathbf{0}_\zeta$.*

Proof. Suppose L is a locally testable language over Σ with window length n and permitted factors $S \subsetneq \Sigma^n$. Consider the generally testable language L' with the same window length and the same permitted factors as locally testable language L. Then $L \subseteq L'$ and, by the properties of a language norm, $\|L\| \leq \|L'\|$; meanwhile $\|L'\| = 0$ from the preceding lemma. □

Regular languages are dense in the upper quotient space

We have now seen that all finite and locally testable languages belong to $\mathbf{0}_\zeta$. On the other hand:

Lemma 52. *Regular languages are dense in the upper quotient space $\mathcal{N}_\zeta$.*

Proof. Let $r \in [0, 1]$. The claim is that for all $\epsilon > 0$ there exists a regular language L such that $|\|L\| - r| < \epsilon$. If $\epsilon \geq \min\{r, 1-r\}$, either $\emptyset$ or Σ^* satisfies the claim, Q.E.D. So we assume that $\epsilon < \min\{r, 1-r\}$. Then $r < r + \epsilon < 1$. Let integers n and q be such that $r < q\alpha^{-n} \leq r + \epsilon \leq (q+1)\alpha^{-n}$, and $0 < q < \alpha^n$. From this we have

$$0 < q\alpha^{-n} - r < \epsilon. \tag{20}$$

Let language $S_\epsilon \subseteq \Sigma^n$ have cardinality q. Consider the right ideal $S_\epsilon\Sigma^*$, which is a disjoint union of the q right word ideals $w\Sigma^*$ with $w \in S_\epsilon$. Note that each of these is a 1-word ideal section $J_{F,\mathbf{v}}$, where $F = \{w\}$ for $w \in S_\epsilon$ and $\mathbf{v} = (0)$. Therefore by Lemmas 13 and 46,

$$\begin{aligned}\|S_\epsilon\Sigma^*\| &= \sum_{w \in S_\epsilon} \|w\Sigma^*\| \\ &= q\alpha^{-n}\end{aligned}$$

From (20) this means that $|\|S_\epsilon\Sigma^*\| - r| < \epsilon$ as required. Finally, by the Myhill-Nerode Theorem (Nerode, 1958) $S_\epsilon\Sigma^*$ is a regular language, since all but finitely many words in $S_\epsilon\Sigma^*$ can be followed by Σ^*. □

This means that the linear, context-free, context-sensitive, and recursively enumerable languages are all dense in the upper quotient space. We still do not know where all these families lie in the lower Besicovitch topological spaces, but we conjecture that the regular languages are indeed also dense in the Besicovitch topology $(2^{\Sigma^*}, \tau_\zeta)$.

Non-r.e. languages are dense in both quotient spaces

We can show fairly simply that the non-recursively enumerable languages are ubiquitous in the Besicovitch topological spaces. Because d_ζ is a strict pseudo-metric, the $\sim$ equivalence classes are uncountable. We present the following without their (uncomplicated) proofs due to space limitations.

Lemma 53. *The single element of the class $\mathbf{0}_\zeta$ is uncountable in 2^{Σ^*} and contains a non-r.e. language.*

Corollary 54. *Every $\sim$ equivalence class contains a non-r.e. language.*

4 Conclusion

We have attempted to improve upon previous definitions of *distance* between languages in a language space. After considering previous work by Vianu (1977) which defined a language distance using the density of their symmetric set difference, we progressed to a new adaptation of a pseudometric inspired by Besicovitch (1932). In a language space, the Besicovitch pseudometric was developed which is essentially the upper density of the set-difference between languages. By lifting to the quotient space $\mathcal{Q}_\zeta$ using Besicovitch equivalence, a natural metric topology was developed and shown to be perfect but not compact. Another step of lifting brought us a compact "upper" quotient space $\mathcal{N}_\zeta$ homeomorphic to the unit interval. The ideals of this upper space were studied, also invoking the notion of *word ideal* defined herein. In the last section it was shown that neither the finite nor locally testable languages are dense in $\mathcal{N}_\zeta$. Finally, the regular languages were shown to be dense in $\mathcal{N}_\zeta$, and the non-r.e. languages were shown to be dense in both $\mathcal{Q}_\zeta$ and $\mathcal{N}_\zeta$.

References

J. Berstel. 1973. Sur la densité asymptotique de langages formels. In *International Colloquium on Automata, Languages and Programming (ICALP, 1972)*, pages 345–358. North-Holland.

A. S. Besicovitch. 1932. *Almost Periodic Functions*. The University Press.

V. G. Bodnarchŭk. 1965. The metrical space of events, part I. *Kibernetika*, 1(1):24–27.

C. S. Calude, H. Jürgensen, and L. Staiger. 2009. Topology on words. *Theoretical Computer Science*, 410:2323–2335.

G. Cattaneo, E. Formenti, L. Margara, and J. Mazoyer. 1997. A shift-invariant metric on $s^{\mathbb{Z}}$ inducing a non-trivial topology. In I. Privara and P. Rusika, editors, *Mathematical Foundations of Computer Science 1997*, volume 1295 of *LNCS*, pages 179–188. Springer-Verlag.

A. Dincă. 1976. The metric properties on the semigroups and the languages. In A. Mazurkiewicz, editor, *Mathematical Foundations of Computer Science 1976*, volume 45 of *LNCS*, pages 260–264. Springer-Verlag.

M. Gehrke. 2009. Stone duality and the recognisable languages over an algebra. In A. Kurz, M. Lenisa, and A. Tarlecki, editors, *Algebra and coalgebra in computer science*, volume 5728 of *LNAI*, pages 236–250. Springer.

D. Genova and N. Jonoska. 2006. Topological properties of forbidding-enforcing systems. *Journal of Automata, Languages and Combinatorics*, 11(4):375–398.

J. M. Howie. 1995. *Fundamentals of Semigroup Theory*. Oxford University Press.

J. Kozik. 2005. Conditional densities of regular languages. *Electronic Notes in Theoretical Computer Science*, 14.

J. Kozik. 2006. *Decidability of relative density in Chomsky hierarchy of languages*. Ph.D. thesis, Jagiellonian University, Cracow, Poland.

S. Marcus. 1966. *Introduction mathématique à la linguistique structurale*. Dunod.

S. Marcus. 1967. *Algebraic Linguistics; Analytical Models*. Academic Press.

E. Nelson. 1980. Categorical and topological aspects of formal languages. *Mathematical Systems Theory*, 13:255–273.

A. Nerode. 1958. Linear automaton transformations. *Proceedings of the American Mathematical Society*, 9(4):541–544.

A. Salomaa and M. Soittola. 1978. *Automata-theoretic aspects of formal power series*. Springer, Berlin.

V. Vianu. 1977. The Bodnarchŭk metric space of languages and the topology of the learning space. In J. Gruska, editor, *Mathematical Foundations of Computer Science 1977*, volume 53 of *LNCS*, pages 537–542. Springer-Verlag.

H. Walter. 1975. Topologies on formal languages. *Mathematical Systems Theory*, 9:142–158.

S. Yu. 1997. Regular languages. In G. Rozenberg and A. Salomaa, editors, *Handbook of formal languages*, volume 1, pages 41–110. Springer, Berlin.

Individuation Criteria, Dot-types and Copredication: A View from Modern Type Theories*

Stergios Chatzikyriakidis
LIRMM
University of Montpellier 2
stergios.chatzikyriakidis@lirmm.fr

Zhaohui Luo
Dept. of Computer Science
Royal Holloway, Univ. of London, U.K.
zhaohui.luo@hotmail.co.uk

Abstract

In this paper we revisit the issue of copredication from the perspective of modern type theories. Specifically, we look at: a) the counting properties of dot-types, and b) the case of a complex dot-type that has remained unsolved in the literature, i.e. that of *newspaper*. As regards a), we show that the account proposed in (Luo, 2010) for dot-types makes the correct predictions as regards counting. In order to verify this, we implement the account in the Coq proof-assistant and check that the desired inferences follow. Then, we look at the case of b), the case of a dot-type which is both resource and context sensitive. We propose a further resource sensitive version of the dot-type, in effect a linear dot-type. This along with local coercions can account for the behaviour attested.

1 Copredication: Dot Types and Individuation Criteria

One of the issues that should be taken care of when giving an account of co-predication, concerns cases of coordination like the one shown below:

(1) John picked up and mastered three books

In the above sentence, the CN book is used in its physical sense (PHY) with respect to the predicate picked up, while for the predicate mastered it is rather used in its informational content sense (INFO). A theory of co-predication should be able to take care of these facts. This is true for the account by means of the dot-types proposed by (Luo, 2010; Luo, 2012b). However, besides capturing this behaviour of dot objects, there is an additional property that has to be captured. The account provided must also make the correct predictions as regards individuation and counting. Let us explain. Consider the following sentences:

(2) John picked up three books

(3) John mastered three books

(4) John picked up and mastered three books

The first example (2) is true in case John picked three distinct physical objects. Thus, it is compatible with a situation where John picked up three copies of the same book. (3) is true in case three distinct informational objects are mastered but does not impose any restrictions on whether these three informational objects should be different physical objects or not. To the contrary, (4) is only compatible with an interpretation where three distinct physical objects as well as three distinct informational objects is involved.[1]

Another issue pertaining to dot types concerns cases of what Retoré (2014) calls rigid and flexible coercions in co-predication cases. These cases in contrast to cases like *Book* where both senses can

* This work is partially supported by the research grants from Leverhulme, Royal Academy of Engineering, the CAS/SAFEA International Partnership Program for Creative Research Teams as well as the ANR ContInt Polymnie project in France.

[1] This is basically an issue of how complex objects, i.e. dot-types, are individuated and stems from the work of (Asher, 2008; Asher, 2011).

Proceedings of the 14th Meeting on the Mathematics of Language (MoL 14), pages 39–50, Chicago, USA, July 25–26, 2015. ©2015 Association for Computational Linguistics

be coordinated, involve examples where if one of the senses is used the other one cannot be used anymore:

(5) Liverpool is spread out and voted (last Sunday).

(6) # Liverpool voted and won (last Sunday).

Perhaps a better example for such cases is Pustejovsky's *newspaper* examples. The CN *newspaper* is associated with three senses: a) physical object, b) informational object and c) institution. It is a strange fact that whereas senses a) and b) can appear together in a coordinated structure, sense c) cannot appear with any of the other two (examples taken from (Antunes and Chaves, 2003)):

(7) # That newspaper is owned by a trust and is covered with coffee.

(8) # The newspaper fired the reporter and fell off the table.

(9) # John sued and ripped the newspaper.

Pustejovsky's proposal (Pustejovsky, 1995) to treat newspaper as a composite dot object does not explain the above facts. Likewise, the proposal of using (ordinary) dot-types in (Luo, 2010) has a similar problem: one would consider *newspaper* to be a subtype of the dot-type INST • (PHY • INFO), which would not disallow the above bad examples. The picture gets complicated in the light of examples like the following, in which it seems that the institutional sense can be used together with one of the two other senses in some cases:

(10) The newspaper you are reading is being sued by Mia.

As far as we know, no satisfactory account has been provided to these questions yet. In this paper, following earlier work on dot-types in MTTs (Luo, 2010; Luo, 2012b; Xue and Luo, 2012) and coordination (Chatzikyriakidis and Luo, 2012), we take up the challenge of providing an account that correctly predicts the individuation criteria in cases of co-predication while it furthermore provides a first look at capturing the behaviour of problematic cases like *newspaper*.

2 Formal Semantics in Modern Type Theories: a Brief Introduction

The term Modern Type Theories (MTTs) refers to type theories studied and developed within the tradition of Martin-Löf, which include predicative type theories such as Martin-Löf's type theory (Nordström et al., 1990) and impredicative type theories such as CIC_p as implemented in Coq (The Coq team, 2007) and UTT (Luo, 1994). Formal semantics in Modern Type Theories (MTT-semantics for short) was first studied by Ranta in his pioneering work (Ranta, 1994).[2] It has been further developed in the last several years, including the crucial employment of the theory of coercive subtyping (Luo, 1999; Luo, Soloviev and Xue, 2012) among other developments and made MTT-semantics a viable and full-blown alternative to the traditional Montagovian framework. In this paper, we use one such modern type theory, UTT with Coercive Subtyping (Luo, 1994; Luo, 1999), whose application to linguistic semantics was first discussed in (Luo, 2010).

Two features of MTTs are worth being mentioned, both important for being a foundational language for linguistic semantics. The first is that an MTT has a consistent internal logic according to the propositions-as-types principle (Curry and Feys, 1958; Howard, 1980).[3] For instance, the higher-order logic is embedded in UTT and it is essentially used when we employ UTT for linguistic semantics (just like how higher-order logic is used in Montague's semantics.)

The second feature of MTTs is that it has rich type structures, which have been recognised by many researchers as very useful in formal semantics. In this

[2] Potentially, even further back, with the work of Sundholm (Sundholm, 1986; Sundholm, 1989), but Ranta (Ranta, 1994) was the first systematic study of formal semantics in a modern type theory.

[3] Having such an internal logic is a basic requirement for a type theory to be employed for linguistic semantics and we need to be careful to keep the internal logic to be consistent when trying to extend an existing type theory to do linguistic semantics, for otherwise, we could be in a muddle situation if the basic requirement is violated. For instance, the framework of Type Theory with Records (TTR) (Cooper, 2011) is based on set theory and, as a consequence, TTR does not have such an internal logic based on the propositions-as-types principle (to see this, it suffices to note that TTR's $a : A$ is just the set-theoretical membership relation $a \in A$ and undecidable).

section, we shall briefly discuss some of these distinctive features of MTTs, specifically the ones most relevant to this paper.

2.1 Type Many-sortedness and CNs as Types

The domain of individuals in MTTs is multi-sorted and not single-sorted as in Church's simple type theory (Church, 1940). Instead of using one coarse-grained domain of entities, like it is done in the Montague Semantics (MS) (Montague, 1974), MTTs contain many types that allow one to make fine-grained distinctions between individuals and further use those different types to interpret subclasses of individuals. For example, one can find $John : [\![man]\!]$ and $Mary : [\![woman]\!]$, where $[\![man]\!]$ and $[\![woman]\!]$ are different types.

A further difference closely related to type many-sortedness concerns the interpretation of CNs. In MS, CNs are interpreted as predicates of type $e \to t$, whereas in MTTs CNs are interpreted as *types*. Thus, in MTTs, CNs *man, human, table* and *book* are interpreted as types $[\![man]\!]$, $[\![human]\!]$, $[\![table]\!]$ and $[\![book]\!]$, respectively. (Such types may be defined by means of type constructors such as Σ etc – see below.) Then, individuals are interpreted as being of one of the types used to interpret CNs. Such interpretations of CNs as types would not work without a proper subtyping mechanism that extends MTTs – coercive subtyping provides us with such a framework.[4]

2.2 Rich Typing

Type structures in MTTs are very rich. They can be used to represent collections of objects (or constructive sets, informally) in a model-theoretic sense, although they are syntactic entities in MTTs. Elaborating on the expressiveness of typing structures of MTTs, we briefly mention the following type structures:

- Dependent sum types (Σ-types $\Sigma(A, B)$ which have product types $A \times B$ as a special case). Σ-types have been used to interpret intersective and subsective adjectives without the need of resorting to meaning postulates. The inferences follow directly from typing (Ranta, 1994; Chatzikyriakidis and Luo, 2013). Note that subtyping is essential for the Σ-type to work (Luo, 2012b).
- Dependent product types (Π-types $\Pi(A, B)$, which have arrow-types $A \to B$ as a special case). These are basic dependent types that, together with universes (see below), provide polymorphism among other things. To give an example, verb modifying adverbs are typed by means of dependent Π-types (together with the universe CN of common nouns) (Luo, 2012b; Chatzikyriakidis, 2014).
- Disjoint union types ($A + B$). Disjoint union types have been proposed to give interpretations of privative adjectives (Chatzikyriakidis and Luo, 2013).
- Universes. These are types of types, basically collections of types. Typical examples of universes in MTT-semantics include, among others, the universe $Prop$ of logical propositions as found in impredicative type theories and the universe CN of (the interpretations of) common nouns (Luo, 2012b). Further uses of universes can be seen in (Chatzikyriakidis and Luo, 2012) where the universe $LType$ of all linguistic types is used in order to deal with coordination.
- Dot-types ($A \bullet B$). These are special types introduced to study co-predication (Luo, 2012b). It is worth mentioning that coercive subtyping is essentially employed in the formulation of dot-types.[5]

2.3 MTT-semantics is Both Proof-theoretic and Model-theoretic

It has been noted recently (Luo, 2014) that one of the advantages of MTT-semantics as compared to traditional Montagovian approaches is that MTT-semantics can be seen as being both model-theoretic and proof-theoretic. NL semantics can first be represented in an MTT in a model-theoretic way and then

[4] See (Luo, 1999; Luo, Soloviev and Xue, 2012) for the formal details of coercive subtyping. Also see (Luo, 2012a) and the next section for further argumentation on interpreting CNs as types.

[5] See (Bassac et al., 2010) for another proposal of using coercions to deal with co-predication.

these semantic representations can be understood inferentially in a proof-theoretic way (Luo, 2014).

In particular, since MTTs are proof-theoretically specified, it is not surprising that many proof assistants implement MTTs. Perhaps, the most advanced of these proof-assistants is the Coq proof-assistant (The Coq team, 2007). Coq is a state-of-the-art proof assistant that has produced a number of impressive results. Some of these include a complete mechanized proof of the four colour theorem (Gonthier, 2005), the odd order theorem (Gonthier et al., 2013) as well as CompCert, a formally verified compiler for C (Leroy, 2013). Because Coq has a powerful reasoning ability and that it implements an MTT, a new avenue of research is opened up – to use Coq as an NL reasoner. This has been attempted in (Chatzikyriakidis and Luo, 2014a; Chatzikyriakidis and Luo, 2014b) with a number of promising results as regards NL inference. In this paper, we also exemplify the way proof-assistants can be used to help in checking the inferences that semantic accounts give rise to.

3 CNs as Types and Individuation Criteria

As already discussed in our introduction to MTTs, CNs are interpreted as types in MTTs. This proposal has also some nice consequences concerning what Geach (1962) has called the criterion of identity, which is pretty much the individuation criterion that we have been referring to in this paper. Intuitively, a CN determines a concept that beside having a criterion of application to be employed to determine whether the concept applies to an object, it further involves a criterion of identity, to be employed to determine whether two objects of the concept are the same. It has been argued that CNs are distinctive in this as other lexical terms like verbs and adjectives do not have such criteria of identity (cf. the arguments in (Baker, 2003)). There seems to be a close link between the constructive notion of a set (Type) and criteria of identity/individuation. This is because, in constructive mathematics, a set is a 'preset', which involves its application criterion, together with an equality, which further gives its criterion of identity determining whether two objects of the set are the same (Bishop, 1967; Beeson, 1985). Modern type theories such as Martin-Löf's type theory (Martin-Löf, 1975; Martin-Löf, 1984) were originally developed for the formalisation of constructive mathematics, where each type is associated with such an equality or criterion of identity. The identification of CNs as types thus provides CNs their criteria of application and identity. We cannot go into the details of how this is to be achieved formally. but the interested reader is directed to (Luo, 2012a) for a detailed exposition of the CNs as Types idea.

In order to proceed, firstly we have to discuss the existing account of dot-types as this was given by (Luo, 2010; Luo, 2012b; Xue and Luo, 2012). Specifically, we have to see whether this account predicts the counting criteria correctly in examples like (4) repeated below:

(11) John picked up and mastered three books

As we have said, the only possible interpretation of (11) we receive is one where three distinct physical as well as informational objects are involved. The sentences cannot be interpreted as involving three distinct informational objects but one physical object or vice versa as involving three distinct physical objects but one informational object. The question is whether this account captures that. First of all, let us say something about coordination, since this would be needed in discussing the examples in a compositional manner. The approach we suggest for coordination, based on earlier work in (Chatzikyriakidis and Luo, 2012) involves a type universe of linguistic types, $LType$:[6]

(12) $\Pi A : LType.\ A \to A \to A$

As regards typing the above is a natural way to encode coordination. However, we need a way to further encode the semantics of coordination in each case. For this paper, we show this for VP coordination only. In order to define VP coordination, we first define an auxiliary object AND:

(13) $AND : \Pi A{:}LType.\ \Pi x, y{:}A.\ \Sigma a{:}A.\ \forall p{:}A \to Prop.\ p(a) \supset p(x) \wedge p(y).$

The auxiliary entities read as follows: for any type A in $LType$ and forall $x, y{:}A$, $AND(A, (x, y))$ is a

[6]See (Chatzikyriakidis and Luo, 2012) for more details on the universe $LType$, its motivation as well as (some of) its introduction rules.

pair (a, f) such that forall $p{:}A \rightarrow Prop$, $f(p)$ is a proof that $p(a)$ implies both $p(x)$ and $p(y)$. Then, *and* is defined as the first projection $\pi 1$ of the auxiliary object:

(14) $and = \lambda A{:}LType.\lambda x, y, z{:}A.\pi_1(AND(A, x, y))$

With these in mind, let us now look at the existing proposal as regards dot-types and its proper formalization as this was provided in (Luo, 2010). The whole idea of forming a dot-type is informally based on the fact that to form a dot-type $A \bullet B$, its constituent types A and B should not share common parts/components. For example, the following two cases cannot be dot-types since they both share components:

(15) PHY • PHY

(16) PHY • (PHY • INFO)

Definition 3.1 (components) *Let T : $Type$ be a type in the empty context. Then, $\mathcal{C}(T)$, the set of components of T, is defined as:*

$$\mathcal{C}(T) =_{df} \begin{cases} \text{SUP}(T) & \text{if the NF of } T \text{ is not } X \bullet Y \\ \mathcal{C}(T_1) \cup \mathcal{C}(T_2) & \text{if the NF of } T \text{ is } T_1 \bullet T_2 \end{cases}$$

where $\text{SUP}(T) = \{T' \mid T \leq T'\}$.

The rules for the dot-types are given in Figure 1, as given in (Luo, 2012b). The notion of dot-type captures copredication in a nice way: it is both formal and suitable for MTT-semantics. The question is whether this account gives us correct individuation criteria. In order to test this, we check it against the Coq proof-assistant (The Coq team, 2007), based on the formal development as considered in (Luo, 2011). In effect, we define in Coq the dot-type PHY • INFO and define $Book$ to be the Σ-type that encodes Pustejovksy's qualia structure; as a consequence, $Book$ is a subtype of PHY • INFO. We further define *mastered* and *picked up* to be of type INFO $\rightarrow Prop$ and PHY $\rightarrow Prop$, respectively, and further provide a tactic to enhance automation, the details of which are out of the scope of this paper. Lastly, the quantifier $three$ is defined.[7]

[7] $Three$ is defined as follows: forall A of type CN and given a predicate $P{:}A \rightarrow Prop$, there exist three elements, x,y and z, that are different, which are true of P.

```
Load LibTactics.
Definition CN:=Set.
Parameter Man Human:CN.
Parameter John:Man.
Axiom mh:Man->Human.
Coercion mh:Man>->Human.
(* Phy dot Info *)
Parameter Phy Info : CN.
Record PhyInfo:CN:=mkPhyInfo{phy:>
Phy;info:>Info}.
(*Book as Sigma-type with PhyInfo &
BookQualia*)
Parameter Hold:Phy->Info->Prop.
Parameter R:PhyInfo->Prop.
Parameter W:Human->PhyInfo->Prop.
Record BookQualia (A:PhyInfo):Set:=
  mkBookQualia {Formal:Hold A A;
                Telic:R A;
                Agent:exists
                h:Human, W h A }.
Record Book:Set:=mkBook{Arg:>
PhyInfo;Qualia:BookQualia Arg}.
Ltac AUTO:=cbv delta;intuition;try
repeat congruence;jauto;intuition.
Parameter mastered:Human->Info->Prop.
Parameter picked_up:Human->Phy->Prop.
Unset Implicit Arguments.
Parameter AND: forall A:Type, forall
x y:A, sigT(fun a:A=>forall p:A->
Prop,p(a)->p(x) /\p(y)).
Definition and:= fun A:Type, fun x
y:A=>projT1(AND A x y).
Definition Three:=fun(A:CN)(P:A->
Prop)=>exists x:A,P x/\(exists y:A,
P y/\(exists z:A,P z/\x<>y/\y<>z/\
x<>z)).
```

With these in line, let us see whether the correct predictions are being made with respect to individuation criteria. What we need to capture is the following entailment:

(17) John picked up and mastered three books ⇒ John picked up three books and John mastered three books

Basically, what we need to be able to get is a situation where three distinct informational as well as

Formation Rule

$$\frac{\Gamma\ valid \quad \langle\rangle \vdash A : Type \quad \langle\rangle \vdash B : Type \quad \mathcal{C}(A) \cap \mathcal{C}(B) = \emptyset}{\Gamma \vdash A \bullet B : Type}$$

Introduction Rule

$$\frac{\Gamma \vdash a : A \quad \Gamma \vdash b : B \quad \Gamma \vdash A \bullet B : Type}{\Gamma \vdash \langle a, b\rangle : A \bullet B}$$

Elimination Rules

$$\frac{\Gamma \vdash c : A \bullet B}{\Gamma \vdash p_1(c) : A} \qquad \frac{\Gamma \vdash c : A \bullet B}{\Gamma \vdash p_2(c) : B}$$

Computation Rules

$$\frac{\Gamma \vdash a : A \quad \Gamma \vdash b : B \quad \Gamma \vdash A \bullet B : Type}{\Gamma \vdash p_1(\langle a, b\rangle) = a : A} \qquad \frac{\Gamma \vdash a : A \quad \Gamma \vdash b : B \quad \Gamma \vdash A \bullet B : Type}{\Gamma \vdash p_2(\langle a, b\rangle) = b : B}$$

Projections as Coercions

$$\frac{\Gamma \vdash A \bullet B : Type}{\Gamma \vdash A \bullet B <_{p_1} A : Type} \qquad \frac{\Gamma \vdash A \bullet B : Type}{\Gamma \vdash A \bullet B <_{p_2} B : Type}$$

Coercion Propagation

$$\frac{\Gamma \vdash A \bullet B : Type \quad \Gamma \vdash A' \bullet B' : Type \quad \Gamma \vdash A <_{c_1} A' : Type \quad \Gamma \vdash B = B' : Type}{\Gamma \vdash A \bullet B <_{d_1[c_1]} A' \bullet B' : Type}$$

where $d_1[c_1](x) = \langle c_1(p_1(x)), p_2(x)\rangle$.

$$\frac{\Gamma \vdash A \bullet B : Type \quad \Gamma \vdash A' \bullet B' : Type \quad \Gamma \vdash A = A' : Type \quad \Gamma \vdash B <_{c_2} B' : Type}{\Gamma \vdash A \bullet B <_{d_2[c_2]} A' \bullet B' : Type}$$

where $d_2[c_2](x) = \langle p_1(x), c_2(p_2(x))\rangle$.

$$\frac{\Gamma \vdash A \bullet B : Type \quad \Gamma \vdash A' \bullet B' : Type \quad \Gamma \vdash A <_{c_1} A' : Type \quad \Gamma \vdash B <_{c_2} B' : Type}{\Gamma \vdash A \bullet B <_{d[c_1,c_2]} A' \bullet B' : Type}$$

where $d[c_1, c_2](x) = \langle c_1(p_1(x)), c_2(p_2(x))\rangle$.

Figure 1: The rules for dot-types.

physical objects are involved. We formulate this as a theorem to be proven in Coq:

```
Theorem DT:(Three Book)(and(PhyInfo
->Prop)(picked_up John)(mastered
John))->(Three Book)(picked_up
John)/\(Three Book)(mastered John).
```

This can be proven in Coq.[8] Indeed, what we need with respect to examples like (22), as Gotham (2012) mentions, is an interpretation were the two objects are double distinct, both informationally and physically. Gotham (2012) shows this in discussing the account as proposed by (Asher, 2011), which provides weaker semantics for this example. In effect, Asher's (2011) account predicts situations where three informational and one physical object are involved (or vice versa) to be possible. The idea developed is roughly as follows: In every situation like (22) the hearer has to option to choose between the physical and the informational individuation criterion. If the former is chosen, then a situation where three physical objects but one informational object are involved is possible. If the hearer chooses the latter criterion, then a situation where three distinct informational objects but only one physical object is involved is possible. If this is true, one can indeed use (22) to refer to let us say one informational and three physical objects (or vice versa), then the double-distinct judgments should be the result of some pragmatic strengthening and thus should be cancelable. This is however not the case as the examples below show (taken from Gotham, 2012):

(18) John picked up and mastered three books, but he didn't pick up three books.

(19) John picked up and mastered three books; in fact, he picked up exactly one book.

(20) John picked up and mastered three books, but he didn't master three books.

(21) John picked up and mastered three books; in fact, he mastered exactly one book.

Most interestingly, what we can further prove is the entailment that from John picked up and mastered three books, it follows that John picked up three physical objects and mastered three informational objects. In Coq notation:

```
Theorem DT:(Three Book)(and(PhyInfo
->Prop)(picked_up John)(mastered
John))->(Three Phy)( picked_up John)
/\(Three Info)(mastered John).
```

This can be proven as well.[9]

It seems in this respect, that the account gives the correct predictions as regards individuation criteria and counting. This can be seen as an advantage compared to approaches like Asher's (2011), which gives the correct results after some additional assumptions on accommodation are made (which really complicate the account), while they further make it too permissive as to allow the following (see (Gotham, 2014)):

(22) # Fred picked up and mastered a stone.

On the other hand, the claim made by (Gotham, 2014) that the dot-type account as this is given by (Luo, 2010) cannot capture the facts, is shown to be incorrect on the basis of what we have presented here. Gotham's account predicts the correct results as well, but we believe at the expense of additional complications (e.g. the introduction of $R - compressible$ pluralities), that the present account does not introduce.

Thus, the account proposed for dot-types is not only formally sound but also gives the correct results with respect to counting and individuation criteria without the need of additional machinery. We take this to be a clear advantage over the other accounts. On a more general level, it seems that using the rich typing structures that MTTs have to offer, provides us with considerable advantages over problematic issues in lexical semantics.

4 The Case of *newspaper*: a Proposal for Linear Dot-types

Cases like *book* or *lunch*, being subtypes of dot-types, seem to have clear properties that are captured

[8]Those that wish to prove this on their own, the tactics to prove both of the examples are: $compute, intro, destruct$ AND, $case\ a\ with\ (ThreeBook)$, $AUTO$, $AUTO$.

[9]In order to prove this, one has to add an additional axiom in Coq that deals with equalities under subtyping. In general, when $X <_c Y$, we do not have $x \neq_X y \implies (x \neq_Y y)$ unless c is injective. For the atomic types like Book and PHY, the equality on a subtype coincides with that of the supertype and so we can axiomatically assume this. See Appendix A for the Coq code and some explanation.

with the existing formalization given for dot-types. There is however a more problematic case, famously exemplified by the word *newspaper*, which seem to require a different, more restrictive treatment. First of all, *newspaper* is associated with three rather than two senses, i.e. institution (23), informational object (24) and physical object (25) as the examples below illustrate:[10]

(23) The newspaper was sued on moral grounds.

(24) He read the newspaper.

(25) He picked up the newspaper.

Now, when it comes to the use of two different senses in the context of the same sentence, a number of strange restrictions appear. The physical object sense can be used along with the informational sense, in the same way as in the case of *book*, but the organizational sense (newspaper as an institution) cannot be used copredicatively with any of the other two senses (examples from (Antunes and Chaves, 2003)):[11]

(26) # That newspaper is owned by a trust and is covered with coffee.

(27) # The newspaper fired the reporter and fell off the table.

(28) # John sued and ripped the newspaper.

Similar words with multiple senses that further involve similar restrictions are also discussed in (Retoré, 2014). There, a multi-sorted higher order logic is used[12] and every word is associated with a kind of basic type along with a number of coercions that can coerce this basic type into additional types. So in the case of book one gets the principal lambda-term $\lambda x.const(x){:}v \rightarrow t$ where v stands for event and two optional lambda-terms, $Id{:}v \rightarrow v$ and $f_a{:}v \rightarrow a$ where a stands for type *artifact*, a subtype of physical objects. The optional terms are declared as rigid, meaning that if one of the coercions is used, the other one cannot and vice versa. For the case of dot-types like *book* the optional lambda terms are dubbed as flexible, meaning that the coercions can be used simultaneously. This is indeed an interesting account. However, the exact nature of the rigid and flexible coercions are not defined formally, and it is rather unclear how such a specification can be made. Furthermore, for cases like *newspaper*, such an approach will not work. This is because, in the case of the coercion from $f_a{:}a \rightarrow i$ (artifact to informational object), this has to be defined as both rigid and flexible at the same time. Flexible, because we want this to be possible with the physical sense, while rigid because we want this not to be possible with the organizational sense. Furthermore, the account is based on the idea that there is always a principal lambda term. For example, in the case of *Book* the physical sense is chosen. How is this sense chosen is something that it is not explained. The question of why the physical rather than the informational aspect is chosen as the principal sense is something that is left unanswered.

The data with respect to *newspaper* get further complicated. As we have seen, the organizational aspect cannot be used with any of the other two aspects. However, this is not without exceptions. There are cases this restriction seems to disappear, allowing the organizational aspect to appear with any of the two other senses:

[10] An anonymous reviewer asks whether there are other analogous cases with three-way polysemy. Words similar to newspaper also exist, e.g. *magazine, journal*. Other cases with more than two meaning are mentioned in (Retoré, 2014) but it is not clear whether they constitute examples of a similar phenomenon. For example, the case of Liverpool, mentioned in (Retoré, 2014) as having the senses *Place*, *Town* and *People*. It is not however clear whether the justification for these types is well-founded. For example the sense *Town* and *People* could be very well reduced into one sense. Unfortunately, this needs discussion that we cannot perform in this paper. However, this is an extremely important question and the range of examples that are similar to newspaper should be investigated in order to end up with a fuller classification of dot-types according to their properties.

[11] As an anonymous reviewer notes, all these examples involve a conjunction of the organizational and the physical aspect. He further asks what happens in case we have a conjunction of the informational and the organizational aspect. These cases are also infelicitous, e.g. *# He mastered and sued the newspaper* or *# That newspaper is owned by a trust and is very badly written*, so the pattern described in the paper is not violated.

[12] The meta-language for the system in (Retoré, 2014) is Girard's system F rather than the simply typed λ-calculus as in Church's simple type theory (Church, 1940) as used by Montague.

(29) The newspaper you are reading is being sued by Mia.

However, if one looks at the examples that allow this kind of constructions, it seems that they are of a specific kind. Most specifically all these cases involve some kind of modification, e.g. a relative clause as in the above example, or adjectival modification as in (30):

(30) The most provocative newspaper of the year has been sued by the government.

(31) The newspaper he just grabbed from the newsstand is doing well in the stock market.

The pattern seems to be the following: the organizational aspect cannot be used with any of the other two aspects, unless one aspect is taking part in a modified CN construction. In case this happens the organizational aspect can be used along the other aspects. The account as proposed in (Pustejovsky, 1995) for *newspaper* cannot deal with these phenomenon and as far as we know, no formal account has been proposed for these cases. This is what we want to discuss here. The original account of dot-types in (Luo, 2010) among others will face similar problems. The dot-type $\text{INST} \bullet (\text{PHY} \bullet \text{INFO})$ will suffer the problem of predicting examples (23)-(25) to be ok contrary to fact. In what follows, we discuss a solution to this extent by proposing to treat these cases by extending the dot-type system to further include resource sensitive dot-types, i.e. linear dot-types.

Linear Dot-types: a Tentative Proposal. It is clear from what we see from the data that we are dealing with a situation where the dot-type is resource sensitive, in the sense of linear logic (Girard, 1987) or Lambek calculus (Lambek, 1958). For example, in linear logic, the rules of weakening and contraction are not available and this has a number of consequences. One of them is that one is has to use an assumption exactly once. An assumption, once used, is not re-usable anymore. It seems that this idea, is quite close to what we need for the *newspaper* case.[13] We need an additional version of the dot-type, more specifically a linear version of the dot-type. This version will be closed related to the tenser product in linear logic and the usual dot-type, one of the important feature being that if one of its components has been used, the other one cannot be used any more.

Let us represent this linear dot-type as $A \ominus B$. We can further have combinations of regular and linear dot-types. In the case of newspaper what we need is the type $\text{INST} \ominus (\text{PHY} \bullet \text{INFO})$. With this type, we can take care of examples like (23) to (25) (these are also taken care of with a regular dot-type), while at the same time excluding examples (26-28) (that would be predicted to be ok with a regular dot-type).

Note that the examples like (29) can be accounted for without employing the linear version of dot-types. For instance, the semantics of (29) can be given as $sue(n)$ where $n : \Sigma(Newpaper, read)$ and $sue : Inst \to Prop$, because we have $\Sigma(Newpaper, read) < Newspaper < \text{INST} \bullet (\text{PHY} \bullet \text{INFO}) < \text{INST}$. The question of course is when do we use a linear dot-type and when a regular dot-type. In order to solve this problem, one can use local coercions, i.e. subtyping assumptions localized in terms (or judgments), as proposed in (Luo, 2010; Luo, 2012b). Local coercions have been used in (Luo, 2011) to deal with cases of homophony and in (Asher and Luo, 2012) to give semantics of linguistic coercions in sophisticated situations. Local coercions are only effective locally for some terms (expressions in type theory). They may be introduced into terms by the following rule (intuitively, the coercions declared locally are only effective in the expressions in the scope of the keyword **in**):

$$\frac{\Gamma,\ A <_c B \vdash J}{\Gamma \vdash \textbf{coercion}\ A <_c B\ \textbf{in}\ J}$$

where J is any of the following four forms of judgement:

$$k : K,\quad k = k' : K,\quad K\ kind,\ \text{ and }\ K = K'.$$

For instance, with $J \equiv k : K$, we have

$$\frac{\Gamma,\ A <_c B \vdash k : K}{\Gamma \vdash \textbf{coercion}\ A <_c B\ \textbf{in}\ k : K}$$

[13]This is based on the fact that in case the organizational aspect is used, no other aspect can be used any more within the same context. This is a kind of resource sensitivity.

In the case of *newspaper*, what we need is to consider two local coercions: $Newspaper$ < INST • (PHY • INFO) in interpreting the cases where the ordinary dot-type should be used and $Newspaper$ < INST ⊖ (PHY • INFO) in interpreting the cases where the linear dot-type should be used. For example, the following (32) will give a correct interpretation of (29):

(32) **coercion** $Newspaper$ < INST • (PHY • INFO) **in** $[\![(29)]\!]$

while the following would not be accepted:

(33) # **coercion** $Newspaper$ < INST ⊖ (PHY • INFO) **in** $[\![(26)]\!]$

We believe that this gives a satisfactory account of a problem that as far as we know has not received a treatment up to now.[14]

However, it has to be kept in mind that we have not formally treated the linear dot-type $A \ominus B$. One of the reasons for this is that, in order to do this, we need to formally study how to incorporate coercive subtyping into a resource sensitive logical system. Put in another way, one needs to study an MTT augmented with resource sensitive contextual segments and its coercive subtyping extension. We leave this as future work.

5 Conclusion

We have discussed dot-types with respect to their counting criteria and have shown that the MTT account proposed captures the fact correctly. The proof-assistant Coq was used in order to verify that the correct inferences are predicted. The account was shown not only to produce the correct results but to do so without resorting to serious extra complications of the original account (actually none is needed). Furthermore, the case of *newspaper* was discussed and a solution based on the introduction of linear dot-types combined with local coercions was provided. The issue of introducing linear dot-types in a formal way presupposes a linear version of type theory that at the moment we do not have. Thus, we leave this issue as a subject of future research. A related piece of work is that the second author has recently developed Lambek dependent types (Luo, 2015), with the motivation of studying a uniform basis for NL analysis: from automated syntactic analysis to logical reasoning in proof assistants based on MTT-semantics.

[14] Another solution that has been proposed to us by an anonymous reviewer, is the following: One basically assumes that the organizational aspect is just a different lexical entry, in effect as we understand it (even though the reviewer has not phrased it in this way) a case of homonymy. We have thought of this possibility. In case this idea is put forth within the framework discussed in this paper, one will use the account by (Luo, 2011) for cases of homonymy like *bank* where local coercions are used to either use $Bank < Institution$ or the $Bank < Riverside$ sense. In the case of newspaper, we would use either $Newspaper$ < INST or $Newspaper$ < PHY • INFO.

References

S. Antunes and R.P Chaves. On the licensing conditions of co-predication. In *Proc of the 2th Inter. Workshop on Generative Approaches to the Lexicon (GL 2007)*, 2003.

N. Asher. A type driven theory of predication with complex types. *Fundamenta Informaticae 84 (2)*, 151-183, 2012.

N. Asher. *Lexical Meaning in Context: a Web of Words*. Cambridge University Press, 2012.

N. Asher and Z. Luo. Formalisation of coercions in lexical semantics. *Sinn und Bedeutung 17, Paris*, 2012.

Mark C Baker. *Lexical categories: Verbs, nouns and adjectives*, volume 102. Cambridge University Press, 2003.

C. Bassac, B. Mery, and C. Retoré. Towards a type-theoretical account of lexical semantics. *Journal of Logic, Language and Information*, 19(2), 2010.

M.J. Beeson. *Foundations of Constructive Mathematics*. Springer-Verlag, 1985.

E. Bishop. *Foundations of Constructive Analysis*. McGraw-Hill, 1967.

S. Chatzikyriakidis. Adverbs in a modern type theory. In N. Asher and S. Soloviev, editors, *Proceedings of LACL2014. LNCS 8535*, pages 44–56, 2014.

S. Chatzikyriakidis and Z. Luo. An account of natural language coordination in type theory with coercive subtyping. In Y. Parmentier and D. Duchier, editors, *Proc. of Constraint Solving and Language Processing (CSLP12). LNCS 8114*, pages 31–51, Orleans, 2012.

S. Chatzikyriakidis and Z. Luo. Adjectives in a modern type-theoretical setting. In G. Morrill and J.M Nederhof, editors, *Proceedings of Formal Grammar 2013. LNCS 8036*, pages 159–174, 2013.

S. Chatzikyriakidis and Z. Luo. Natural language reasoning using proof-assistant technology: Rich typing and beyond. In *Proceedings of EACL2014*, 2014.

S. Chatzikyriakidis and Z. Luo. Natural Language Inference in Coq. *Journal of Logic, Language and Information*, 23(4):441–480, 2014.

A. Church. A formulation of the simple theory of types. *J. Symbolic Logic*, 5(1), 1940.

R. Cooper. Type theory and semantics in flux. In R. Kempson, T. Fernando and N. Asher, editors, *Handbook of the Philosophy of Science*. Elsevier, 2011.

The Coq Team. *The Coq Proof Assistant Reference Manual (Version 8.1), INRIA*, 2007.

H.B. Curry and R. Feys. *Combinatory Logic*, volume 1. North Holland Publishing Company, 1958.

P.T. Geach. *Reference and Generality: An examination of some Medieval and Modern Theories*. Cornell University Press, 1962.

J.-Y. Girard. Linear logic. *Theoret. Comput. Sci.*, 50, 1987.

G. Gonthier. A computer checked proof of the four colour theorem, 2005.

Georges Gonthier, Andrea Asperti, Jeremy Avigad, Yves Bertot, Cyril Cohen, François Garillot, Stéphane Le Roux, Assia Mahboubi, Russell OConnor, Sidi Ould Biha, et al. A machine-checked proof of the odd order theorem. In *Interactive Theorem Proving*, pages 163–179. Springer, 2013.

Gotham, M.: Numeric quantification in copredication. UCL Working Papers in Linguistics pp. 1–20 (2012)

M. Gotham. *Copredication, quantification and individuationy*. PhD thesis, University College London, 2014.

W. A. Howard. The formulae-as-types notion of construction. In J. Hindley and J. Seldin, editors, *To H. B. Curry: Essays on Combinatory Logic*. Academic Press, 1980.

J. Lambek. The mathematics of sentence structure. *The American Mathematical Monthly*, 65(3), 1958.

X. Leroy. The compcert c verified compiler: Documentation and users manual. http://compcert.inria.fr/man/manual.pdf, 2013.

Z. Luo. *Computation and Reasoning: A Type Theory for Computer Science*. Oxford University Press, 1994.

Z. Luo. Coercive subtyping. *Journal of Logic and Computation*, 9(1):105–130, 1999.

Z. Luo. Type-theoretical semantics with coercive subtyping. *Semantics and Linguistic Theory 20 (SALT20), Vancouver*, 2010.

Z. Luo. Contextual analysis of word meanings in type-theoretical semantics. *Logical Aspects of Computational Linguistics (LACL'2011). LNAI 6736*, 2011.

Z. Luo. Common nouns as types. In D. Bechet and A. Dikovsky, editors, *Logical Aspects of Computational Linguistics (LACL'2012). LNCS 7351*, 2012.

Z. Luo. Formal semantics in modern type theories with coercive subtyping. *Linguistics and Philosophy*, 35(6):491–513, 2012.

Z. Luo. Formal Semantics in Modern Type Theories: Is It Model-theoretic, Proof-theoretic, or Both? *Invited talk at Logical Aspects of Computational Linguistics 2014 (LACL 2014), Toulouse. LNCS 8535*, pages 177–188, 2014.

Z. Luo. A Lambek calculus with dependent types. *Types for Proofs and Programs (TYPES 2015), Tallinn*, 2015.

Z. Luo, S. Soloviev, and T. Xue. Coercive subtyping: theory and implementation. *Information and Computation*, 223:18–42, 2012.

P. Martin-Löf. An intuitionistic theory of types: predicative part. In H.Rose and J.C.Shepherdson, editors, *Logic Colloquium '73*, 1975.

P. Martin-Löf. *Intuitionistic Type Theory*. Bibliopolis, 1984.

R. Montague. *Formal Philosophy*. Yale University Press, 1974. Collected papers edited by R. Thomason.

B. Nordström, K. Petersson, and J. Smith. *Programming in Martin-Löf's Type Theory: An Introduction*. Oxford University Press, 1990.

J. Pustejovsky. *The Generative Lexicon*. MIT, 1995.

A. Ranta. *Type-Theoretical Grammar*. Oxford University Press, 1994.

C. Retore. The montagovian generative lexicon Tyn: a type theoretical framework for natural language semantics. In R. Matthes and A. Schubert, editors, *Proc of TYPES2013*, 2013.

G. Sundholm. Proof theory and meaning. In D. Gabbay and F. Guenthner, editors, *Handbook of Philosophical Logic III: Alternatives to Classical Logic*, pages 471–506. Reidel, 1986.

G. Sundholm. Constructive generalized quantifiers. *Synthese*, 79(1):1–12, 1989.

T. Xue and Z. Luo. Dot-types and their implementation. *Logical Aspects of Computational Linguistics (LACL 2012). LNCS 7351*, 2012.

A Some notes on a Coq proof

In order to prove:

```
Theorem DT:(Three Book)(and(PhyInfo
->Prop)(picked_up John)(mastered
John))->(Three Phy)( picked_up John)
/\(Three Info)(mastered John)
```

We first need to introduce the axiom in order to deal with the subtyping equality problem. We introduce this in a new local context:

```
Section Book.
Variable PHY:forall x:Book, forall
y:Book, not(x=y:>Book)-> not(x=y:>Phy).
Variable INFO:forall x:Book, forall
y:Book, not(x=y:>Book)-> not(x=y:>Info).
```

We show the proof of:

```
Theorem DT:(Three Book)(and(PhyInfo
->Prop)(picked_up John)(mastered
John))->(Three Phy)( picked_up John)
```

We use the following:

```
compute.intro.destruct AND1.case a with
(Three Book). AUTO. AUTO.destruct H0.
destruct H0.destruct H2.destruct H2.
destruct H3.destruct H3.destruct H4.
destruct H5.exists x0.AUTO.exists x1.
AUTO.
```

The interested reader can check for himself/herself for the other cases.

Lexical Semantics and Model Theory: Together at Last?

András Kornai
HAS Institute of Computer Science and Automation
11-13 Kende utca
H-1111 Budapest, Hungary
andras@kornai.com

Marcus Kracht
Bielefeld University
Postfach 10 01 31
33501 Bielefeld, Germany
marcus.kracht@uni-bielefeld.de

Abstract

We discuss the model theory of two popular approaches to lexical semantics and their relation to transcendental logic.

1 Introduction

Recent advances in formal and computational linguistics have brought forth two classes of theories, algebraic conceptual representation (ACR) and continuous vector space (CVS) models. Together with Montague grammar (MG) and its lineal descendants (Discourse Representation Theory, Dynamic Predicate Logic, etc.) we now have three broad families of semantic theories competing in the same space. MG and related theories fit well with most versions of transformational and post-transformational grammar and retain a strong presence in theoretical linguistics, but have long been abandoned in computational work as too brittle (Landsbergen, 1982). As we have argued elsewhere, MG-like theories fail not just as performance grammar but, perhaps more surprisingly, on competence grounds as well (Kornai et al., 2015). Nevertheless, MG will be our starting point, as it is familiar to virtually all linguists.

From an abstract point of view we should distinguish between a framework for compositionality and a commitment to a particular brand of semantics. While we still want to uphold the idea of compositionality, we are less enthusiastic about the dominance of standard first order models, even if suitably intensionalized, in explaining or representing meanings. Luckily, other choices can be made, though they come with a different conception of meaning. The main difference between ACR, CVS, and the standard MG treatment is in fact the choice of model structures: both ACR and CVS aim at modeling 'concepts in the head' rather than 'things in the world', and thus clash strongly with the ostensive anti-psychologism of MG. How can we make sense of such theories after Lewis (1970) without being attacked for promulgating yet another version of markerese? The answer proposed in this paper is that we divest model theory from the narrow meaning it has acquired in linguistics, as being about formulas in some first- or higher-order calculus, and interpret natural language expressions either directly in the models, the original approach of Montague (1970a), or through some convenient knowledge representation language, still composed of formulas, but without the standard logical baggage. The main novelty is that the formulas themselves will be very close to the models, though not quite like in Herbrand models for reasons that will become clear as we develop the theory.

Section 2 provides a brief justification for the enterprise, and sketches as much of ACR and CVS as we will need for Section 3, where essential properties of their models are discussed. Our focus will be on CVS, and we shall discuss the challenge of compositionality, which appears to be nontrivial for CVSs. ACR graphs are simple discrete structures, very attractive for representing meaning (indeed, they have a long history in Knowledge Representation), but more clumsy for syntax. CVS representations, finite dimensional vectors over $\mathbb{R}$, are primarily about distribution (syntactic cooccurrence), and

Proceedings of the 14th Meeting on the Mathematics of Language (MoL 14), pages 51–61,
Chicago, USA, July 25–26, 2015.

meaning, especially the linear structures that encode analogy such as *king:queen = man:woman* will arise in them chiefly as a result of probabilistic regularities (Arora et al., 2015). We take the view that CVS models 'concepts in the head' and to understand how these can be similar across speakers we need to invoke 'concepts in the world' as described by ACR. Section 4 discusses the challenge posed by changing to mentalist semantics. If meanings are in the head, we are losing, or so it appears, the objectivity of meanings. However, we think that this is not so. Instead, our working hypothesis of this paper is what we call 'One Reality': meanings describe a common reality so that anything that is true of the world must be compatible with anything else that is true. The section explores some immediate consequences of this hypothesis. We close with some speculative remarks in Section 5.

2 Out with the old, in with the new

Classical MG (Montague, 1970b; Montague, 1973) provides a translation from expressions of natural language into (higher order) predicate logic. Predicate logic itself is just a technical device, a language, to represent the actual meanings, which are thought to reside in models. Thus, already at the inception, formal semantics differentiated two kinds of "semantics": the abstract level, consisting of linguistic objects (here: expressions of simple type theory), and the concrete level, represented by a model. In what is to follow we shall investigate the effects of making two changes. One is to replace the simple type theory by radically different kinds of semantics, and the second to uphold the idea that the semantics is not just about some model, but about reality, and as such cannot be arbitrarily fixed.

Let us briefly recall how a Montague-style semantics looks like. Following (Kracht, 2011), a grammar consists of a finite signature (F, Ω) of function symbols ($\Omega : F \to \mathbb{N}$ assigns an arity to the symbols), together with an interpretation that interprets each function symbol f as an $\Omega(f)$-ary function on the space of signs, (see also Hodges 2001). Further down we shall meet only two kinds of functions: constants, where $\Omega(f) = 0$, representing the lexicon, and binary functions ($\Omega(f) = 2$), representing syntax proper.) We may take as signs either pairs (w, m), where w is a word over the alphabet and m its meaning; or we may take them as triples (w, c, m), where c is an additional component, the *category* (Kracht, 2003). In the best of all cases, the action of f on the signs is independent in each of the components. The independence of the string action from the meanings is exactly Chomsky's famous principle of the *autonomy of syntax* while the independence of the meaning action from the words is the principle of *compositionality*. If these are granted, each function symbol f then gives rise to a *pair* of functions (f^ε, f^μ), where f^ε is an $\Omega(f)$-ary function on strings and f^μ an $\Omega(f)$-ary function on meanings. Further, given any constant term t over this signature, "unfolding" (the homomorphic operation denoted by $\spadesuit$) it into a sign means

$$(f(t_1, t_2))^\spadesuit = (f^\varepsilon(t_1^\spadesuit, t_2^\spadesuit), f^\mu(t_1^\spadesuit, f_2^\spadesuit))$$

and for 0-ary f, simply $f^\spadesuit() = (f^\varepsilon(), f^\mu())$. Omitting obvious brackets this is simply $f^\spadesuit = (f^\varepsilon, f^\mu)$. A constant therefore is defined by its two components, the string (f^ε) and the meaning (f^μ).

Effectively, we can now view not only the terms as elements of an algebra (called the *term algebra*), but also the strings together with the functions f^ε $(f \in F)$, and the meanings together with the functions f^μ $(f \in F)$. Expressions and meanings thus become algebras, and there are then two homomorphisms from the algebra of terms: one to the algebra of strings and another to the algebra of meanings. Both algebras may have additional functions, of course. This will play a role in the case of CVS models which have the structure of a vector space, whose natural operations enter into the definition of the functions.

When we say 'out with the old' we will not dwell much on the inadequacies of the standard MG treatment, except to summarize some of the well known issues. Technical inadequacies, ranging from narrow issues of proposing invalid readings and missing valid ones to more far-reaching problems as provided e.g. by hyperintensionals (Pollard, 2008) are not viewed as fatal – to the contrary, these provide the impetus for further developments. A more general, systemic issue however is the chronic *lack of coverage*. The problem is not so much that the pioneering examples from *Every man loves a woman such that she loves him* to *John seeks a unicorn and Mary seeks it* could hardly be regarded examples of ordinary language as the alarming lack of progress in this regard – forty years have passed, and best of breed implementations such as CatLog (Morrill, 2011) and grammar fragments such as Jacobson (2014) still cover only a few dozen constructions. An equally deep, and perhaps even more critical, problem is the continuing disregard for *information*. No matter how we look at it, well over 80% of the information carried by sentences comes from the lexicon, with only 10-15% coming from compositional structure (Kornai, 2010). By putting lexical semantics front and center, we will address both these issues.

One of the biggest challenges for MG is the disambiguation. In the standard picture, readings correspond to parse terms. Thus, a string w has as many readings as there are parse terms t such that $t^{\varepsilon} = w$. Unfortunately, scholars in the MG tradition have spent little effort on building grammatical models of natural language that could serve as a starting point of disambiguation in the sense Montague urged, and the disambiguation in terms of parse terms is more a promissory note than an actual algorithm. This is particularly clear as we come to effects of *contextuality*, restated from Frege by Janssen (2001) as follows: 'Never ask for the meaning of a word in isolation, but only in the context of a sentence'.

In other words, what a particular word means in a sentence can be determined only by looking at the context, since the context selects a particular reading. The standard MG picture handles lexical ambiguity by invoking separate lexical entries (that is, 0-ary function symbols) for each sense a word may have, e.g. for pen_1 'writing instrument' and pen_2 'enclosed area for children or cattle'. When we say *The box is in the pen* we clearly have pen_2 in mind, and when we say *The pen is in the box* it is pen_1. Strict adherence to MG orthodoxy demands that we bite the bullet and claim that the false readings are actually wonderful to have, since a smaller playpen could really be delivered in a box, and even for a large cattle pen an artist like Christo could always come by and box up the entire thing. Yet somehow the claim rings false, both from a cognitive standpoint, since the odd readings do not even enter our mind when we hear the sentence unless we are specifically primed, and from the computational standpoint, since it is common knowledge (at least since Bar-Hillel (1960) where the box/pen example originates) that the bulk of the effort e.g. in machine translation is to disambiguate the word meanings. According to Bar-Hillel, the average English word is 3-way ambiguous, so a sentence of length 15 will require over 14 million disambiguated options. However much our computational resources have grown since 1960, and they have actually grown more than 14 million-fold, this is still unrealistic.

Another part of the theory that remained, for the past forty years, largely unspecified, is the mapping g that would *ground* elements of the mathematical model structure in reality (as opposed to the 'valuation' that is built into the model structure). For a mathematical theory, such as the theory of groups, there is no need for g as such in that there are no groups "in the world". All objects in mathematics that have group structure (e.g. the symmetries of some geometrical figure) can be built directly from sets (since a symmetry is a function, and functions are sets), so restricting attention to model structures that are sets is entirely sufficient for doing mathematics. Here we must give some thought to what we consider 'ground truth', a notion that is already problematic for proper names without referents such as *Zeus*.

The abstract structure outlined above does not require the meanings to be anything in particular. All that is required is that we come up with an algebra of meanings into which the terms can be mapped so that certain equations, the meaning postulates, come out true. Our exercise consists therefore in throwing out the old semantics and bringing in the new, here CVS and ACR, and see where this leads us. When we say 'in with the new' this is something of an exaggeration – both ACR and CVS theories go back to the late 1960s and early 1970s, and are thus as old as MG, except both suffered a long hiatus during the 'AI Winter'. Algebraic conceptual representation (ACR) begins with Quillian (1969) and Schank (1972), who put the emphasis on associations (graph edges) between concepts (graph nodes). Quillian only used one kind of (directed) edge, while Schank used several – the ensuing proliferation of link types is famously criticized in Woods (1975). For a summary of the early work see Findler (1979), for modern treatments see Sowa (2000), Banarescu et al. (2013).

Continuous vector space (CVS) representation was first developed by (Osgood et al., 1975), whose interest is also with association between concepts, which they directly measured by asking informants to rate the strength of the association on a 7-point scale. From such data, Osgood and his coworkers proceeded by data reduction via principal component analysis (PCA), obtaining vectors that were viewed as directions in semantic space. In the modern version, which has taken computational linguistics by storm in the past five years, the associations are mined from cooccurrence data in large corpora (Schütze, 1993), but data reduction by PCA or similar techniques is still a central part of establishing the mapping from the vocabulary V to $\mathbb{R}^n$.

Importantly, both ACR and CVS are essentially type free. They assume that the representation of the whole utterance is not any different from the representation of the constituents, down to the lexical entries: in ACR every meaning is a graph, and in CVS a vector. As said above, there are two types of function symbols. Those of arity 0 constitute the *conceptual dictionary*. The remaining function symbols are of arity 2. On the string side they are interpreted as concatenation, giving rise to a CFG. For ACR, the meanings of the parts are combined by ordinary substitution operations, graph rewriting and adjunction. For CVS, several combination operations have been proposed, including vector addition (Mitchell and Lapata, 2008), coordinatewise (weighted) multiplication (Dinu and Lapata, 2010), function application (Coecke et al., 2010) and substitution into recurrent neural nets (Socher et al., 2013). For a summary, see Baroni (2013). Here we will use $\otimes$ to denote any composition operation, as tensorial products have long been suggested in this area (Smolensky, 1990).

A key point is that $\otimes$ itself may be parametrized, more similar to the 'type-driven' versions of MG (Klein and Sag, 1985) than to the classic variant which has a single composition operation, function application. Berkeley Construction Grammar (CxG, see Goldberg 1995) has long urged a full theory of constructional meanings, and Kracht (2011) makes clear that languages must employ many, many simple constructions, if as above compositionality and autonomy of syntax are assumed.

3 The structure of CVS and ACR model structures

Recall that the functions f^ε and f^μ $(f \in F)$ impose an algebraic structure both on the set of exponents and the set of meanings, respectively. There may be additional structure on the meanings, which we may take advantage of. For example, if meanings are vectors, we additionally have scalar multiplication and addition, which can be used in calculations, but which also have their own semantic relevance. Indeed, it has been observed by Mikolov et al. (2013) that in an analogy $a : b = c : d$ we can calculate v_d approximately as $v_a - v_b + v_c$. Or, what is the same, we expect $v_a - v_b = v_c - v_d$.

The currently best performing Context Vector Grammar (CVG, see Socher et al. 2013) uses what looks like a single binary function $\otimes$, however it is parametrized by part of speech. CVGs work on ordered pairs $(\vec{v}, X)$ where $\vec{v}$ contributes the semantics, and X is some part of speech category (including nonterminals such as NP). In our notation $(\vec{v}, X)$ combines with $(\vec{w}, Y)$ by two square matrices L_{XY}, R_{XY} and a bias $\vec{b}_{XY}$ that depend on X and Y (but not on $\vec{v}$ or $\vec{w}$) to yield $\vec{v} \otimes \vec{w} = \tanh(L\vec{v} + R\vec{w} + \vec{b})$ (dropping the parts of speech) where the squishing function $\tanh$ is applied coordinatewise.

Since $\tanh$ is strictly monotonic, we have $x = y$ iff $\tanh(x) = \tanh(y)$, so the last step of squishing can be ignored in the kind of equational deduction that we will deal with. As an example, consider the `gram3-comparative` task. It is an accident of English that comparative is sometimes denoted by the suffix *-er* and sometimes by the prefix *more* written as a separate word. Ideally, the semantics should support equations such as

$$\vec{big} \otimes \vec{er} - \vec{nice} \otimes \vec{er} = \vec{big} - \vec{nice} \qquad (1)$$

or, equivalently,

$$\tanh(L\vec{big} + R\vec{er} + \vec{b}) - \tanh(L\vec{nice} + R\vec{er} + \vec{b}) = \vec{big} - \vec{nice}$$

In reality both the matrix and the vector coefficients are small enough for $\tanh(x) = x$ to be a reasonable approximation, so we have

$$L\vec{big} - L\vec{nice} = \vec{big} - \vec{nice} \qquad (2)$$

or, what is the same, $(L - I)(\vec{big} - \vec{nice}) = 0$ not just for *big* and *nice* but for every pair of adjective vectors $\vec{u}, \vec{v}$. This is possible only if $\langle A \rangle$, the subspace generated by the adjectives, is contained in Ker$(L - I)$. Since L does not even need to be defined outside $\langle A \rangle$, and must coincide with I within $\langle A \rangle$, the simplest assumption is $L = I$ everywhere. Now, R and b are fixed for the comparative task, so $R\vec{er} + \vec{b}$ is some constant vector $\vec{c}$ on $\langle A \rangle$, so that we finally get

$$\forall \vec{x} \in \langle A \rangle : \vec{x} \otimes \vec{er} = \vec{x} + \vec{c} \quad (3)$$

and obviously if (3) holds the analogical requirement in (1) is satisfied. The same argument can be made (with different constant $\vec{c}$) for every derivational and inflexional suffix such as the *-ly* of the `gram1-adjective-to-adverb` or the *-ing* of the `gram5-present-participle` Google task. Further, the same must hold for every case where a fixed formative is used to derive a higher constituent, such as PP[from] from a base NP and a prefix *from*, or NP from a base N and the prefix *the*. Remarkably, just as PP[from] can differ from PP[by] only by a fixed offset, the difference between the constant for *from* and that for *by*, NP[every] and NP[some] can also differ only in a fixed offset irrespective of what the base N was.

This shows how analogies can help in identifying the functions for certain derivations. However, more can be achieved. Consider the case of two synonymous expressions e and e'. Retracing their respective parses, assuming that the result vectors are the same we derive further constraints. Consider *the mayor's hat* and *the hat of the mayor* which should get the same vector assigned compositionally through two different routes. If $\vec{m}$ and $\vec{h}$ are the vectors for *mayor* and *hat*, we have some $\vec{m} + \vec{c_1}$ for *the mayor* and $\vec{h} + \vec{c_1}$ for *the hat*. If the *'s* possessive construction is defined by matrices L_1, R_1 and bias $\vec{b_1}$, and the *of*-possessive by $L_2, R_2, \vec{b_2}$, the fact that these mean the same will be expressed, again ignoring the squishing, by

$$L_1(\vec{m} + \vec{c_1}) + R_1\vec{h} + \vec{b_1} = L_2(\vec{h} + \vec{c_1}) + R_2(\vec{m} + \vec{c_1}) + \vec{b_2} \quad (4)$$

By collecting like terms together, this means

$$(L_1 - R_2)\vec{m} + (R_1 - L_2)\vec{h} + \vec{c_4} = \vec{0} \quad (5)$$

for some constant $\vec{c_4}$ and for all noun vectors $\vec{m}, \vec{h}$. This of course requires $L_1 = R_2, L_2 = R_1$ and $\vec{c_4} = 0$, meaning that the two constructions differ only in the order they take the possessor and possessed arguments. Also, if instead of *((the hat) of (the mayor))* we had chosen the structure *(the (hat of (the mayor)))* the matrices would be the same.

To summarize, all productive derivational and inflectional processes will have the output differ from the input by some constant $\vec{c}$ that depends only on the construction in question, and the same goes for all 'syntactic' processes such as forming a PP or NP whose output differs from its input only by the addition of some fixed grammatical formative, including the formation of modal verb complexes (*must go, will eat, ...*) by a fixed auxiliary. Note that such processes crosslinguistically often end up in the morphology, cf. Romanian *-ul* 'the' or Hungarian *-val/vel* 'with'.

An important consequence of what we said so far is that the effects of fixed formatives, be they attached morphologically or by a supporting clitic or full word, are commutative. This explains how even closely related languages like Finnish and Hungarian can have different conventional suffix orders (e.g. between case endings and possessive endings), as it takes no effort to rejigger the semantics with a change of inflection order. Also, a good number of bracketing paradoxes (Williams, 1981; Spencer, 1988) simply disappear: in light of commutative semantics brackets are not at all called for, and the 'paradox' is simply a by-product of an overly detailed (context free) descriptive technique.

The less productive a process, the less compelling the argument we made above, since it depends on some identity holding not just for a handful of vectors but for an entire subspace generated by the part of speech class of the input. For example the morphologically still perceptible relatedness of latinate prefixes and stems (Aronoff, 1976) as in *commit, remit, permit, submit, compel, repel, impel, confer, refer, infer, ...* will hardly allow for computing separate vectors for *con-, re-, ...* on the one hand and *pel, mit, sume, fer, ceive, ...* on the other as we have too many unknowns for too few equations. Or consider *bath:bathe, sheath:sheathe, wreath:wreathe, teeth:teethe, safe:save, strife:strive, thief:thieve, grief:grieve, half:halve, shelf:shelve, serf:serve, advice:advise, ...* where the relationship between the noun and the verb is quite transparent, yet the set on which the rule applies is almost lost among the much larger set of nouns that can be 'verbed' by zero affixation or stress shift alone.

This is not to say that suppletive forms, such as found in irregular plurals or strong verbs are outside the scope of our finding, for clearly if plural formation is the addition of a single fixed $\vec{c}$ in all regular cases, $\vec{horses} = \vec{horse} + \vec{c}$, we must also have $\vec{oxen} = \vec{ox} + \vec{c}$ since the analogy *horse:horses=ox:oxen* is intact. But given their paucity, derivational forms may still be sensitive to order of affixation, so that something like the Mirror Principle (Baker, 1985) may still make sense.

Looking at the 882 L and R matrices (25 by 25 dimensions) in the CVG instance available as part of the Stanford Dependency Parser, we note that over half (55% for L, 53% for R) of the variance in this set is explained by the first 25 eigenmatrices, so the structure is likely considerably simpler than the full CVG model allows for. We tested this hypothesis by grammars CVG$^{(k)}$ constructed from the Socher et al. (2013) CVG$^{(882)}$ by replacing all 882 L and R matrices by approximations based on the first k eigenmatrices (middle column of Table 1). The case $k = 1$ corresponds to the earlier RNN (Socher et al., 2011) with a single global $\otimes$, and gets only 81.0% on the WSJ task. As we increase the number of coefficients kept, we obtain results closer and closer to the original CVG$^{(882)}$: at $k = 100$ we are already within 1% of the full result.

k	no I	I first
1	81.02	
5	82.85	84.59
25	86.88	89.32
50	88.50	90.07
100	89.47	90.24
200	90.08	90.32
882	90.36	90.36

Table 1 Parsing performance as a function of the number of coefficients kept in $\otimes$ definitions

As Socher et al. (2013) already observe, the diagonal of the L (resp. R) matrix is dominant for left- (resp. right-)headed endocentric constructions, so we also experimented with keeping only $k - 1$ of the eigenmatrices and replacing the kth by I before finding the best approximations (right column of Table 1). With this choice of basis, the phenomenon is even more marked: it is sufficient to keep the top 24 (plus the coefficient for I) to get within 1% of the original result.

By limiting k we can limit the actual information content of $\otimes$, which would otherwise grow quadratically in d. Given that the 882 matrix pairs were already abstracted on the basis of sizeable corpora (63m words from the Reuters newswire, see Turian et al. 2010), direct numerical investigation of the 882 $\otimes$ operators to detect this simpler structure faces stability issues. In fact, it is next to impossible to guess, based strictly on an inspection of the eigenmatrices, that replacing the least one by I would be advantageous – for this we need to have a more model-based strategy, to which we now turn.

We speak about distributions in two main senses: discrete (class-level) and continuous (item-level). The distinction is reflected in the notation of generative grammar as between preterminals and terminals, and in the practice of language modeling as between states and emissions of Hidden Markov Models (HMMs). In generative grammar, the class-level distribution is typically conceived of in 0-1 terms: either a string of preterminals is part of the language or it is not – weighted grammars that make finer distinctions only became popular in the 1980s, decades after the original work on constituency (Wells, 1947; Harris, 1951; Chomsky, 1957). The standard (unweighted) grammar already captures significant generalizations such that A+N (adjective followed by noun) is very likely in English, while N+A is more likely in French. However, as (Harris, 1951) already notes,

> All elements in a language can be grouped into classes whose relative occurrence can be stated exactly. However, for the occurrence of a particular member of one class relative to a particular member of another class, it would be necessary to speak in terms of probability, based on the frequency of that occurrence in a sample.

Retrofitting generative rules such as $N \rightarrow AN$ achieves very little, in that it is not clear which adjective will go with which noun. As (Kornai, 2011) noted, HMM transition probabilities tend to stay in a relatively narrow range of $10^{-4} - 10^{-1}$ (the low values typically coming from smoothing) while emissions can span 8-9 orders of magnitude – this is precisely why n-gram HMMs remain a viable alternative to PCFGs to this day. CVS models capture a great deal of the distributional nuances because the vectors encode not just an estimate of unigram probabilities

$$\log(p(w)) = \frac{1}{2d}||\vec{w}||^2 - \log Z \pm o(1) \quad (6)$$

but also a cooccurrence estimate

$$\log p(w, w') = \frac{1}{2d}||\vec{w} + \vec{w}'||^2 - 2\log Z \pm o(1) \quad (7)$$

for some fixed Z (Arora et al., 2015). For unigrams, the GloVe dictionary (Pennington et al., 2014) actually shows a Pearson correlation of 0.393 with the Google 1T frequencies and 0.395 with the BNC. While these are not bad numbers, (especially considering that G1T and BNC only correlate to 0.882), clearly a lot more need to be done before (7) becomes realistic. Table 2 shows some frequent, rare, and nonexistent A+N combinations together with their Google 1T frequency; the right-hand side of eq. (7); the scalar product of the GloVe word vectors; and their cosine angles.

A-N pair	freq	(6) rhs	$\langle , \rangle$	cos
popular series	95k	-3.153	13.95	0.39
popular guidance	127	-3.158	2.80	0.08
popular extent	0	-3.175	6.78	0.23
rapid development	299k	-3.137	20.40	0.50
rapid place	182	-3.165	7.88	0.25
rapid percent	0	-3.115	11.30	0.24
private student	134k	-3.121	16.79	0.37
rare student	989	-3.133	5.30	0.13
cold student	0	-3.121	4.58	0.10

Table 2 Cooccurrence predictors for frequent, rare, and nonexistent adjective+noun combinations

Evidently, GloVe captures a great deal of the distribution, clearly ranking the frequent above the rare/nonexistent both in unnormalized (scalar product) and normalized (cosine) terms, while (7) largely obscures this. All of these predictors fare badly when it comes to comparing rare to nonexistent forms. (Of course Google 1T 'nonexistence' only means 'below the cutoff' but here this is as good as nonexistence since such pairs don't participate in the training.) It is reasonable to conclude that embeddings model the high- to mid-range of the distribution quite well, but fail on very rare data, which call for a corrective term in the Arora et al. estimate in Eq. (7).

Remarkably, word similarity measures based on definitional similarity do nearly as well on semantic world similarity tasks as those based on distributions (Recski and Ács, 2015). These definitions, common to ACR models, manifest no distributional similarity between definiendum and definiens, compare *rascal* to *a child who behaves badly but whom you still like*. Yet when we compare *rascal* to *imp* 'a child who behaves badly, but in a way that is funny' the similarity becomes evident: both *rascal* and *imp* are defined as 'children behaving badly'. There are many idiosyncratic traits to these words, for example both *little rascal* and *little imp* are plausible, but *??old imp* is not, even though *old rascal* is. More often than not, these differences in distribution have to do with accidents of history rather than any semantic difference to speak of – this is especially clear on the case of exact synonyms like *twelve* and *dozen*.

Here we simply assume that observable distribution is the result of two factors: pure syntax, as expressed by the system of lexical (part of speech) categories such as N, and their projections such as NP, and pure semantics, expressed by their conceptual representations. The manner these two factors combine is not transparent, we hope to address the issue in a follow-on paper.

4 One Reality

Let us return to the question posed above concerning 'real' meanings: the challenge is not so much to encode meanings into some clever abstract language but to actually account for their successful use in conversation. If we believe in a common reality about which we talk to each other, meanings have to have a property that allows them to be merged in a particular way: anything that is true of the world must be compatible with anything else that is true. Yet, reality is not given to us in one fell swoop but rather needs to be explored. Despite the fact that we think of the one real model as the justification of our way of talking, we can only hypostatise its existence and take it from there. The constructed models of reality must form a family of models each approaching the single one. This can be explored in two ways. One way is to insist that any language – even an abstract one – is already equipped with a realist interpretation, and that leads to what is known as Robinson Consistency and the so-called Joint Embedding Property. The second approach considers only the constructed models as given and constructs reality out of them. This leads us to inverse systems of models, or dually, direct systems of algebras.

The models of choice, we argue here, are ACR representations, in essence graphs with colored edges. Some additional markup may be necessary on the nodes (to govern the loci of substitution/adjunction operations) and some additional constraints (in particular limiting out-degrees) may hold, but on the whole such structures are well understood. CVS models may stand in various relations to one another, in a manner far more complex than the alternative relations familiar from Kripke-style models. For example, embeddings I_p and I_q created from the same raw data by PCA but keeping a different number of dimensions $p < q$ are in an *extension of* relation which we can state directly on the corresponding models as $\mathcal{M}_p < \mathcal{M}_q$ where $<$ means 'can be embedded in'.

In ACR, there are many cases when one model structure can be embedded in the other, central among these being the case of the smaller structure simply containing fewer *existents* than the larger one. (The term 'existent' is a bit awkward, but helps to avoid non-Meinongian ontological commitments: in a model whose base elements are graphs or vectors corresponding to *mountain* and *gold*, $I(gold) \otimes I(mountain)$ is an 'existent'.) Moreover, if $\mathcal{K}, \mathcal{L}$ are isomorphic substructures of $\mathcal{M}$, the isomorphy between $\mathcal{K}$ and $\mathcal{L}$ can be extended to an automorphism of $\mathcal{M}$, making model structures *homogeneous* in the sense of Fraïssé (1954).

For the graph structures to actually be *models* they must satisfy certain requirements. The requirements are sine qua non because the models are models of something, namely, in first approximation, external reality. If models are about external reality then it follows that there can be only one. As such however it is not to be found in anyone's head. Instead, we picture the acquisition of the model structure as a process that walks through a number of smaller model structures, expanding them as new information comes in. The process of expansion by necessity produces substructures of one bigger structure. Thus, the classes of model structures must satisfy what is known as the *amalgamation property* that for each $\mathcal{K}, \mathcal{L}, \mathcal{M}$ where we have k and l embeddings of $\mathcal{M}$ into $\mathcal{K}$ and $\mathcal{L}$ respectively, we have some $\mathcal{N}$ and embeddings k' and l' of $\mathcal{K}$ and $\mathcal{L}$ into $\mathcal{N}$ such that the following diagram commutes:

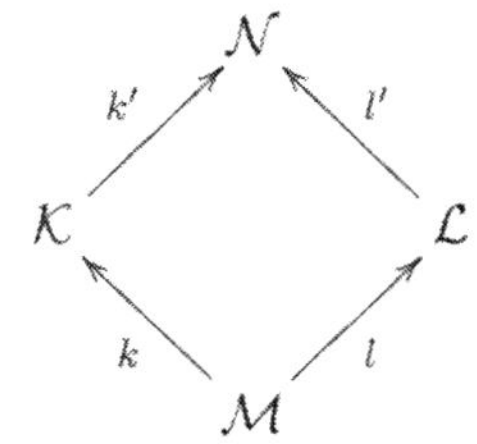

On the logical side we expect a joint consistency in the spirit of Robinson' theorem: if T_1 and T_2 are two theories such that the intersection is consistent and there is no formula φ such that $T_1 \vdash \varphi$ while $T_2 \vdash \neg\varphi$, then $T_1 \cup T_2$ is consistent. Assuming that the world is consistent, we expect this behaviour. Let U be the intersection of T_1 and T_2. Suppose our database is U. Then after some steps of learning we may end up in T_1 or in T_2. However, both states cannot be in conflict by deriving one of them a formula and the other its negation. So, they are jointly consistent.

This property makes perfect sense for lexical entries, where extending a model $\mathcal{M}$ with new entries to build $\mathcal{K}$ or $\mathcal{L}$ can be amalgamated to produce $\mathcal{N}$. What this means, in naive terms, is that the lexicon harbors no contradictions. To see that this is already a non-empty requirement, consider the lexical entry for *cancer* which will, under the ACR theory, contain an IS_A link to *incurable*. When (hopefully soon) a cure is found, this means that the lexical entry itself will have to be revised, just as gay marriage forced the revision of 'between a man and a woman'. More significant are the contradictory cases, for instance when in one extension we learn that Colonel Mustard killed Mr. Boddy, and in another we learn that Professor Plum did. Admitting model structures that harbor internal contradictions (as in paraconsistent logic) clashes with the use of a single model; an alternative that suggests itself is to allow for much richer embeddings, e.g. ones that contain propositional attitude clauses: Miss Scarlet believes that Colonel Mustard killed Mr. Boddy, while Mrs. Peacock believes it's Professor Plum.

As the last example shows, there is an additional complicating factor at play. Even if we assume the model to be a model of a single reality, this grounding model may vary from person to person as in the 'lifelong' DRT of (Alberti, 2000). Communication may reveal that this is the case, but the remedy is not simple. Differences may arise about facts of the matter as well as over meanings, hence they may concern either the grounding model itself ('reality') or the map g that grounds the meanings (Kracht, 2011). Thus, the fact that language is shared among a group of individuals in and of itself calls for a different approach in model theory. This must be left for another occasion, however.

If we believe in a single and unique model structure, we will assume that any model we build must be embeddable into the one existing model. Thus, we must have the *joint embedding property* (JEP) for the family of 'candidate' models of reality. This property requires that for any two models $\mathcal{K}, \mathcal{L}$ the existence of an $\mathcal{N}$ in which both can be embedded. Such a consistency requirement must be made if we insist that all semantics is about a single external reality.

However, suppose the real model is unknown, even unknowable. Then, if we want to understand what it means to talk about real objects appeal to external reality is futile if all we have is appearances. This is where Kant saw the need of a logic he called *transcendental*. The transcendental object is so to speak the limit of approximation made by our inquiry. It is our construction of reality, which rationalises our previous models as being about *something*. In this connection it is rather interesting to note the proposal by Achourioti and van Lambalgen (2011) concerning the transcendental logic. The authors propose that what Kant had actually in mind was what is nowadays called an *inverse system*. This is a family of models $\mathcal{M}_s$ indexed by a poset $(S, \leq)$ such that for all s, t there is r such that $s, t \leq r$, together with maps $h_{st} : \mathcal{M}_s \to \mathcal{M}_t$ for $s \geq t$ satisfying $h_{tr} \circ h_{st} = h_{sr}$. Even if there is no unique model structure, the system of model structures itself is *objective* in Kant's sense (that is *about an object*) if it has the structure of an inverse system; and the transcendental object itself can somehow be imagined as a member of the inverse limit of that system. What is interesting to note is that the formulae for which the transition from the inverse system to the inverse limit is what is known as *geometrical formulae*, having the form $\forall\overline{x}(\varphi(\overline{x}) \to \exists\overline{y}\theta(\overline{x}, \overline{y}))$.

5 Conclusions

As usual, model theory does not solve many outstanding problems, but brings a great deal of much needed clarity in organizing the minor variants one could conceive of. As long as we stay with a purely deductive apparatus, we have to figure out whether natural deduction, Beth tableaux, Hilbert systems, sequent calculi, or some new combination of the above is what we use, and this inevitably gets mixed up with other design choices we have within the ACR/CVS world. (Also, for reasons shrouded in the mists of history, proof theory somehow has a very bad reputation within linguistics.)

This paper has taken the first, rather tentative steps towards understanding the structure of the new model structures. We have seen that operations of inflectional morphology, whether realized by actual inflection or by function words, amount to a shift by a constant vector. Second, we have seen that data can be pooled across semantically equivalent but syntactically different constructions such as the *of* and *'s* possessives. Third, we have seen that the numerical limitations of the current model make it impossible to explore the low frequency tail of the distribution where many phenomena of great linguistic interest, such as causativization and other forms of predicate decomposition, are to be found. Even so, our results in Table 1 make clear that the actual complexity of construction operations is considerably less than the POS-pair assumption built into CVGs would suggest.

The use of existents provides, for the first time we

believe, a reasonable framework to approach both standard and Meinongian ontology on equal footing. This is not to say that one has to be committed to some higher plane of ideal existence where the entirety of Meinong's Jungle is present, to the contrary, all one needs is a notion of finitely generated models, and a compositional semantics that is willing to interpret $a \otimes b$ based on the interpretation of a and b. Similarly, the key property of Fraïssé homogeneity is the one at stake in the entire philosophical debate surrounding *inverted qualia* (see Byrne (2014) for a summary). What is clear is that automorphisms mapping one synonym on another can be extended to automorphisms of the whole lexicon, but from here on one may take several paths depending on one's philosophical predilections.

Many questions remain open, and perhaps more importantly, many questions can be meaningfully asked for the first time. The traditional riddle of *class meanings* (how nouns designate 'things', adjectives 'qualities', and verbs 'actions') is now amenable to empirical work relating the vectors of pronouns, proadjectives and other pro-forms to the center of gravity of $\langle N \rangle, \langle A \rangle, \ldots$. On the pure semantics side, we may begin to see how, by finite mechanism, humans are capable of infinite comprehension, learning of (transcendental) objects.

Acknowledgments

We thank Gábor Recski (HAS Research Institute for Linguistics) for performing the PCA on the $\otimes$ operators of the Socher et al. (2013) CVG.

References

Gábor Alberti. 2000. Lifelong discourse representation structure. *Gothenburg Papers in Computational Linguistics*.

Mark Aronoff. 1976. *Word Formation in Generative Grammar*. MIT Press.

Sanjeev Arora, Yuanzhi Li, Yingyu Liang, Tengyu Ma, and Andrej Risteski. 2015. Random walks on context spaces: Towards an explanation of the mysteries of semantic word embeddings. *arXiv:1502.03520v1*.

Mark Baker. 1985. *Incorporation: a theory of grammatical function changing*. MIT.

Laura Banarescu, Claire Bonial, Shu Cai, Madalina Georgescu, Kira Griffitt, Ulf Hermjakob, Kevin Knight, Philipp Koehn, Martha Palmer, and Nathan Schneider. 2013. Abstract meaning representation for sembanking. In *Proceedings of the 7th Linguistic Annotation Workshop and Interoperability with Discourse*, pages 178–186, Sofia, Bulgaria, August. Association for Computational Linguistics.

Yehoshua Bar-Hillel. 1960. A demonstration of the nonfeasibility of fully automatic high quality translation. In *The present status of automatic translation of languages*, volume Advances in Computers I, pages 158–163.

Marco Baroni. 2013. Composition in distributional semantics. *Language and Linguistics Compass*, 7(10):511–522.

Alex Byrne. 2014. Inverted qualia. In Edward N. Zalta, editor, *The Stanford Encyclopedia of Philosophy*. Summer 2014 edition.

Noam Chomsky. 1957. *Syntactic Structures*. Mouton, The Hague.

Bob Coecke, Mehrnoosh Sadrzadeh, and Stephen Clark. 2010. Mathematical foundations for a compositional distributional model of meaning. *arXiv:1003.4394v1*.

Georgiana Dinu and Mirella Lapata. 2010. Measuring distributional similarity in context. pages 1162–1172.

Nicholas V. Findler, editor. 1979. *Associative Networks: Representation and Use of Knowledge by Computers*. Academic Press.

Roland Fraïssé. 1954. Sur l'extension aux relations de quelques propriétés des ordres. *Ann. Sci. Ecole Norm. Sup*, 71:361–388.

Adele E. Goldberg. 1995. *Constructions: A Construction Grammar Approach to Argument Structure*. University of Chicago Press.

Zellig Harris. 1951. *Methods in Structural Linguistics*. University of Chicago Press.

Wilfrid Hodges. 2001. Formal features of compositionality. *Journal of Logic, Language and Information*, 10:7–28.

Pauline Jacobson. 2014. *Compositional Semantics*. Oxford University Press.

T.M.V. Janssen. 2001. Frege, contextuality and compositionality. *Journal of Logic, Language and Information*, 10(1):115–136.

Ewan Klein and Ivan Sag. 1985. Type-driven translation. *Linguistics and Philosophy*, 8:163–201.

András Kornai, Judit Ács, Márton Makrai, Dávid Nemeskey, Katalin Pajkossy, and Gábor Recski. 2015. Competence in lexical semantics. To appear in Proc. *SEM-2015.

András Kornai. 2010. The algebra of lexical semantics. In Christian Ebert, Gerhard Jäger, and Jens Michaelis, editors, *Proceedings of the 11th Mathematics of Language Workshop*, LNAI 6149, pages 174–199. Springer.

András Kornai. 2011. Probabilistic grammars and languages. *Journal of Logic, Language, and Information*, 20:317–328.

Marcus Kracht. 2003. *The Mathematics of Language*. Mouton de Gruyter, Berlin.

Marcus Kracht. 2011. *Interpreted Languages and Compositionality*, volume 89 of *Studies in Linguistics and Philosophy*. Springer, Berlin.

Jan Landsbergen. 1982. Machine translation based on logically isomorphic montague grammars. In *Proceedings of the 9th conference on Computational linguistics-Volume 1*, pages 175–181. Academia Praha.

D. Lewis. 1970. General semantics. *Synthese*, 22(1):18–67.

Tomas Mikolov, Wen-tau Yih, and Zweig Geoffrey. 2013. Linguistic regularities in continuous space word representations. In *Proceedings of NAACL-HLT-2013*, pages 746–751.

Jeff Mitchell and Mirella Lapata. 2008. Vector-based models of semantic composition. In *Proceedings of ACL-08: HLT*, pages 236–244, Columbus, Ohio. Association for Computational Linguistics.

Richard Montague. 1970a. English as a formal language. In R. Thomason, editor, *Formal Philosophy*, volume 1974, pages 188–221. Yale University Press.

Richard Montague. 1970b. Universal grammar. *Theoria*, 36:373–398.

Richard Montague. 1973. The proper treatment of quantification in ordinary English. In R. Thomason, editor, *Formal Philosophy*, pages 247–270. Yale University Press.

Glynn Morrill. 2011. CatLog: A categorial parser/theorem-prover. In *Type Dependency, Type Theory with Records, and Natural-Language Flexibility*.

Charles E. Osgood, William S. May, and Murray S. Miron. 1975. *Cross Cultural Universals of Affective Meaning*. University of Illinois Press.

Jeffrey Pennington, Richard Socher, and Christopher Manning. 2014. Glove: Global vectors for word representation. In *Conference on Empirical Methods in Natural Language Processing (EMNLP 2014)*.

Carl Pollard. 2008. Hyperintensions. *Journal of Logic and Computation*, 18(2):257–282.

M. Ross Quillian. 1969. The teachable language comprehender. *Communications of the ACM*, 12:459–476.

Gábor Recski and Judit Ács. 2015. Mathlingbudapest: Concept networks for semantic similarity. In *Proceedings of the 9th International Workshop on Semantic Evaluation (SemEval 2015)*, pages 543–547, Denver, Colorado, June. Association for Computational Linguistics.

Roger C. Schank. 1972. Conceptual dependency: A theory of natural language understanding. *Cognitive Psychology*, 3(4):552–631.

Hinrich Schütze. 1993. Word space. In SJ Hanson, JD Cowan, and CL Giles, editors, *Advances in Neural Information Processing Systems 5*, pages 895–902. Morgan Kaufmann.

Paul Smolensky. 1990. Tensor product variable binding and the representation of symbolic structures in connectionist systems. *Artificial intelligence*, 46(1):159–216.

Richard Socher, Cliff Chiung-Yu Lin, and Christopher D Manning. 2011. Parsing natural scenes and natural language with recursive neural networks. In *Proc. 28th ICML*.

Richard Socher, John Bauer, Christopher D. Manning, and Andrew Y. Ng. 2013. Parsing with compositional vector grammars. In *The 51st Annual Meeting of the Association for Computational Linguistics (ACL 2013)*.

J.F. Sowa. 2000. *Knowledge representation: logical, philosophical, and computational foundations*. MIT Press.

Andrew Spencer. 1988. Bracketing paradoxes and the English lexicon. *Language*, 64:663–682.

Joseph Turian, Lev Ratinov, and Yoshua Bengio. 2010. Word representations: a simple and general method for semi-supervised learning. In *Proceedings of the 48th Annual Meeting of the Association for Computational Linguistics*, pages 384–394. Association for Computational Linguistics.

Michiel van Lambalgen and Theodora Achourioti. 2011. A Fromalization of Kant's Transcendental Logic. *The Review of Symbolic Logic*, 4:254 – 289.

Roulon S. Wells. 1947. Immediate constituents. *Language*, 23:321–343.

Edwin Williams. 1981. On the notions 'lexically related' and 'head of a word'. *Linguistic Inquiry*, 12:245–274.

William A. Woods. 1975. What's in a link: Foundations for semantic networks. *Representation and Understanding: Studies in Cognitive Science*, pages 35–82.

A Frobenius Model of Information Structure in Categorical Compositional Distributional Semantics

Dimitri Kartsaklis **Mehrnoosh Sadrzadeh**

Queen Mary University of London
School of Electronic Engineering and Computer Science
Mile End Road, London E1 4NS, UK
{d.kartsaklis;m.sadrzadeh}@qmul.ac.uk

Abstract

The categorical compositional distributional model of Coecke et al. (2010) provides a linguistically motivated procedure for computing the meaning of a sentence as a function of the distributional meaning of the words therein. The theoretical framework allows for reasoning about compositional aspects of language and offers structural ways of studying the underlying relationships. While the model so far has been applied on the level of syntactic structures, a sentence can bring extra information conveyed in utterances via intonational means. In the current paper we extend the framework in order to accommodate this additional information, using Frobenius algebraic structures canonically induced over the basis of finite-dimensional vector spaces. We detail the theory, provide truth-theoretic and distributional semantics for meanings of intonationally-marked utterances, and present justifications and extensive examples.

1 Introduction

Distributional models of meaning, in which a word is represented as a high dimensional vector of contextual statistics in a metric space, provide a convincing framework for lexical semantics that has been found useful in a number of natural language processing tasks (Schütze, 1998; Landauer and Dumais, 1997; Manning et al., 2008). Despite their success at the word level, the underlying hypothesis of these approaches does not naturally scale up to phrases or sentences due to the infinite capacity of language to produce new meanings from a finite vocabulary and a set of grammar rules.

Coecke et al. (2010) provide a solution to the problem by noticing that the category of finite-dimensional vector spaces and linear maps is homomorphic to a grammar expressed as a pregroup (Lambek, 2008); specifically, both share compact closed structure (Kelly, 1972). In practice this means that any grammatical derivation based on the type-logical identities of the individual words in a sentence can be translated to a (multi-)linear map which, when applied on the vectorial representations of the words therein, results in a sentence vector. The grammatical type of a word determines the vector space in which this word lives. Taking nouns to be simple vectors in a basic vector space N, an adjective, for example, becomes a linear map $N \to N$, or equivalently, a matrix in $N \otimes N$; furthermore, a transitive verb is a bi-linear map $N \otimes N \to S$, living in $N \otimes S \otimes N$. Composition takes the form of tensor contraction, which is a generalization of matrix multiplication to higher order tensors.

In general, the model resembles a quantitative linear-algebraic version of the formal semantics approach (Montague, 1970), in the sense that syntax strictly guides the semantic composition. Interestingly, syntax seems to co-exist with a distinct structural layer, the purpose of which is to optimize the message that an utterance conveys. This aspect is known as *information structure*, and at the phrase or sentence level is expressed as a distinction between a theme part (information that is generally agreed to be known to both of the interlocutors) and a rheme part—information that is new for the addressee. The exact relation that holds between syntactical and information structure is an interesting and controversial topic. For example, a theme does not have to comprise a valid grammatical constituent

Proceedings of the 14th Meeting on the Mathematics of Language (MoL 14), pages 62–74,
Chicago, USA, July 25–26, 2015. ©2015 Association for Computational Linguistics

in the strict sense of the term, as it is evident in the following example:

(1) Q: Do you need anything?
A: [I would like]$_T$ [some tea]$_R$

The distinction between a theme and rheme is denoted by the presence of a boundary that can be expressed by phonological, morphological or even syntactical means, depending on the language. Furthermore, the presence of such boundaries suggest the existence of a distinct composition operator related to information structure and different than the one that would be normally used for syntax.

In this paper we extend the categorical model of Coecke et al. (2010) in a way to accommodate an information structure layer of composition. In order to achieve this, we model intonational boundaries (the devices for defining information structure in English) by using the multiplication part of the Frobenius algebra that is canonically induced over any vector space with fixed basis, in order to endow equal contribution of the theme and rheme on the vectorial representation of a sentence, thus putting emphasis on the appropriate part. The resulting model can be seen as containing two types of composition operators: the usual tensor contraction for accommodating syntax, and the Frobenius multiplication for accommodating information structure. We discuss the implications in terms of the resulting vectorial representations for phrases and sentences, and provide connections with existing models from the current literature of compositional distributional semantics. Various examples demonstrate the potential of the model.

2 Categorical compositional distributional semantics

The categorical model of Coecke et al. (2010) assigns semantic representations to phrases and sentences of language, based on their grammatical structure and the semantics of individual words. In its most abstract form, this model can be expressed in terms of a structure-preserving passage between grammar and meaning:

$$\mathcal{F}\colon \text{Grammar} \to \text{Meaning}$$

Given a sequence of words $\mathrm{w}_1 \cdots \mathrm{w}_n$, its categorical meaning is defined to be:

$$[\![\mathrm{w}_1 \cdots \mathrm{w}_n]\!] := \mathcal{F}(\alpha)([\![\mathrm{w}_1]\!], \cdots, [\![\mathrm{w}_n]\!]) \qquad (1)$$

Here, α is derived from the grammatical relationships amongst the words in the sequence. This notion can be formalised in a coherent way, if both the grammar and the meaning are expressed in a high level logical structure, referred to by *compact closure*. Lambek's pregroup algebras (Lambek, 2008) and vector space distributional semantics are examples of compact closed structures. Stipulating that the grammar is expressed in a pregroup algebra and that the meaning of words are vectors constructed using the distributional hypothesis (Harris, 1968), Eq. 1 gets a more concrete form:[1]

$$\overrightarrow{\mathrm{w}_1 \cdots \mathrm{w}_n} := \mathcal{F}(\alpha)(\overrightarrow{\mathrm{w}}_1 \otimes \cdots \otimes \overrightarrow{\mathrm{w}}_n) \qquad (2)$$

In the proceeding subsections we make these notions precise and provide intuitions and examples.

2.1 Pregroup grammars

A pregroup grammar is a pregroup algebra, linked to the vocabulary of a language via the notion of a type dictionary. We define these structures below.

A pregroup algebra is a partially ordered monoid where each element has a left and a right adjoint. It is denoted by a tuple $(P, \leq, \cdot, 1, (-)^l, (-)^r)$, where $(P, \leq)$ is a partially ordered set, and $\cdot$ is a monoid multiplication with 1 as its unit. For each element $p \in P$ there are $p^l, p^r \in P$, referred to by p's *left* and *right* adjoints, satisfying the following four inequalities:

$$p \cdot p^r \leq 1 \leq p^r \cdot p \qquad p^l \cdot p \leq 1 \leq p \cdot p^l$$

When a pregroup algebra is generated over a base set $\mathcal{B}$, it is denoted by $P(\mathcal{B})$. Given the vocabulary of a language Σ and a set of its basic grammatical types $\mathcal{B}$, a pregroup grammar is a relation $D \subseteq \Sigma \times P(\mathcal{B})$ that assigns grammatical types from the pregroup algebra $P(\mathcal{B})$ to the words of the vocabulary Σ. Such a pregroup grammar is denoted by $P(\mathcal{B}, \Sigma)$.

As an example, suppose $\mathcal{B} = \{n, s\}$, where n stands for a well-formed noun phrase and s for a well-formed sentence. Suppose further that $\Sigma = \{\text{Mary, snores, likes, musicals}\}$. The pregroup dictionary consists of the following set:

$$\{(\text{Mary}, n), (\text{snores}, n^r s), (\text{likes}, n^r s n^l), (\text{musicals}, n)\}$$

[1]One can translate the types from other type logics, such as the syntactic calculus and CCG to pregroups and carry on with the same calculations. There is also recent work that directly assigns vector semantics to CCG (Lewis and Steedman, 2013).

One says that a sequence of words $w_1 w_2 \cdots w_n$ for $w_i \in \Sigma$ forms a grammatical sentence, according to a pregroup grammar $P(\mathcal{B}, \Sigma)$, whenever we have:

$$t_1 \cdot t_2 \cdot \ldots \cdot t_n \leq s$$

for $(w_i, t_i) \in D$. The above inequality is often referred to by *grammatical reduction*. For example, 'Mary likes musicals' is a grammatical sentence, since we have the following reduction:

$$n \cdot n^r \cdot s \cdot n^l \cdot n \leq 1 \cdot s \cdot 1 = s \tag{3}$$

2.2 Distributional models

The only piece of information provided by a derivation like the one in Eq. 3 is whether the sentence in question is well-formed or not. Furthermore, we are unable to distinguish between words of the same type. Distributional models of meaning provide a solution to these problems by following the *distributional hypothesis* (Harris, 1968), which states that semantically similar words must appear in similar contexts. Hence, the semantic representation of a word can be given in terms of its distributional behaviour in a large corpus of text. In its simplest form, a word vector is comprised by numbers that show how many times the target word co-occurs with every other word in a selected subset of the vocabulary (usually the most frequent content-bearing words). This allows the representation of words as points in some high dimensional space, where semantic relatedness can be measured (usually by cosine distance) and evaluated. For a concise introduction to distributional models, see (Turney and Pantel, 2010). We will now proceed to show how the quantitative approach of distributional models can be combined with the compositional model of Section 2.1 into a unified account.

2.3 Categorical generalization

The theory of categories generalises algebraic constructions to categorical ones (Mac Lane, 1971). Herein, instead of sets, functions or relations, one has objects A, B and morphisms $f \colon A \to B$. The generalised binary operation over these is referred to by a product. Posing different conditions on the objects, morphisms, or the product results in different kinds of categories. A *monoidal category* has a product with a unit I, that is $A \otimes I \cong I \otimes A \cong A$. These categories are generalisations of partially ordered monoids: elements of the partial order become objects of the category and the partial orderings between them become morphisms. Furthermore, *compact closed categories* are generalisations of pregroups, where the adjunction inequalities correspond to the following ϵ and η morphisms:

$$\epsilon^r : A \otimes A^r \to I \qquad \eta^r : I \to A^r \otimes A$$
$$\epsilon^l : A^l \otimes A \to I \qquad \eta^l : I \to A \otimes A^l$$

These maps needs to adhere to four axioms, referred to as *yanking equations*, which ensure that all relevant diagrams commute.

The importance of the theory of categories for this paper is that finite-dimensional vector spaces and linear maps also form a compact closed category, denoted by **FVect**. Herein, objects are vector spaces, morphisms are linear maps, and the product is the tensor product between vector spaces whose unit is the scalar field of the vector spaces, in our case, real numbers ($\mathbb{R}$). In the presence of a fixed basis (which is the case we are interested in) the adjoints become identity, that is we have $V^r \cong V^l \cong V$, for a vector space V spanned by $\{\overrightarrow{v}_i\}_i$. As a result the four ϵ and η maps reduce to two:

$$\epsilon \colon V \otimes V \to \mathbb{R} \qquad \eta \colon \mathbb{R} \to V \otimes V$$

The ϵ map takes the inner product of two vectors and the η map produces a diagonal matrix. The fact that both pregroup algebras and vector spaces form compact closed categories allows us to develop a structure preserving passage between the two mathematical structures, thus enabling us to bridge the grammatical structure to distributional semantics.

2.4 From grammar to distributions

A structure preserving passage from grammatical structures (in the form of a pregroup grammar) to semantics (in the form of vector spaces) is given by a map denoted as follows:

$$\mathcal{F} : P(\Sigma, \mathcal{B}) \to \textbf{FVect} \tag{4}$$

This is a strongly monoidal passage, which means that it has the following compositional properties for juxtapositions of types in a pregroup grammar:

$$\mathcal{F}(1) = \mathbb{R} \tag{5}$$
$$\mathcal{F}(p \cdot q) = \mathcal{F}(p) \otimes \mathcal{F}(q) \tag{6}$$
$$\mathcal{F}(p^r) = \mathcal{F}(p^l) = \mathcal{F}(p) \tag{7}$$
$$\mathcal{F}(p \leq q) = \mathcal{F}(p) \to \mathcal{F}(q) \tag{8}$$

On the level of basic types we assign a vector space to each basic type, that is, $\mathcal{F}(n) = N$ and $\mathcal{F}(s) = S$. As a result of the above assignments, words that have simple types, for example noun phrases, will become vectors in vector space N. Words that are functions of one argument become matrices, e.g. intransitive verbs with type $n^r \cdot s$ are elements of $N \otimes S$; and words that are functions of two arguments, e.g. transitive verbs with type $n^r \cdot s \cdot n^l$, become tensors of order 3, living in $N \otimes S \otimes N$ for the specific case. The grammatical reductions are translated to compositions of morphisms, and in particular ϵ-maps.

$$\begin{aligned}\overrightarrow{\text{Mary snores}} &= \mathcal{F}(n \cdot n^r \cdot s)(\overrightarrow{\text{Mary}} \otimes \overline{\text{snores}}) \\ &= (\epsilon^r_N \otimes 1_S)(\overrightarrow{\text{Mary}} \otimes \overline{\text{snores}})\end{aligned}$$

A simple computation shows that the above is equal to $\overrightarrow{\text{Mary}} \times \overline{\text{snores}}$; similarly, for the meaning of a transitive sentence we obtain:

$$\overrightarrow{\text{Mary likes musicals}} = \overrightarrow{\text{Mary}} \times \overline{\text{likes}} \times \overrightarrow{\text{musicals}}$$

Note that tensor contraction (in spaces with fixed basis) is associative, so there is no need to keep track of brackets in the above. The situation is similar to pregroups, where the monoid multiplication is again associative.

2.5 Frobenius algebras

Compact closed categories on their own do not have much structure: there is a binary operation and the maps ϵ and η. The expressive power of these categories can be increased using Frobenius algebras. We define these below.

Given a compact closed category $\mathcal{C}$, an object $X \in \mathcal{C}$ has a Frobenius structure on it if there exist the following morphisms:

$$\Delta \colon X \to X \otimes X \qquad \iota \colon X \to I$$
$$\mu \colon X \otimes X \to X \qquad \zeta \colon I \to X$$

These have to satisfy certain conditions, the most important to us being the *Frobenius condition*:

$$(\mu \otimes 1_X) \circ (1_X \otimes \Delta) = \Delta \circ \mu = (1_X \otimes \mu) \circ (\Delta \otimes 1_X)$$

Vector spaces with fixed basis do have such structures over them, generally referred to by *copying* and *merging*. For $\overrightarrow{v} \in V, \overline{w} \in V \otimes V$, we have that $\Delta(\overrightarrow{v}) \in V \otimes V$ is a diagonal matrix whose diagonal elements are weights of $\overrightarrow{v}$, and $\mu(\overline{w}) \in V$ is a vector consisting only of the diagonal elements of $\overline{w}$.

These structures have been used in previous work to encode lower dimensional verb matrices into higher dimensional tensors (Kartsaklis et al., 2012; Kartsaklis et al., 2014) and to pass the information around sentences with relative clauses by copying and merging (Sadrzadeh et al., 2013; Sadrzadeh et al., 2014).

2.6 Graphical calculus

In the presence of higher order tensor product spaces, calculations can become quite complex. The formalism of compact closed categories and Frobenius structures is complete with regard to a graphical calculus (Selinger, 2011) that simplifies the computations to a great extend. We briefly overview the main components of this language.

Objects are depicted by lines and morphisms by boxes. Tensor products between objects and morphisms are given by juxtaposition of their diagrams, while composition of morphisms amounts to connecting outputs to inputs. Examples are as follows:

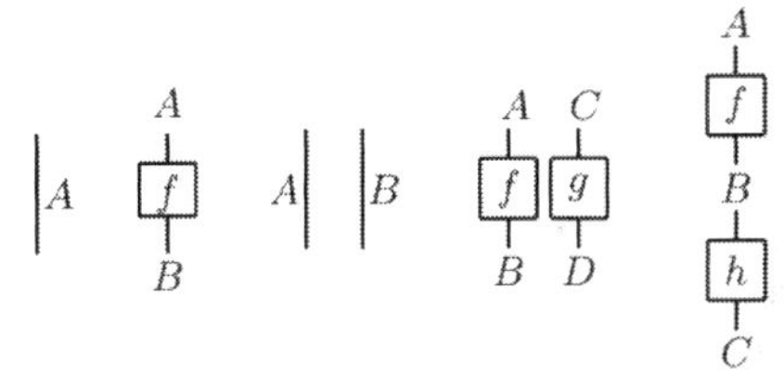

The ϵ maps are depicted by cups, η maps by caps, and yanking by their composition and straightening of the strings. For instance:

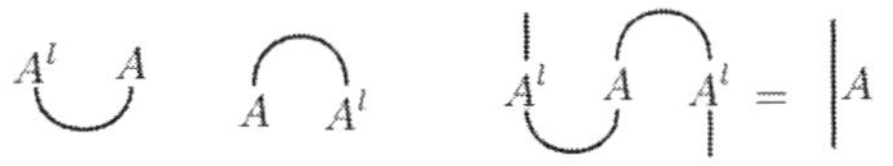

The diagrams corresponding to the Frobenius morphisms are as follows:

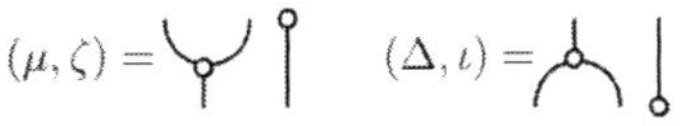

with the Frobenius condition being depicted as:

The defining axioms guarantee that any picture of a Frobenius computation can be reduced to a normal form (so-called a "spider") that only depends on the number of input and output strings of the nodes:

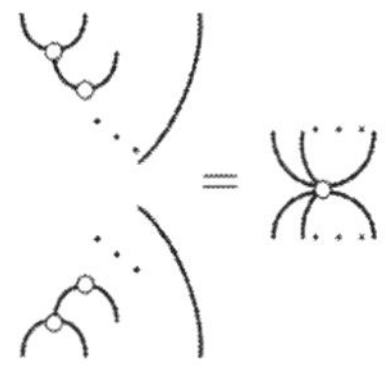

Elements within the objects (for the case of vector spaces, vectors) are depicted by morphisms from the unit. These are shown by triangles with a number of strings emanating from them. The number of strings denotes the order of the tensor; for instance, the diagrams for $\overrightarrow{v} \in V, \overrightarrow{v'} \in V \otimes W$, and $\overrightarrow{v''} \in V \otimes W \otimes Z$ are as follows:

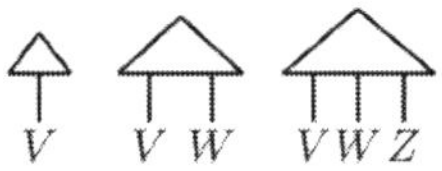

3 Information structure and intonation

The term *information structure* collectively refers to techniques that aim to enhance the communication between two interlocutors in order to optimize the conveyed message for the benefit of the addressee (Chafe, 1976). One such technique, for example, is to emphasize a particular part of the utterance that is important for the listener by changing the spoken pitch:

(2) Q: What does Mary like?
A: Mary likes MUSICALS

The emphasis imposes a specific information structure to the uttered sentence, essentially splitting it in two parts: The part in upper-case above is what Steedman (2000) calls *rheme*—the information that the speaker wishes to make common ground for the listener; the rest of the sentence, i.e. what the listener already knows, is called *theme*. The question in (2) puts the listener in a specific attentional state, in the context of which an answer such as:

(3) A: #MARY likes musicals

will be *infelicitous*, that is, not compatible with that state.

The distinction between theme (or topic) and rheme (or comment) has great significance from an information structure point view, since it defines a generic shape for the sentence that directly reflects the attentional needs of the addressee. A further dimension that can be found in both rheme and theme distinguishes between the *focus*, that is, the specific word that receives most of the intonational emphasis, and the background, which consists of the rest of the words in the specific text segment. Note that in contrast to rheme/theme distinction, focus and background seem to operate at the lexical level.[2]

Furthermore, we should point out that although the examples we use in this paper are mainly based on question/answer dialogues, this is not by any means the only case where the presence of a specific information structure can be useful. For example, consider the dialogue:

(4) –I think Mary likes jazz.
–Mary likes MUSICALS.

Information structure can be expressed in different ways that may vary from language to language. In English, for example, the means for defining information structure is *intonation*: variations of spoken pitch, the purpose of which is to emphasize parts of the utterance that might be important for the conveyed message, as we saw above in our examples. However, in other languages such as Japanese or Cantonese, the intonational boundaries can be also specifically marked by morphological devices, e.g. special particles (Féry and Krifka, 2008). Finally, the position of a text segment in a sentence can also be an indication of its information-structural role. In English, for example, themes tend to appear at the beginning of a clause.

In this paper we concentrate on the sentence-level distinction between rheme and theme.

4 Grammar and intonation

The presence of a distinct layer of information structure that seems to co-exist with the grammatical structure of a sentence, poses the interesting question regarding the exact relationship that holds between those different structural aspects. For example, although the text segment "Mary likes" forms a perfectly acceptable theme, most linguists would agree that it does not also comprise a valid grammatical constituent. In spite of this claim, though, it is interesting to note that a number of categorial grammars, including Combinatory Categorial Grammar (CCG) (Steedman, 2001), treat text segments like the above as possible syntactic constituents. Consider the following ditransitive sentence:

[2] Actually many authors use the term *focus* as a synonym for *rheme*; the definitions we give in this paper follow (Steedman, 2000).

(5) John gave Mary a flower

In CCG, this sentence has a number of different syntactic derivations, two of them are the following:

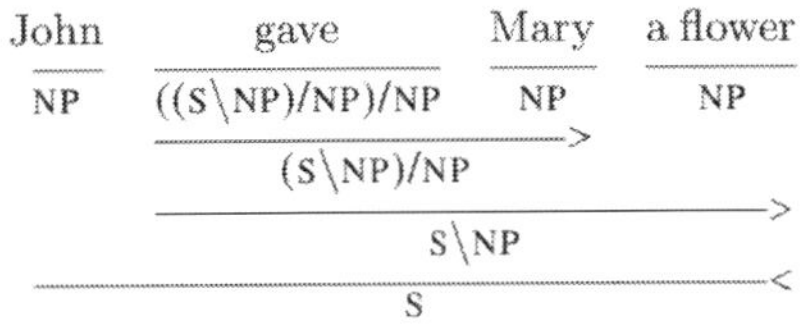

(9)

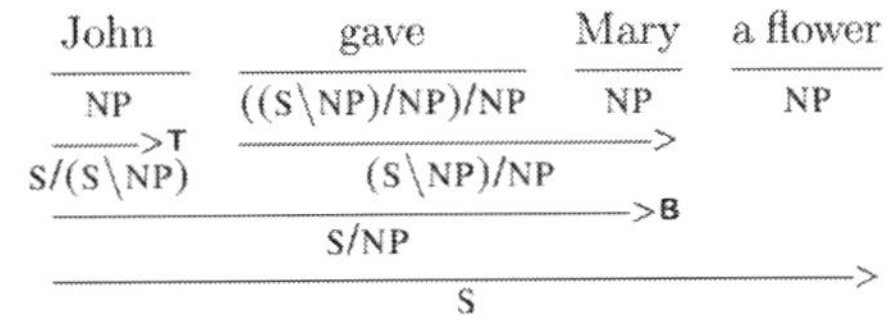

(10)

Note that (9) proceeds by first composing the part corresponding to the verb phrase ("gave Mary a flower"); later, in the final step, the verb phrase is composed with the subject 'John'. The situation is reversed for (10), where the use of type-raising and composition rules of CCG allow the construction of the fragment "John gave Mary" as valid grammatical text constituent, which is later combined with the direct object of the sentence ('a flower'). Steedman (2000) argues that this form of different syntactic derivations that one can get even for very simple sentences when using CCG (some times referred to with the somewhat belittling term "spurious readings"), actually serve to reflect variations in information structure. Each one of the above derivations subsumes a different intonational pattern, distinguishing the rheme from the theme when the sentence is used for answering different questions: (9) answers to "Who gave Mary a flower?", whereas (10) to "What did John give to Mary?".

In other words, the claim here is that (a) surface structure and information structure coincide; and (b) the role of information structure is to provide a particular interpretation of the surface structure. Let us define this important idea in a precise way, since it will be the cornerstone of the model presented in this paper:

Postulate 4.1 *Intonational boundaries in an utterance determine the intended syntactic structure.*

In our grammatical formalism, pregroup grammars, variations in a grammatical derivation similar to above are only implicitly assumed, since the order of composition remains unspecified. This fact is apparent in the pregroup derivation of the example sentence, where both (9) and (10) are subsumed into the following *reduction diagram*:

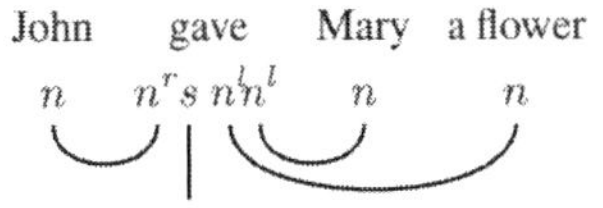

(11)

Furthermore, it is directly reflected in our semantic space through the functorial passage, via the fact that tensor contraction is associative:

$$\overrightarrow{\text{John}} \times (\overline{\text{gave}} \times \overrightarrow{\text{Mary}} \times \overrightarrow{\text{flower}}) = (\overrightarrow{\text{John}} \times \overline{\text{gave}} \times \overrightarrow{\text{Mary}}) \times \overrightarrow{\text{flower}} \quad (12)$$

Eq. 12 constitutes a natural manifestation of the *principle of combinatory transparency* (Steedman, 2001): no matter in what order the various text constituents are combined, the semantic representation assigned to the sentence is always the same; in other words, information structure should not affect semantic conditions. Note, however, that even in the strict setting of formal semantics this is not always the case. Consider the behaviour of the following sentence under the presence of the focus-sensitive particle 'only':

(6) a. John only gave Mary A FLOWER
b. John only gave MARY a flower

The use of different intonational focus clearly changes the semantic value of the sentence: (6a) is true if the only thing that John gave to Mary was a flower (but he might have given things to other girls as well), while (6b) is true if the only person who got a flower from John was Mary.

In the more relaxed and quantitative setting of a compositional distributional model of meaning, the idea of having vectorial representations of words and sentences that reflect intonational patterns seems even more legitimate. This concept is aligned with the distributional nature of such models: given a text corpus containing information structure annotations (of any kind), we would assume that the co-occurrence vector of a word under focus (say, $\overrightarrow{\text{BOOK}}$) would slightly differ from that of the vector representing the normal use of the word ($\overrightarrow{\text{book}}$).[3] Furthermore, we would expect that, after the composition, this difference would be also reflected in

[3] In the trivial case, this would be true by the presence of intonational markers in the immediate context of a word under focus, as opposed to its normal use.

the vector representing the meaning of the entire sentence. From the next section we start working towards imposing this behaviour on the categorical model of Coecke et al. (2010).

5 Intonation in pregroups

Traditionally, a notational system describing intonation consists of markings that indicate pitch accents and boundaries. Using the notation of Pierrehumbert and Hirschberg (1990), for example, we get the following for our example sentence:[4,5]

(7) [MARY]$_R$ [likes MUSICALS]$_T$
H* L L+H* LH%

The prosody starts with a sharp pitch accent (H*) that puts the focus on 'Mary', and continues with a rapid fall to low pitch (L boundary) that signifies a transition from rheme to theme. Within theme now, the focus goes to 'musicals' which gets the less rapidly rising pitch L+H*, whereas the boundary LH% expresses a rising continuation that marks the end of theme. In the case that theme precedes the rheme, we have the following pattern:

(8) [MARY likes]$_T$ [MUSICALS]$_R$
L+H* LH% H* LL%

As mentioned earlier, this paper mainly addresses the rheme/theme aspect of information structure, which is directly related to boundary markings. We start by representing an intonational boundary using a special token $\triangleright$, for which the following relations hold:

$$\textit{theme} \triangleright \textit{rheme} \quad \text{or} \quad \textit{rheme} \triangleleft \textit{theme} \tag{13}$$

Naturally, $\triangleright$ is equivalent to LH%, while $\triangleleft$ corresponds to L in the Pierrehumbert and Hirschberg (1990) notation. It is very important to emphasize at this point that the above introduced tokens are far from an ad-hoc means for achieving a goal. Recall from our discussion in Section 3 that while in English the means of imposing information structure is purely phonological, this is not necessarily the case for other languages. As a concrete case, in Buli (a Gur language spoken in Ghana), the rheme is preceded by a *focus marker*, which again can be interpreted as an information-structural boundary since it separates the theme from rheme; this is shown in the following example (Fiedler et al., 2006):

(9) Q: What did the woman eat?
A: ɔ̀ ŋɔ̀b **kà** **túé**
3sg eat (FM) beans

To formalize this mixing of syntactical and information structure in the context of a pregroup grammar, we add the two boundary markers to the vocabulary and introduce two new atomic types:

$$\text{Theme: } \theta \qquad \text{Rheme: } \rho \tag{14}$$

An intonation pregroup grammar then will have the following form:

$$P(\Sigma \cup \{\triangleright, \triangleleft\}, \{n, s, \theta, \rho\})$$

For the case of a simple transitive sentence, we get the following boundary types, based on the fact that now the boundary (and not the verb) becomes the head of our sentence:

$$\triangleright : \theta^r \cdot s \cdot \rho^l \qquad \triangleleft : \rho^r \cdot s \cdot \theta^l \tag{15}$$

The type dictionary changes accordingly: a transitive verb such as 'like' will be assigned two more types $n^r \cdot \theta$ and $\theta \cdot n^l$ depending whether it produces a left-hand theme or a right-hand theme in the sentence; similarly, nouns will be assigned the extra type ρ. For the two cases of Eq. 13, we obtain the following derivations:

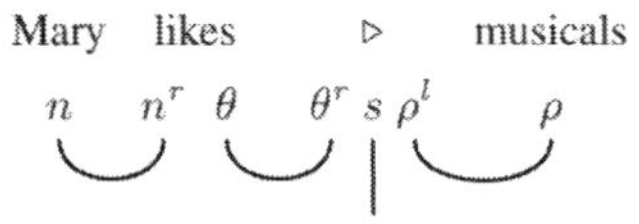

(16)

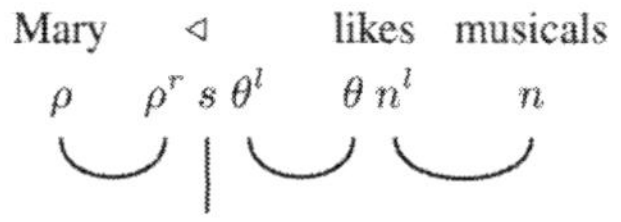

(17)

After transferring this to **FVect** via our functor in Eq. 4, and extending its action on atomic types by defining $\mathcal{F}(\theta) = \Theta$ and $\mathcal{F}(\rho) = P$, we get the obvious semantic counterpart:

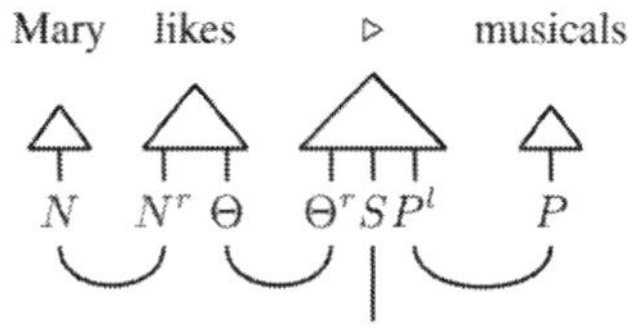

(18)

There are some important observations based on the derivation in (18) above. Firstly, our simple sentence now is given in terms of a theme and a rheme,

[4]Example taken from (Steedman, 2000).

[5]From now on we explicitly mark themes and rhemes in our examples for clarity.

as required, both of which contribute equally to its construction. Additionally, note that our verb is not any more a function of two arguments (of a subject and an object) as in the canonical case, but of a single noun: it takes as input a subject in order to return a theme. Hence, in contrast to a typical case of a transitive verb, the semantic representation of which requires a tensor of order 3, in this case the corresponding linear map takes the form $N \rightarrow \Theta$, which can be canonically represented by a *matrix* $N \otimes \Theta$.

The question of how to properly model intonation in compositional distributional semantics is evidently epitomized in choosing an appropriate form for the tensor of the $\triangleright$ token in (18). In order to provide an answer to this, we first need to examine the concepts of rheme and theme from a semantic point of view.

6 A semantic truth-theoretic argument

We use as an example the following simple case:

(10) Q: Who does John like?
A: [John likes]$_\text{T}$ [MARY]$_\text{R}$

From an extensional point of view, the semantic value of the theme can be seen as a set of *alternative* options (Rooth, 1992), each one of which may be used as a response to the given question:

$$[\![\text{John (might) like}]\!] = \{x|\text{John (might) like } x\}$$

As a consequence, the role of the rheme now is to *restrict* the set of alternatives to a specific choice (Steedman, 2000). Note that this action of restricting the available choices is responsibility of the intonational boundary; indeed, the boundary can be seen as a binary operator that performs the merging of the theme with the rheme, restricting the alternatives set of the theme to a specific response:

(11) [John likes]$_\text{T}$ $\triangleright$ [MARY]$_\text{R}$:=
$\triangleright$([John likes]$_\text{T}$,[MARY]$_\text{R}$)

This is what Diagram (18) shows; in our multilinear setting, the boundary becomes a bi-linear map $\Theta \otimes P \rightarrow S$ that performs the required "restriction". Now, what is the most appropriate way to model this operation in the extensional setting discussed above? Note that by simply checking if rheme is contained in the alternatives set is not sufficient; this would return *true* or *false* as an answer to a question that expects a person. A more appropriate choice then is to model the boundary by using set intersection: we take the meaning of rheme to be a singleton that contains the answer, and the meaning of the sentence to be the intersection of rheme with theme:

$$\{\text{Mary}\} \cap \{x|\text{John (might) like } x\} \quad (19)$$

The answer will be again the singleton {Mary} if Mary is included in the set of people who John potentially likes, and the empty set otherwise. Thus we have achieved our goal: the theme set has been restricted according to the provided response. We generalize this argument to an arbitrary pair of rheme and theme (with S_theme denoting theme's corresponding alternative set) as follows:

$$\triangleright(\text{rheme}, \text{theme}) = \{\text{rheme}\} \cap S_\text{theme} \\ \begin{cases} \text{rheme} & \text{if rheme} \in S_\text{theme} \\ \varnothing & o.w. \end{cases} \quad (20)$$

6.1 From sets to vector spaces

We transfer the above reasoning to vector spaces, by encoding sets and relations in vectorial forms. The vectorial form of a set is a vector space (let it be $N = \{n_i\}_i$) whose basis vectors are the elements of the set. For the sake of demonstration (and this will become clear as the section reads on), we define our sentence space to be a one dimensional space where the origin denotes falsity and everything else denotes truth. One can take this to be a dimension in any vector space; here we take it to be in N and denote its basis vector with a basis vector of N. Furthermore, a binary relation such as $likes(x, y)$ can be represented as an adjacency matrix $\mathbf{W}$ in which W_{ij} is 1 if the pair (i, j) is contained in the relation and 0 otherwise. Note that this matrix is isomorphic to a tensor in $N \otimes S \otimes N$, since our sentence space is one-dimensional.

Let us apply categorical composition to compute a vectorial representation for the theme of our sentence, "John likes".

$$(\epsilon_N^r \otimes 1_S)\left(\overrightarrow{n_3} \otimes \left(\sum_{ij} W_{ij}\overrightarrow{n_i} \otimes \overrightarrow{n_j}\right)\right) = \\ \sum_{ij} W_{ij}\langle\overrightarrow{n_3}|\overrightarrow{n_i}\rangle\overrightarrow{n_j} = \sum_{ij} W_{ij}\delta_{3i}\overrightarrow{n_j} = \sum_j W_{3j}\overrightarrow{n_j} \quad (21)$$

Hence the vectorial representation of "John likes" becomes indeed the subset of all individuals who might be liked by the person denoted by vector $\overrightarrow{n_3}$, and can be seen as the semantic value of the theme of our sentence. The next step is to compose this theme with the rheme 'Mary'; in other words, we must decide an appropriate type of composition for our intonational boundary. Let us first try again standard categorical composition:

$$(1_S \otimes \epsilon^l_N)\left(\left(\sum_j W_{3j}\overrightarrow{n_j}\right) \otimes \overrightarrow{n_1}\right) =$$
$$\sum_j W_{3j}\langle n_j | n_1\rangle = \sum_j W_{3j}\delta_{j1} = W_{31} \qquad (22)$$

Note that this corresponds to a set membership test; the result is 1 if Mary is included in the set of alternative responses and 0 otherwise. However, as noted before, in information structure terms a more appropriate operation would be to take the intersection of the singleton $\{\text{Mary}\}$ with the set of alternatives. Interestingly, set intersection now corresponds to element-wise vector multiplication (in this work denoted by symbol $\odot$) and the vector space equivalent of Eq. 19 becomes:

$$\left(\sum_j W_{3j}\overrightarrow{n_j}\right) \odot \overrightarrow{n_1} = \begin{cases} \overrightarrow{n_1} & \text{if } W_{31} = 1 \\ \overrightarrow{0} & o.w. \end{cases} \qquad (23)$$

The result is now 'Mary', if Mary is included in the set of valid answers, and the zero vector otherwise. The fact that the meaning of our sentence becomes an element of the noun space demonstrates clearly that, in information structure terms, there is a necessity for a shared vector space between sentences and nouns (or noun phrases)—a direct consequence of the fact that now the meaning of a sentence is mainly focused on a specific noun or noun phrase therein. Furthermore, since a sentence is now expressed as a merging of a theme and a rheme, it is also required that $\Theta = S = P$ (and equal to what we took to be N in the preceding). In the next section we encode the above reasoning in the abstract form of compact closed categories and then present an instantiation in vector spaces.

7 Intonation in compact closed categories with Frobenius structure

The point with regard to shared spaces is accomplished by the following types assignment:

$$\mathcal{F}(x) = W \qquad \forall x \in \{n, s, \theta, \rho\} \qquad (24)$$

As a consequence of the above, the vector spaces assigned to transitive verbs are computed as follows:

$$\mathcal{F}(n^r \cdot \theta) = \mathcal{F}(\theta \cdot n^l) = W \otimes W$$

Furthermore, boundaries are assigned to the following vector space:

$$\mathcal{F}(\theta^r \cdot s \cdot \rho^l) = \mathcal{F}(\rho^r \cdot s \cdot \theta^l) = W \otimes W \otimes W$$

We have now arrived at a central point of this paper. As the semantic representation of a boundary, we assign the following morphism:

$$(1_W \otimes \mu_W \otimes 1_W) \circ (\eta_W \otimes \eta_W) \qquad (25)$$

Note that the above is indeed an element in $W \otimes W \otimes W$:

$$\triangleright, \triangleleft : I \cong I \otimes I \xrightarrow{\eta_W \otimes \eta_W} W \otimes W \otimes W \otimes W \qquad (26)$$
$$\xrightarrow{1_W \otimes \mu_W \otimes 1_W} W \otimes W \otimes W$$

The reasoning behind our assignment will become clear in a moment. For now, we proceed to a formal definition:

Definition 7.1 *The meaning vector of a sentence expressed in information structure terms is given by:*

$$(1_W \otimes \mu_W \otimes 1_W) \circ (\eta_W \otimes \eta_W)(\overrightarrow{theme} \otimes \overrightarrow{rheme}) \qquad (27)$$

when theme precedes the rheme, or as follows in the opposite case.

$$(1_W \otimes \mu_W \otimes 1_W) \circ (\eta_W \otimes \eta_W)(\overrightarrow{rheme} \otimes \overrightarrow{theme}) \qquad (28)$$

These vectors are depicted as follows:

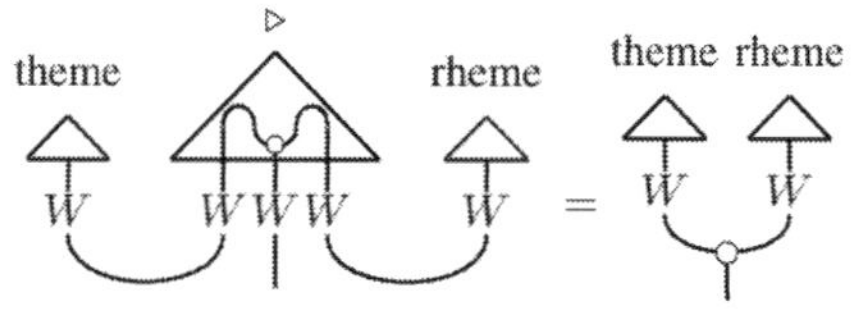

(29)

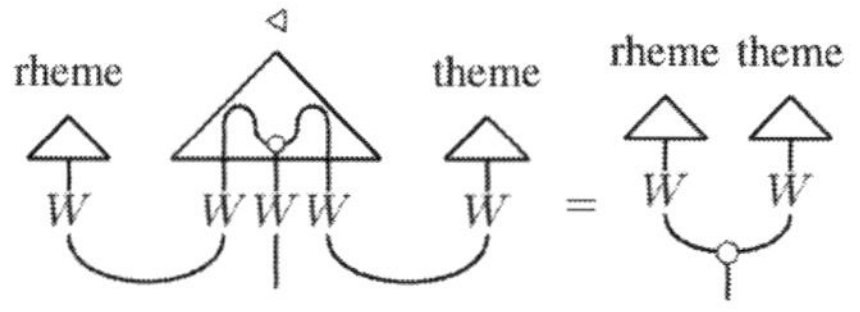

(30)

Note that the normal forms at the right-hand side of the diagrams above are direct applications of the Frobenius condition. Furthermore, either the theme or rheme here might correspond to large text constituents, i.e. phrases or even sentences. In this case, the proposed framework guarantees that an appropriate vector will be created for them based on categorical composition.

8 Vector space instantiation

Our justification for using the semantic form of Eq. 25 for the boundary comes from the fact that it produces normal forms as below:

$$\mu(\overrightarrow{\text{theme}} \otimes \overrightarrow{\text{rheme}}) \quad \mu(\overrightarrow{\text{rheme}} \otimes \overrightarrow{\text{theme}}) \tag{31}$$

This is exactly how element-wise vector multiplication is defined from a categorical perspective:

$$\mu(\vec{v_1} \otimes \vec{v_2}) = \vec{v_1} \odot \vec{v_2} =$$ 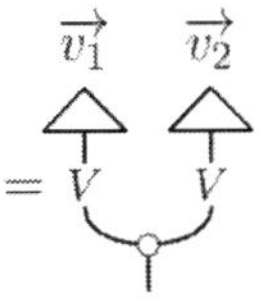 (32)

As a result, the linear algebraic instantiations of Definition 7.1 become as follows:

$$\overrightarrow{\text{rheme}} \odot \overrightarrow{\text{theme}} \qquad \overrightarrow{\text{theme}} \odot \overrightarrow{\text{rheme}}$$

We stress again the fact that rheme and theme can have complex structures, and their vector meanings will reflect this strutter. For simple transitive sentences[6] of the form "subject verb ▷ object" or "subject ◁ verb object", we get linear algebraic meanings as follows:

$$(\overrightarrow{\text{subj}} \times \overline{\text{verb}}) \odot \overrightarrow{\text{obj}} \qquad \overrightarrow{\text{subj}} \odot (\overline{\text{verb}} \times \overrightarrow{\text{obj}})$$

As an example of a composed theme, consider:

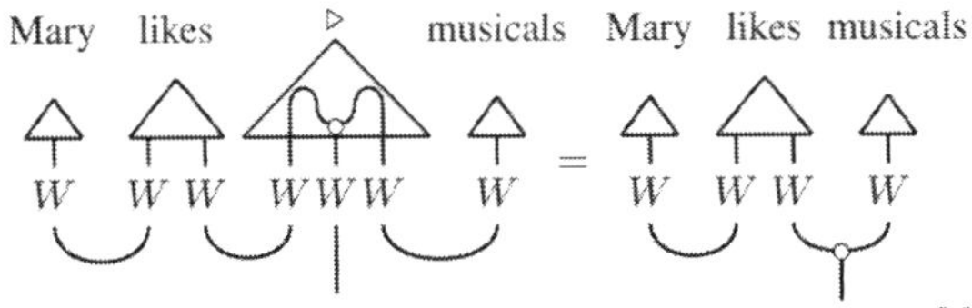

(33)

A vector is computed for the theme 'Mary likes' according to the rules of the grammar, and then this vector is element-wise multiplied with the vector of the rheme (which, in this example, is just the distributional vector of the word).

9 Interpretation

The transition from the set-theoretical framework to high dimensional real vector spaces poses the question what is the role of element-wise vector multiplication in the latter setting. Compositional models based on element-wise vector addition or multiplication are usually referred to as *vector mixture* models—a term that emphasizes on the *equal contribution* of each word to the final result, which produces a kind of average of the input vectors. Note that this behaviour stands in direct contrast with the categorical compositional approach, in which the type-logical identities of words strictly depend on their grammatical role. Due to their simplicity, vector mixture models have been studied extensively (Mitchell and Lapata, 2008), demonstrating steady and reasonably good performance in a number of tasks.

The significance of the Frobenius operators for our model (as opposed to some other form of combinatory mechanism) is that their concrete manifestation in a vector space setting imposes exactly this vector mixture behaviour, in the form of element-wise vector multiplication. In other words, the result is a combination of two compositional approaches, vector mixtures and categorical models, in a unified framework: while categorical composition is still applied to compute vectorial representations for a theme and a rheme, the two parts contribute equally to the final result via element-wise multiplication imposed by the Frobenius operators. This puts the necessary focus on the appropriate part of the sentence, reflecting the variation in meaning intended by the intonational pattern.

To what extent the notion of a rheme as a means for restricting the theme applies in **FVect**? Note that, from a geometric perspective, element-wise vector multiplication acts as a scaling of the basis; for example, $\binom{x}{y} \odot \binom{2.0}{0.5}$ transforms the vector space in which the first vector lives so that the units on the x-axis are doubled and the units on the y-axis are halved.[7] Furthermore, a zero value in one vector would completely eliminate the corresponding component in the other. Hence, the concept of restricting the theme has now taken a new quantitative form, generalizing appropriately our initial intuition (motivated by set intersection) to the multi-dimensional, real-valued setting of **FVect**.

10 Relation to previous work

How does the above derivations correlate to the premises of the original framework, in which 'likes' is a transitive verb with type $n^r \cdot s \cdot n^l$? Note that another application of the Frobenius condition on the normal form of Diagram (33) will give us:

[6] We provide more complicated examples later in Sect. 11.

[7] Of course we can think of a similar scaling taking place on the two axes of the second vector by factors x and y.

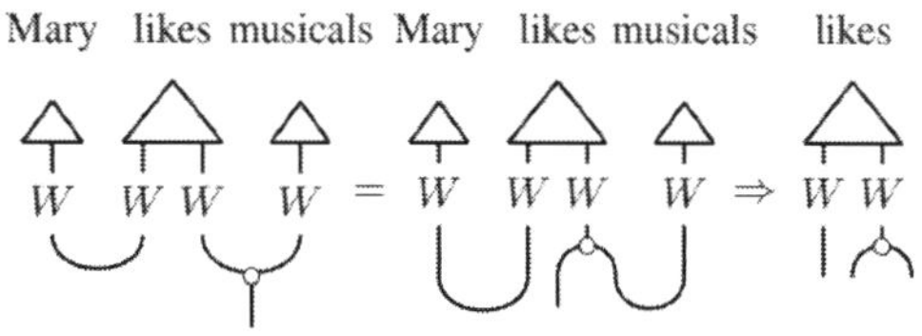

(34)

In other words, the semantic representation of word 'likes' can be still regarded as a bi-linear map, faithfully encoded in a tensor of order 3, as required by the framework. In this case, the tensor of 'likes' in **FVect** is seen as created by applying the morphism $1_W \otimes \Delta_W$ on a matrix representing the verb 'likes'. The limitation, of course, is that now the middle wire carrying the result (the sentence vector space) cannot be any more differentiated from the two argument wires (the noun vector spaces), since it is produced by copying one of them.

Note that these are the Frobenius models of Kartsaklis et al. (2014), referred to as Copy-Subject and Copy-Object, and originally used as a means for faithfully encoding a verb matrix to a tensor of order 3, thus restoring the functorial relation between the semantic representation and the grammatical type. The present theory[8] offers an alternative more complete account that goes far beyond providing a convenient way to expand a matrix to a cube.

11 Covering complex intonational patterns

So far we examined simple cases of intonation, in which our sentence consisted of a single rheme and a theme. In this section we turn our attention to some more interesting examples.

11.1 Multiple rhemes

We will first examine the case of a sentence with more than one rhemes. Imagine the following question/answer dialogue:

(12) Q: Who likes whom?
A: [JOHN]$_R$ [likes]$_T$ [MARY]$_R$

In our pregroup notation, this introduces two distinct intonational boundaries in the sentence. The derivation takes the following form:

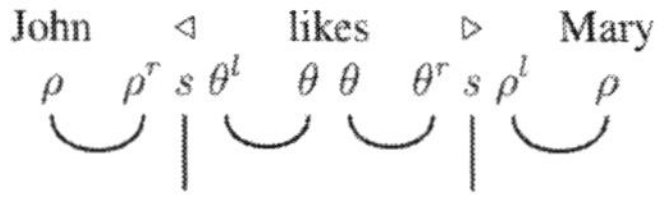

(35)

[8] An early account of which also appears in the doctoral thesis of the first author (Kartsaklis, 2015).

Note that the type of 'likes' now becomes $\theta \cdot \theta$; in other words, the theme is not any more a function (no adjoint is present in the type), but a *higher order* atomic entity. This is directly reflected in **FVect** where we get:

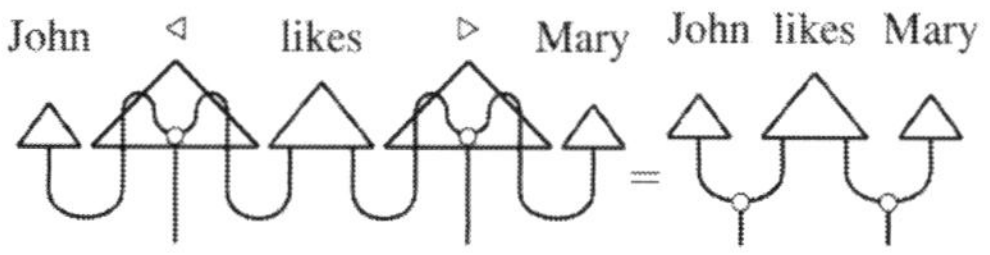

(36)

The result of this computation is now a matrix and not a vector. Indeed, if we follow the linear algebraic calculations we get:

$$(\mu_W \otimes \mu_W)(\overrightarrow{\text{John}} \otimes \overline{\text{likes}} \otimes \overrightarrow{\text{mary}}) = (\overrightarrow{\text{John}} \otimes \overrightarrow{\text{Mary}}) \odot \overline{\text{likes}} \quad (37)$$

The behaviour above follows the premises of the proposed model: Since our theme is a matrix, the calculations follow naturally, producing another matrix as the rheme (the tensor product of the two individual rhemes) that *restricts* as required the theme via element-wise multiplication. Note that this means that a sentence with one rheme would not be comparable with a sentence with two rhemes, since it would live in a different space. That is again not surprising: the shape of theme defines the shape of the sentence vector space, and only themes of the same order can be compared to each other.

11.2 Relational words as rhemes

We have conveniently avoided to discuss until now the case in which the rheme is not a noun phrase, but a relational word as below:

(13) Q: How does John feel about Mary?
A: [John]$_T$ [LIKES]$_R$ [Mary]$_T$

In pregroups we model such a situation by the following derivation:

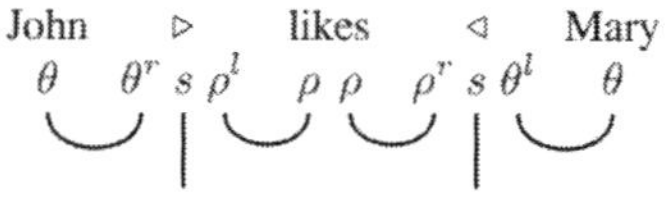

(38)

Note that this time the verb becomes a higher order rheme, getting the type $\rho \cdot \rho$. However, when this is transferred to **FVect** the symmetry of the category and the commutativity of the Frobenius algebra means that the vector of the sentence becomes equal to that of Example (12). In general, problems due to commutativity of the Frobenius operators can be resolved if one moves to non-commutative versions of Frobenius algebras. Piedeleu et al. (2015) explore such constructions in the context of language by elevating the categorical model of Coecke et al. (2010) to an open quantum system setting, in which words are represented as mixed states.

11.3 Nested rhemes

Consider the following case:

(14) What was the book Mary wrote about?
[Mary wrote a book about]$_T$ [ART]$_R$

The interesting point here is that the intonational boundary is placed in a position that constitutes a glaring violation of the grammatical structure, which in the normal case has the following form:

(39)

For cases like these we should recall that our framework is entirely built on the assumption of Postulate 4.1: in the context of information structure, intonational boundaries *determine* the intended syntactical structure. For our case, we get:

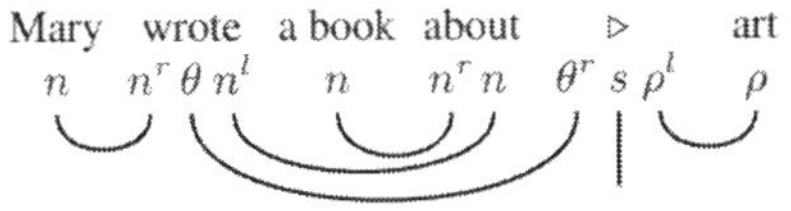

(40)

The linear-algebraic result follows trivially as the usual element-wise composition of the theme with the rheme.

11.4 Rheme in the middle of sentence

In many cases a noun phrase can serve as the rheme while being placed in the middle of the sentence, splitting the theme into two parts:

(15) Did Mary write an essay about art?
[Mary wrote]$_T$ [A BOOK]$_R$ [about art]$_T$

In these cases, the left-hand intonational boundary gets the type $\theta^r \cdot \rho \cdot \rho^l$, as below:

(41)

In other words, a new rheme is produced that is used as input to the right-hand intonational boundary. In **Fvect** we get the following interaction:

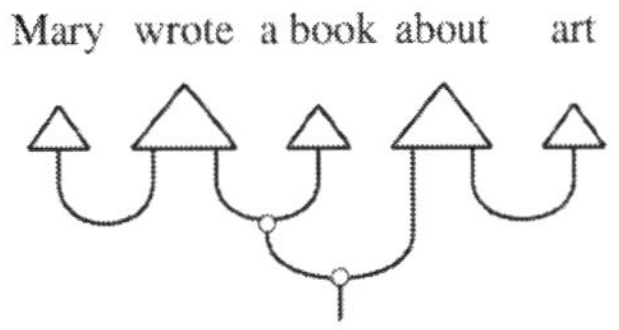

(42)

By application of the spider equality (Section 2.6) we get the normal form below, which computes a meaning for the sentence as the element-wise multiplication of the vectors composed for the two themes with the vector of the rheme:

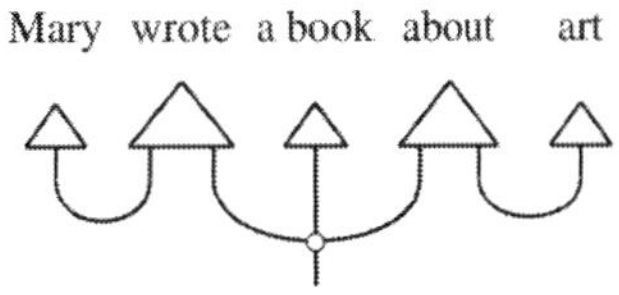

(43)

12 Conclusion and future work

The present paper provides a first account of intonation and information structure for the emerging field of categorical compositional distributional semantics. In a more generic level, it lays the groundwork for a model capable of accommodating two different types of composition over a distributional setting. An experimental evaluation is deferred for the future, preferably in the context of a question-answering task. There is also a lot of interesting work to be done on the theory side. At the current stage, for example, the semantic value of intonational boundaries is given by direct assignment of a specific morphism—a common practice in the past for the relevant literature. A future direction, then, more aligned with the categorical nature of the model, would be to embed the appropriate translation into the functorial passage itself. This challenging goal requires novel theoretical contributions that will elevate the concept of a pregroup grammar to a new entity equipped with Frobenius structure.

Finally, the categorical compositional model of Piedeleu et al. (2015) is very relevant to our interests, since it can accommodate a variety of non-commutative Frobenius algebras the linguistic intuition of which in relation to this work remains to be explored.

Acknowledgements

We would like to thank the three anonymous reviewers for their comments, and Bob Coecke for great discussions on the paper. Financial support from AFOSR is gratefully acknowledged by the authors.

References

Wallace L. Chafe. 1976. Givenness, contrastiveness, definiteness, subjects, topics and point of view. In Charles N. Li, editor, *Subject and Topic*. Academic Press, New York.

B. Coecke, M. Sadrzadeh, and S. Clark. 2010. Mathematical Foundations for a Compositional Distributional Model of Meaning. Lambek Festschrift. *Linguistic Analysis*, 36:345–384.

Caroline Féry and Manfred Krifka. 2008. Information Structure: Notional distinctions, ways of expression. In Piet van Sterkenburg, editor, *Unity and diversity of languages*, pages 123–135. John Benjamins Publishing.

Ines Fiedler, Katharina Hartmann, Brigitte Reineke, Anne Schwarz, and Malte Zimmermann. 2006. Subject focus in West African languages. In *International Conference of Information Structure*, Potsdam.

Z. Harris. 1968. *Mathematical Structures of Language*. Wiley.

Dimitri Kartsaklis, Mehrnoosh Sadrzadeh, and Stephen Pulman. 2012. A unified sentence space for categorical distributional-compositional semantics: Theory and experiments. In *Proceedings of 24th International Conference on Computational Linguistics (COLING 2012): Posters*, pages 549–558, Mumbai, India, December. The COLING 2012 Organizing Committee.

Dimitri Kartsaklis, Mehrnoosh Sadrzadeh, Stephen Pulman, and Bob Coecke. 2014. Reasoning about meaning in natural language with compact closed categories and Frobenius algebras. *arXiv preprint arXiv:1401.5980*.

Dimitri Kartsaklis. 2015. *Compositional Distributional Semantics with Compact Closed Categories and Frobenius Algebras*. Ph.D. thesis, University of Oxford.

G Maxwell Kelly. 1972. Many-variable functorial calculus (I). In G.M. Kelly, M. Laplaza, G. Lewis, and S. MacLane, editors, *Coherence in categories*, pages 66–105. Springer.

J. Lambek. 2008. *From Word to Sentence*. Polimetrica, Milan.

T. Landauer and S. Dumais. 1997. A Solution to Plato's Problem: The Latent Semantic Analysis Theory of Acquision, Induction, and Representation of Knowledge. *Psychological Review*.

Mike Lewis and Mark Steedman. 2013. Combined distributional and logical semantics. *Transactions of the Association for Computational Linguistics*, 1:179–192.

Saunders Mac Lane. 1971. *Categories for the Working Mathematician*. Number 5 in Graduate Texts in Mathematics. Springer-Verlag.

C.D. Manning, P. Raghavan, and H. Schütze. 2008. *Introduction to Information Retrieval*. Cambridge University Press.

J. Mitchell and M. Lapata. 2008. Vector-based Models of Semantic Composition. In *Proceedings of the 46th Annual Meeting of the Association for Computational Linguistics*, pages 236–244.

Richard Montague. 1970. English as a formal language. In *Linguaggi nella Società e nella Tecnica*, pages 189–224. Edizioni di Comunità, Milan.

Robin Piedeleu, Dimitri Kartsaklis, Bob Coecke, and Mehrnoosh Sadrzadeh. 2015. Open System Categorical Quantum Semantics in Natural Language Processing. In *Proceedings of the 6th Conference on Algebra and Coalgebra in Computer Science (CALCO)*, Nijmegen, Netherlands, June.

Janet Pierrehumbert and Julia Hirschberg. 1990. The meaning of intonational contours in the interpretation of discourse. In Philip Cohen, Jerry Morgan, and Martha Pollack, editors, *Intentions in communication*, pages 271–312. MIT Press, Cambridge MA.

Mats Rooth. 1992. A theory of focus interpretation. *Natural language semantics*, 1(1):75–116.

M. Sadrzadeh, S. Clark, and B. Coecke. 2013. The Frobenius anatomy of word meanings I: subject and object relative pronouns. *Journal of Logic and Computation*, Advance Access, October.

M. Sadrzadeh, S. Clark, and B. Coecke. 2014. The Frobenius anatomy of word meanings II: Possessive relative pronouns. *Journal of Logic and Computation*, June.

H. Schütze. 1998. Automatic Word Sense Discrimination. *Computational Linguistics*, 24:97–123.

Peter Selinger. 2011. A survey of graphical languages for monoidal categories. In Bob Coecke, editor, *New structures for physics*, pages 289–355. Springer.

Mark Steedman. 2000. Information structure and the syntax-phonology interface. *Linguistic inquiry*, 31(4):649–689.

Mark Steedman. 2001. *The Syntactic Process*. MIT Press.

Peter D Turney and Patrick Pantel. 2010. From frequency to meaning: Vector space models of semantics. *Journal of artificial intelligence research*, 37(1):141–188.

A Synopsis of Morphoid Type Theory

David McAllester
TTI-Chicago
mcallester@ttic.edu

Abstract

Morphoid type theory (MTT) is a type-theoretic foundation for mathematics supporting the concept of isomorphism and the substitution of isomorphics. Unlike homotopy type theory (HoTT), which also supports isomorphism, morphoid type theory is a direct extension of classical predicate calculus and avoids the intuitionistic constructs of propositions-as-types, path induction and squashing. Although HoTT is capable of supporting classical inference, MTT's thoroughly classical treatment is expected to be more comfortable for those who take a Platonic or realist approach to the practice of mathematics.

1 Introduction

The central issue in both homotopy type theory (HoTT-Authors, 2013) and morphoid type theory (McAllester, 2014) is isomorphism. The notion of isomorphism in mathematics seems related to the notion of an application programming interface (API) in computer software. An API specifies what information and behavior an object provides. Two different implementations can produce identical behavior when interaction is restricted to that allowed by the API. For example, textbooks on real analysis typically start from axioms involving multiplication, addition, and ordering. Addition, multiplication and ordering define an abstract interface — the well formed statements about real numbers are limited to those that can be defined in terms of the operations of the interface. We can implement real numbers in different ways — as Dedekind cuts or Cauchy sequences. However, these different implementations provide identical behavior as viewed through the interface — the different implementations are isomorphic as ordered fields. The axioms of real analysis specify the reals up to isomorphism for ordered fields. Peano's axioms (the second order version) similarly specify the structure of the natural numbers up to isomorphism.

The general notion of isomorphism is best illustrated by considering dependent pair types. Here we will write a dependent pair type as $\mathbf{PairOf}\,(x\!:\!\sigma,\ y\!:\!\tau[x])$ where the instances of this type are the pairs $\mathbf{Pair}(x, y)$ where x is an instance of the type σ and y is an instance of $\tau[x]$. The type of directed graphs can be written as $\mathbf{PairOf}\,(\mathcal{N}\!:\!\mathbf{type},\ P\!:\!(\mathcal{N} \times \mathcal{N}) \to \mathbf{Bool})$ where $\mathcal{N}$ is a type representing the set of nodes of the graph and P is a binary predicate on the nodes giving the edge relation. Two directed graphs $\mathbf{Pair}(\mathcal{N}, P)$ and $\mathbf{Pair}(\mathcal{M}, Q)$ are isomorphic if there exists a bijection from $\mathcal{N}$ to $\mathcal{M}$ that carries P to Q. Some bijections will carry P to Q while others will not. Two pairs $\mathbf{Pair}(x, y)$ and $\mathbf{Pair}(u, w)$ of a general dependent pair type $\mathbf{PairOf}\,(x\!:\!\sigma,\ y\!:\!\tau[x])$ are isomorphic if there is a σ-isomorphism from x to u that carries y to w. Some σ-isomorphisms from x to u will carry y to w while others will not. This implies that to define isomorphism at general dependent pairs types we need that for any type σ, and for any two isomorphic values x and u of type σ, we can define the full set of σ-isomorphisms from x to u. An interesting special case is the full set of σ-isomorphisms from x to x. This is the symmetry group of x.

Proceedings of the 14th Meeting on the Mathematics of Language (MoL 14), pages 75–86,
Chicago, USA, July 25–26, 2015. ©2015 Association for Computational Linguistics

Both Homotopy type theory (HoTT) and morphoid type theory (MTT) are intended as type-theoretic foundations for mathematics supporting a concept of isomorphism. HoTT is an extension of constructive logic while MTT is an extension of classical predicate calculus. More specifically, HoTT is a version of Martin Löf type theory (Martin-Löf, 1971; Coquand and Huet, 1988; Sambin and Smith, 1998) extended to includes Voevodsky's univalence axiom (HoTT-Authors, 2013). Martin-Löf type theory involves propositions-as-types and path induction, neither of which are used in MTT. To accommodate classical (nonconstructive) inference, HoTT can be extended with a version of the law of the excluded middle. However, even in the classical version propositions continue to be represented as types rather than Boolean-valued expressions. To accommodate classical inference HoTT also includes squashing — a technicality required to allow propositions-types to be treated more like Boolean-valued expressions. In MTT all propositions are Boolean-valued and there is no need for squashing.

Perhaps the most significant difference between HoTT and MTT involves the abstraction barrier imposed on types. In MTT two types are type-isomorphic if there exists a bijection between them. In MTT two types with the same cardinality (number of equivalence classes) cannot be distinguished by well-typed predicates on types. In HoTT, however, types with the same cardinality can still be distinguished by well-typed predicates. In HoTT two types are equivalent only when they have the same higher-order groupoid structure. For example, two graphs fail to be isomorphic unless the node types have the same higher order groupoid structure. This can be interpreted as implementation details of a type leaking from the abstraction barrier on types. This leakage interpretation is discussed more explicitly in section 3.

HoTT allows one to block the leakage of type implementations by squashing types to "sets". A set is a type whose internal groupoid structure is effectively suppressed. One can construct the type of topological space whose point types are sets. In this case we get the familiar notion isomorphism (homeomorphism) where two topological spaces are homeomorphic if there is *any* bijection between their points that identifies their open sets. We can then define the groupoid of topological spaces to be the category consisting of the topological spaces and the homeomorphisms between them. This is the "first order" groupoid of topological spaces. If we take the point types of topological spaces to be first order groupoids rather than sets, and restrict the point bijections to functors, we get the second order groupoid of topological spaces. We can then define a third order groupoid and so on. In HoTT we can even have ω-order groupoids.

In MTT the internal structure of types is approached in a different way. In MTT natural mappings are distinguished from general functions. For example, there is an isomorphism (a linear bijection) from a finite dimensional vector space to its dual. However, there is no natural isomorphism. Although not covered in this synopsis, MTT takes a function to be natural if it can be written as a lambda expression. Lambda expressions (natural functions) have commutation properties not shared by general functions. Two types σ and τ are cryptomorphic (in the sense of Birkoff or Rota) if there exists a pair of natural functions (lambda expressions) $f : \sigma \to \tau$ and $g : \tau \to \sigma$ such that $f \circ g$ and $g \circ f$ are both identity functions (viewed as functions on the isomorphism classes of σ and τ respectively). MTT does not attempt to handle higher order groupoid structure.

This synopsis of MTT is preliminary and many of the features described here go beyond the features covered by soundness proofs in version 4 of (McAllester, 2014). This synopsis should be viewed as a plan, or program, for the next version of (McAllester, 2014).

2 The Core Rules of Morphoid Type Theory

Morphoid type theory starts from the syntax and semantics of classical predicate calculus. In sorted first order logic every term has a sort and each function symbol f specifies the sorts of its arguments and the sort of its value. We write $f : \sigma_1 \times \cdots \times \sigma_n \to \tau$ to indicate that f is a function that takes n arguments of sort $\sigma_1, \ldots, \sigma_n$ respectively and which produces a value of sort τ. The syntax of sorted first order logic can be defined by the following grammar where function and predicate applications must sat-

$$\epsilon \vdash \mathbf{type}_j : \mathbf{type}_i \text{ for } j < i \qquad \frac{\Sigma \vdash \tau : \mathbf{type}_i \quad x \text{ not declared in } \Sigma}{\Sigma;\ x : \tau \vdash x : \tau} \qquad \frac{\Sigma \vdash \tau : \mathbf{type}_i \quad \Sigma \vdash \sigma : \mathbf{type}_i}{\Sigma \vdash (\tau \to \sigma) : \mathbf{type}_i} \qquad \frac{\Sigma \vdash f : \sigma \to \tau \quad \Sigma \vdash e : \sigma}{\Sigma \vdash f(e) : \tau}$$

$$\epsilon \vdash \mathbf{Bool} : \mathbf{type}_i \qquad \frac{\Sigma \vdash \Phi : \mathbf{Bool}}{\Sigma; \Phi \vdash \Phi} \qquad \frac{\Sigma; \Theta \vdash \Theta \quad \Sigma \vdash \Psi}{\Sigma; \Theta \vdash \Psi}$$

$$\frac{\Sigma \vdash \Phi : \mathbf{Bool} \quad \Sigma \vdash \Psi : \mathbf{Bool}}{\Sigma \vdash (\Phi \vee \Psi) : \mathbf{Bool}} \qquad \frac{\Sigma \vdash \Phi : \mathbf{Bool}}{\Sigma \vdash \neg\Phi : \mathbf{Bool}} \qquad \frac{\Sigma;\ x : \tau \vdash \Phi[x] : \mathbf{Bool}}{\Sigma \vdash (\forall x : \tau\ \Phi[x]) : \mathbf{Bool}} \qquad \frac{\Sigma \vdash w : \tau \quad \Sigma \vdash u : \tau}{\Sigma \vdash (w =_\tau u) : \mathbf{Bool}}$$

Figure 1: **Predicate Calculus Expressions**. Here $\mathbf{type}_0, \mathbf{type}_1, \mathbf{type}_2, \ldots$ are distinct constants and ϵ is a constant denoting the empty context. The first two rules of the first row allow us to derive $\epsilon; \alpha : \mathbf{type}_j \vdash \alpha : \mathbf{type}_j$ thereby declaring primitive types. We can then declare additional symbols such as $c : \alpha$ or $P : \alpha \to \mathbf{Bool}$. The requirement of $j < i$ in the first rule is needed to avoid Russel's paradox. The second rule of the second row allows Boolean assumptions to be introduced into contexts.

isfy the sort constraints associated with the function and predicate symbols.

$$\begin{array}{rcl} t & ::= & x \mid c \mid f(t_1, \ldots, t_n) \\ \Phi & ::= & P(t_1, \ldots, t_n) \mid t_1 =_\sigma t_2 \\ & & \mid \Phi_1 \vee \Phi_2 \mid \neg\Phi \mid \forall x : \sigma\ \Phi[x] \end{array}$$

Note that in the above grammar the equality symbol $=_\sigma$ is subscripted with a sort σ to which it applies. The labeling of equality with sorts is important for the treatment of isomorphism.

Given this basic grammar of terms and formulas it is standard to introduce the following abbreviations.

$$\begin{array}{rcl} \Phi \wedge \Psi & \equiv & \neg(\neg\Phi \vee \neg\Psi) \\ \Phi \Rightarrow \Psi & \equiv & \neg\Phi \vee \Psi \\ \exists x : \sigma\ \Phi[x] & \equiv & \neg\forall x : \sigma\ \neg\Phi[x] \\ \exists! x : \sigma\ \Phi[x] & \equiv & \left\{ \begin{array}{l} \exists x : \sigma\ \Phi[x] \\ \wedge \forall x, y : \sigma \\ \quad \Phi[x] \wedge \Phi[y] \Rightarrow x =_\sigma y \end{array} \right. \\ (\exists x : \sigma) & \equiv & \exists x : \sigma\ x =_\sigma x \end{array}$$

We now replace the word "sort" with the word "type". To define the set of well formed terms and formulas we need to specify primitive types and a set of constant and function symbols each with specified argument and value types. In formal type systems this is done with symbol declarations. We write $\Sigma \vdash t : \sigma$ to indicate that the symbol declarations in Σ imply that t is a well-formed expression of type σ. For example we have the following.

$$\left. \begin{array}{l} \alpha : \mathbf{type}; \\ \beta : \mathbf{type}; \\ c : \alpha; \\ f : \alpha \to \beta \end{array} \right\} \vdash \quad f(c) : \beta$$

$$\left. \begin{array}{l} \alpha : \mathbf{type}; \\ c : \alpha; \\ f : \alpha \to \alpha; \\ P : \alpha \to \mathbf{Bool} \end{array} \right\} \vdash \quad P(f(f(c))) : \mathbf{Bool}$$

An expression of the form $\Sigma \vdash \Theta$ is called a *sequent* where Σ is called the context and Θ is called the judgement. The sequent $\Sigma \vdash \Theta$ says that judgement Θ holds in context Σ. We allow a context to contain both symbol declarations and Boolean assumptions. For example we have

$$\left. \begin{array}{l} \alpha : \mathbf{type};\ a : \alpha;\ b : \alpha; \\ f : \alpha \times \alpha \to \alpha; \\ \forall x : \alpha\ \forall y : \alpha \\ \quad f(x, y) =_\alpha f(y, x) \end{array} \right\} \vdash f(a, b) =_\alpha f(b, a)$$

$$\frac{\Sigma \vdash \Phi : \mathbf{Bool} \quad \Sigma \vdash \Psi : \mathbf{Bool} \quad \Sigma; \Phi \vdash \Psi \quad \Sigma; \neg\Phi \vdash \Psi}{\Sigma \vdash \Psi}$$

$$\frac{\Sigma \vdash \Phi : \mathbf{Bool} \quad \Sigma \vdash \Psi : \mathbf{Bool} \quad \Sigma \vdash \Phi}{\Sigma \vdash \Phi \vee \Psi \quad \Sigma \vdash \neg\neg\Phi}$$

$$\frac{\Sigma \vdash \Phi : \mathbf{Bool} \quad \Sigma \vdash \Psi : \mathbf{Bool} \quad \Sigma \vdash \Psi}{\Sigma \vdash \Phi \vee \Psi}$$

$$\frac{\Sigma \vdash \Phi : \mathbf{Bool} \quad \Sigma \vdash \Psi : \mathbf{Bool} \quad \Sigma \vdash \neg\Psi \quad \Sigma \vdash \neg\Phi}{\Sigma \vdash \neg(\Phi \vee \Psi)}$$

$$\frac{\Sigma \vdash \forall x : \tau\ \Phi[x] \quad \Sigma \vdash e : \tau}{\Sigma \vdash \Phi[e]}$$

$$\frac{\Sigma; x : \tau \vdash \Phi[x] : \mathbf{Bool} \quad \Sigma; x : \tau \vdash \Phi[x]}{\Sigma \vdash \forall x : \tau\ \Phi[x]}$$

$$\frac{\Sigma \vdash \exists! x : \sigma\ \Phi[x]}{\Sigma \vdash \mathbf{The}(x : \sigma,\ \Phi[x]) : \sigma \quad \Sigma \vdash \Phi[\mathbf{The}(x : \sigma,\ \Phi[x])]}$$

$$\frac{\Sigma \vdash e : \tau}{\Sigma \vdash e =_\tau e}$$

$$\frac{\Sigma \vdash u =_\tau w}{\Sigma \vdash w =_\tau u}$$

$$\frac{\Sigma \vdash u =_\tau w \quad \Sigma \vdash w =_\tau s}{\Sigma \vdash u =_\tau s}$$

$$\frac{\Sigma \vdash \tau : \mathbf{type}_i \quad \Sigma;\ x : \sigma \vdash e[x] : \tau \quad \Sigma \vdash w =_\sigma u \quad \Sigma, \sigma, \tau \text{ and } e[x] \text{ are pure}}{\Sigma \vdash e[w] =_\tau e[u]}$$

$$\frac{\Sigma \vdash f, g : \sigma \to \tau \quad \Sigma \vdash \forall x : \sigma\ f(x) =_\tau g(x)}{\Sigma \vdash f =_{\sigma \to \tau} g}$$

$$\frac{\Sigma \vdash \tau : \mathbf{type}_i \quad \Sigma \vdash \forall x : \sigma\ \exists y : \tau\ \Phi[x,\ y] \quad \Sigma, \sigma, \tau \text{ and } \Phi[x, y] \text{ are pure}}{\Sigma \vdash \exists f : \sigma \to \tau\ \forall x : \sigma\ \Phi[x, f(x)]}$$

Figure 2: **Predicate Calculus Inference Rules.** The first row is a complete set of rules for Boolean logic. A rule with two conclusions abbreviates two rules each with the same antecedents. The third rule in the second row handles definite descriptions (Hilbert's ι-operator). An expression is "pure" if does not involve any of the constructs introduced in figure 6. The last row gives the axioms of extensionality and choice.

In higher order predicate calculus the type system is extended to include not only primitive types but also function types and we can write, for example, $P(f)$ where we have $f : \sigma \to \tau$ and $P : (\sigma \to \tau) \to$ **Bool**. In the higher order case we can use the following standard abbreviations due to Curry.

$$\sigma_1 \times \sigma_2 \to \tau \quad \equiv \quad \sigma_1 \to (\sigma_2 \to \tau)$$
$$f(a, b) \quad \equiv \quad f(a)(b)$$

This extends in the obvious way to abbreviations of the form $\sigma_1 \times \cdots \times \sigma_n \to \tau$. Without loss of generality we then need consider only single argument functions.

Figure 1 gives a set of inference rules for forming the expressions of higher order predicate calculus. Each rule allows for the derivation of the sequent below the line provided that the sequents above the line are derivable. A rule with no antecedents is written as a single derivable sequent.

Figure 2 gives inference rules for predicate calculus including definite descriptions of the form $\mathbf{The}(x : \sigma,\ \Phi[x])$ (Hilbert's ι-operator) and rules representing the axiom of extensionality and the axiom of choice.

Figure 3 gives inference rules for dependent pair types, subtypes, and existential types. A dependent pair type has the form $\mathbf{PairOf}\,(x : \sigma,\ y : \tau[x])$ and is the type whose instances are the pairs $\mathbf{Pair}(x, y)$ where x is an instance of σ and y is an instance of $\tau[x]$. A subtype expression has the form $\mathbf{SubType}(x : \sigma,\ \Phi[x])$ where $\Phi[x]$ is a Boolean expression. This expression denotes the type whose elements are those elements x in σ such that $\Phi[x]$ holds. We let $\mathbf{PairOf}(x : \sigma,\ y : \tau[x]$ s. t. $\Phi[x, y])$

$$\frac{\begin{array}{l}\Sigma \vdash \sigma : \mathbf{type}_i \\ \Sigma;\ x:\sigma \vdash \tau[x] : \mathbf{type}_i\end{array}}{\Sigma \vdash \mathbf{PairOf}\,(x:\sigma,\ y:\tau[x]) : \mathbf{type}_i} \qquad \frac{\begin{array}{l}\Sigma \vdash \mathbf{PairOf}\,(x:\sigma,\ y:\tau[x]) : \mathbf{type}_i \\ \Sigma \vdash u:\sigma \\ \Sigma \vdash w:\tau[u]\end{array}}{\begin{array}{l}\Sigma \vdash \mathbf{Pair}(u,w) : \mathbf{PairOf}\,(x:\sigma,\ y:\tau[x]) \\ \Sigma \vdash \pi_1(\mathbf{Pair}(u,w)) \doteq u \\ \Sigma \vdash \pi_2(\mathbf{Pair}(u,w)) \doteq w\end{array}} \qquad \frac{\Sigma \vdash p : \mathbf{PairOf}\,(x:\sigma,\ y:\tau[x])}{\begin{array}{l}\Sigma \vdash \pi_1(p):\sigma \\ \Sigma \vdash \pi_2(p):\tau[\pi_1(p)] \\ \Sigma \vdash p \doteq \mathbf{Pair}(\pi_1(p),\ \pi_2(p))\end{array}}$$

$$\Sigma \vdash e \doteq e \qquad \frac{\Sigma \vdash u \doteq w}{\Sigma \vdash w \doteq u} \qquad \frac{\begin{array}{l}\Sigma \vdash u \doteq w \\ \Sigma \vdash w \doteq s\end{array}}{\Sigma \vdash u \doteq s} \qquad \frac{\begin{array}{l}\Sigma \vdash u \doteq w \\ \Sigma \vdash \Theta[u]\end{array}}{\Sigma \vdash \Theta[w]}$$

$$\frac{\begin{array}{l}\Sigma \vdash \tau : \mathbf{type}_i \\ \Sigma;\ x:\tau \vdash \Phi[x] : \mathbf{Bool}\end{array}}{\Sigma \vdash \mathbf{SubType}\,(x:\tau,\ \Phi[x]) : \mathbf{type}_i} \qquad \frac{\begin{array}{l}\Sigma \vdash \mathbf{SubType}\,(x:\tau,\ \Phi[x]) : \mathbf{type}_i \\ \Sigma \vdash e:\tau \\ \Sigma \vdash \Phi[e]\end{array}}{\Sigma \vdash e : \mathbf{SubType}\,(x:\tau,\ \Phi[x])} \qquad \frac{\Sigma \vdash e : \mathbf{SubType}\,(x:\tau,\ \Phi[x])}{\begin{array}{l}\Sigma \vdash e:\tau \\ \Sigma \vdash \Phi[e]\end{array}}$$

$$\frac{\begin{array}{l}\Sigma \vdash \sigma : \mathbf{type}_i \\ \Sigma;\ x:\sigma \vdash \tau[x] : \mathbf{type}_i\end{array}}{\Sigma \vdash (\exists x:\sigma\ \tau[x]) : \mathbf{type}_i} \qquad \frac{\begin{array}{l}\Sigma \vdash (\exists x:\sigma\ \tau[x]) : \mathbf{type}_i \\ \Sigma \vdash w:\sigma \\ \Sigma \vdash e:\tau[w]\end{array}}{\Sigma \vdash e : (\exists x:\sigma\ \tau[x])} \qquad \frac{\begin{array}{l}\Sigma \vdash e : (\exists x:\sigma\ \tau[x]) \\ \Sigma;\ y:\sigma; z:\tau[y] \vdash \Theta[z] \\ y \text{ does not occur free in } \Theta[z]\end{array}}{\Sigma \vdash \Theta[e]}$$

Figure 3: **Pair Types, Subtypes and Existential Types.** Note the use of absolute equality (judgemental equality) $\doteq$ in the rules for pair types. We can have two distinct but isomorphic things — we can have $a =_\sigma b$ with $a \not\doteq b$. It is important that absolute equalities are not Boolean expressions — otherwise the substitution of isomorphics would yield that $a =_\sigma b$ implies $a \doteq b$.

abbreviate

$$\mathbf{SubType}\left(\begin{array}{l} z : \mathbf{PairOf}\,(x:\sigma,\ y:\tau[x]), \\ \Phi[\pi_1(z), \pi_2(z)] \end{array}\right).$$

The type of groups, abbreviated **Group**, can then be written as

$$\mathbf{PairOf}\,(\alpha : \mathbf{type},\ f:(\alpha \times \alpha) \to \alpha \text{ s. t. } \Phi[\alpha, f])$$

where $\Phi[\alpha, f]$ states the group axioms. For example, the group axiom that an identity element exists can be written as

$$\exists x:\alpha\ \forall y:\alpha \quad f(x,y) =_\alpha y \quad \wedge \quad f(y,x) =_\alpha y.$$

The type of topological spaces, denoted **TOP**, can be written as

$$\mathbf{PairOf}\left(\begin{array}{l} \alpha : \mathbf{type}, \\ \mathbf{Open} : (\alpha \to \mathbf{Bool}) \to \mathbf{Bool}, \\ \text{s. t. } \Psi[\alpha, \mathbf{Open}] \end{array}\right)$$

where $\Psi[\alpha, \mathbf{Open}]$ states the topology axioms. Here the open sets of the topological space are represented by predicates. Note that the types **Group** and **TOP** are closed type expressions — these type expressions do not contain free variables.

We should note that subtypes are literally subsets and, for example, we can derive the sequent

$$G : \mathbf{AbelianGroup} \vdash G : \mathbf{Group}.$$

Existential types have the form $\exists x : \sigma\ \tau[x]$ where $\tau[x]$ is a type expression. This is the type whose members are those values v such that there exists a value $u : \sigma$ such that v is in the type $\tau[u]$. Existential types allow one to express the type of permutation groups as distinct from the type of groups. A permutation group is a group whose group elements are permutations of an underlying set. The type of permutation groups, denoted **PermGroup**, has the form

$$\exists\alpha\!:\!\textbf{type}\ \exists P\!:\!(\textbf{Permutation}[\alpha] \to \textbf{Bool})\ \tau[\alpha, P].$$

Here the predicate P represents a set of permutations on the type α and $\tau[\alpha, P]$ is a pair type specifying a group whose elements are the permutations of α satisfying P and where the group operation is functional composition. Again note that **PermGroup** is a closed type expression. The rules for existential types then allow the derivation of the sequent

$$G\!:\!\textbf{PermGroup} \vdash G\!:\!\textbf{Group}.$$

We also have the representation theorem

$$\vdash \forall G\!:\!\textbf{Group}\ \exists H\!:\!\textbf{PermGroup}\ G =_{\textbf{Group}} H.$$

However, the isomorphism relations $=_{\textbf{Group}}$ and $=_{\textbf{PermGroup}}$ are different. Two permutation groups can be group-isomorphic while operating on underlying sets of different sizes. Such permutation groups are group-isomorphic but not permutation-group-isomorphic.

3 Observational Equivalence

Our intention now is to interpret an equation such as $G =_{\textbf{Group}} H$ as stating that G and H are group-isomorphic. The significance of isomorphism arises from the substitution rule of figure 2. The rules of equality — reflexivity, symetry, transitivity and substitution — support the congruence closure algorithm for reasoning about equality. The ability to apply congruence closure to the isomorphism relation should be of great value in automated reasoning.[1]

The core rules define a notion of "observational equivalence" — two closed terms a and b of type σ are observationally σ-equivalent if for every predicate expression $P\!:\sigma \to \textbf{Bool}$ (typable by the core rules) we have $P(a)$ if and only if $P(b)$. We want $=_\sigma$ to be as course as possible subject to the constraint that $a =_\sigma b$ implies that a and b are observationally σ-equivalent.

The desire to be as coarse as possible while staying within observational equivalence motivates the interpretation of type-isomorphism as same-cardinality. Predicates on types that are well formed under the core rules cannot distinguish between types of the same cardinality.

4 Morphoids

The semantics of morphoid type theory is developed within a meta-theory of Platonic mathematics — we adopt the position that it is meaningful to discuss actual (Platonic) mathematical objects.

The semantics of morphoid type theory is based on a class of values called morphoids. A rigorous definition of a class of morphoids for a subset of the language can be found in (McAllester, 2014) version 4. Here we give some intuition for morphoids and state some formal properties.

Morphoids are built from "points". Morphoid points are analogous to the ur-elements of some early versions of set theory. General morphoids are built from points in a manner analogous to the way that sets are constructed from ur-elements.

A morphoid point has the form $\textbf{Point}(i, j)$ where i is called the left index and j is the right index of the point. We define the operations of **Left**, **Right**, inverse and composition on points as follows.

$$\begin{aligned}
\textbf{Left}(\textbf{Point}(i,j)) &= \textbf{Point}(i,i)\\
\textbf{Right}(\textbf{Point}(i,j)) &= \textbf{Point}(j,j)\\
\textbf{Point}(i,j)^{-1} &= \textbf{Point}(j,i)\\
\textbf{Point}(i,j) \circ \textbf{Point}(j,k) &= \textbf{Point}(i,k)
\end{aligned}$$

Here we have that $x \circ y$ is defined only in the case where $\textbf{Right}(x) = \textbf{Left}(y)$.

Every morphoid value is either a Boolean value (**True** or **False**), a point, a morphoid type, a pair of morphoid values, or a morphoid function. The operations of **Left**, **Right**, inverse and composition are defined recursively on all morphoid values where $x \circ y$ is defined when $\textbf{Right}(x) = \textbf{Left}(y)$. We consider each kind of value in turn.

A morphoid type is a set σ of morphoid values satisfying certain properties defined in (McAllester, 2014). A fundamental property is the following.

(M) For $x, y, x \in \sigma$ with $x \circ y^{-1} \circ z$ defined we have $x \circ y^{-1} \circ z \in \sigma$.

The following equations define the morphoid operations on types where $\sigma \circ \tau$ is defined only when

[1] It is tempting to suggest that congruence closure is of great value in subconscious human thought.

σ

$$\left\{\begin{array}{c}\mathbf{Point}(a,A),\ \mathbf{Point}(a,\tilde{A}),\\ \mathbf{Point}(\tilde{a},A),\ \mathbf{Point}(\tilde{a},\tilde{A}),\\ \\ \mathbf{Point}(b,B),\ \mathbf{Point}(b,\tilde{B}),\\ \mathbf{Point}(\tilde{b},B),\ \mathbf{Point}(\tilde{b},\tilde{B}),\\ \\ \mathbf{Point}(c,C),\ \mathbf{Point}(c,\tilde{C}),\\ \mathbf{Point}(\tilde{c},C),\ \mathbf{Point}(\tilde{c},\tilde{C})\end{array}\right\}$$

$\mathbf{Left}(\sigma)$

$$\left\{\begin{array}{c}\mathbf{Point}(a,a),\ \mathbf{Point}(a,\tilde{a}),\\ \mathbf{Point}(\tilde{a},a),\ \mathbf{Point}(\tilde{a},\tilde{a}),\\ \\ \mathbf{Point}(b,b),\ \mathbf{Point}(b,\tilde{b}),\\ \mathbf{Point}(\tilde{b},b),\ \mathbf{Point}(\tilde{b},\tilde{b}),\\ \\ \mathbf{Point}(c,c),\ \mathbf{Point}(c,\tilde{c}),\\ \mathbf{Point}(\tilde{c},c),\ \mathbf{Point}(\tilde{c},\tilde{c})\end{array}\right\}$$

$\mathbf{Right}(\sigma)$

$$\left\{\begin{array}{c}\mathbf{Point}(A,A),\ \mathbf{Point}(A,\tilde{A}),\\ \mathbf{Point}(\tilde{A},A),\ \mathbf{Point}(\tilde{A},\tilde{A}),\\ \\ \mathbf{Point}(B,B),\ \mathbf{Point}(B,\tilde{B}),\\ \mathbf{Point}(\tilde{B},B),\ \mathbf{Point}(\tilde{B},\tilde{B}),\\ \\ \mathbf{Point}(C,C),\ \mathbf{Point}(C,\tilde{C}),\\ \mathbf{Point}(\tilde{C},C),\ \mathbf{Point}(\tilde{C},\tilde{C})\end{array}\right\}$$

σ^{-1}

$$\left\{\begin{array}{c}\mathbf{Point}(A,a),\ \mathbf{Point}(\tilde{A},a),\\ \mathbf{Point}(\tilde{A},a),\ \mathbf{Point}(\tilde{A},\tilde{a}),\\ \\ \mathbf{Point}(B,b),\ \mathbf{Point}(\tilde{B},b),\\ \mathbf{Point}(\tilde{B},b),\ \mathbf{Point}(\tilde{B},\tilde{b}),\\ \\ \mathbf{Point}(C,c),\ \mathbf{Point}(\tilde{C},c),\\ \mathbf{Point}(\tilde{C},c),\ \mathbf{Point}(\tilde{C},\tilde{c})\end{array}\right\}$$

τ

$$\left\{\begin{array}{c}\mathbf{Point}(A,AA),\ \mathbf{Point}(A,\widetilde{AA}),\\ \mathbf{Point}(\tilde{A},AA),\ \mathbf{Point}(\tilde{A},\widetilde{AA}),\\ \\ \mathbf{Point}(B,BB),\ \mathbf{Point}(B,\widetilde{BB}),\\ \mathbf{Point}(\tilde{B},BB),\ \mathbf{Point}(\tilde{B},\widetilde{BB}),\\ \\ \mathbf{Point}(C,CC),\ \mathbf{Point}(C,\widetilde{CC}),\\ \mathbf{Point}(\tilde{C},CC),\ \mathbf{Point}(\tilde{C},\widetilde{CC})\end{array}\right\}$$

$\sigma \circ \tau$

$$\left\{\begin{array}{c}\mathbf{Point}(a,AA),\ \mathbf{Point}(a,\widetilde{AA}),\\ \mathbf{Point}(\tilde{a},AA),\ \mathbf{Point}(\tilde{a},\widetilde{AA}),\\ \\ \mathbf{Point}(b,BB),\ \mathbf{Point}(b,\widetilde{BB}),\\ \mathbf{Point}(\tilde{b},BB),\ \mathbf{Point}(\tilde{b},\widetilde{BB}),\\ \\ \mathbf{Point}(c,CC),\ \mathbf{Point}(c,\widetilde{CC}),\\ \mathbf{Point}(\tilde{c},CC),\ \mathbf{Point}(\tilde{c},\widetilde{CC})\end{array}\right\}$$

Figure 4: **The operations of Left, Right, inverse and composition on point types.**

G

```
Point(a,A)
     \
      Point(b,B)
     /
Point(c,C)
```

$\mathbf{Left}(G)$

```
Point(a,a)
     \
      Point(b,b)
     /
Point(c,c)
```

$\mathbf{Right}(G)$

```
Point(A,A)
     \
      Point(B,B)
     /
Point(C,C)
```

G^{-1}

```
Point(A,a)
     \
      Point(B,b)
     /
Point(C,c)
```

H

```
Point(A,AA)
     \
      Point(B,BB)
     /
Point(C,CC)
```

$G \circ H$

```
Point(a,AA)
     \
      Point(b,BB)
     /
Point(c,CC)
```

Figure 5: **The operations of Left, Right, inverse and composition on abstract morphoid graphs.**

$\mathbf{Right}(\sigma) = \mathbf{Left}(\tau)$.

$$\begin{aligned}
\mathbf{Left}(\sigma) &= \left\{ \begin{array}{l} x_1 \circ x_2^{-1} : x_1, x_2 \in \sigma, \\ \mathbf{Right}(x_1) = \mathbf{Right}(x_2) \end{array} \right\} \\
\mathbf{Right}(\sigma) &= \left\{ \begin{array}{l} x_1^{-1} \circ x_2 : x_1, x_2 \in \sigma, \\ \mathbf{Left}(x_1) = \mathbf{Left}(x_2) \end{array} \right\} \\
\sigma \circ \tau &= \left\{ \begin{array}{l} x \circ y : x \in \sigma, \ y \in \tau, \\ \mathbf{Right}(x) = \mathbf{Left}(y) \end{array} \right\} \\
\sigma^{-1} &= \{x^{-1} : x \in \sigma\}
\end{aligned}$$

A morphoid type whose elements are points is called a point type. Figure 4 gives examples of morphoid point types and examples of the morphoid operations applied to point types. The morphoid closure condition (M) implies that for any morphoid type σ we have that $\mathbf{Left}(\sigma)$ and $\mathbf{Right}(\sigma)$ are equivalence relations (see figure 4). Note that morphoid types are not required to be closed under inverse. This allows types to be directed from left to right. Again consider the types in figure 4. Furthermore, property (M) implies that any morphoid type σ defines a bijection between the equivalence classes of $\mathbf{Left}(\sigma)$ and the equivalence classes of $\mathbf{Right}(\sigma)$.

A morphoid pair is simply a pair of morphoids. The morphoid operations on pairs are defined as follows where again $x \circ y$ is defined only when $\mathbf{Right}(x) = \mathbf{Left}(y)$.

$$\begin{aligned}
\mathbf{Left}(\mathbf{Pair}(x, y)) &= \mathbf{Pair}(\mathbf{Left}(x), \mathbf{Left}(y)) \\
\mathbf{Right}(\mathbf{Pair}(x, y)) &= \mathbf{Pair}(\mathbf{Right}(x), \mathbf{Right}(y)) \\
\mathbf{Pair}(x, y)^{-1} &= \mathbf{Pair}(x^{-1}, y^{-1}) \\
\mathbf{Pair}(x, y) \circ \mathbf{Pair}(z, w) &= \mathbf{Pair}(x \circ z, \ y \circ w)
\end{aligned}$$

The treatment of morphoid functions involves subtleties. In this synopsis we note only that for any two morphoid types σ and τ we can define the type $\sigma \to \tau$ such that for $f \in \sigma \to \tau$ and $x \in \sigma$ we can define the application $f(x)$ so that we have $f(x) \in \tau$. Furthermore, these definitions are such that the elements of the type $\sigma \to \tau$ represents *all* functions from the equivalence classes of σ to the equivalence classes of τ. Details, including the definitions of the morphoid operations on functions, can be found in (McAllester, 2014).

It is possible to prove that morphoids satisfy the following properties where properties (G4), (G5), (G6), and (G9) apply when the compositions are defined.

(G1) For any morphoid x we have that $\mathbf{Left}(x)$, $\mathbf{Right}(x)$ and x^{-1} are also morphoids.

(G2) For any morphoids x and y we have that $x \circ y$ is defined if and only if $\mathbf{Right}(x) = \mathbf{Left}(y)$ and when $x \circ y$ is defined we have that $x \circ y$ is a morphoid.

(G3) $\mathbf{Left}(x^{-1}) = \mathbf{Right}(x)$ and $\mathbf{Right}(x^{-1}) = \mathbf{Left}(x)$

(G4) $\mathbf{Left}(x \circ y) = \mathbf{Left}(x)$ and $\mathbf{Right}(x \circ y) = \mathbf{Right}(y)$.

(G5) $(x \circ y) \circ z = x \circ (y \circ z)$.

(G6) $x^{-1} \circ x \circ y = y$ and $x \circ y \circ y^{-1} = x$.

(G7) $\mathbf{Right}(x) = x^{-1} \circ x$ and $\mathbf{Left}(x) = x \circ x^{-1}$.

(G8) $(x^{-1})^{-1} = x$.

(G9) $(x \circ y)^{-1} = y^{-1} \circ x^{-1}$.

Properties (G1) through (G9) state that the class of morphoids forms a groupoid under the morphoid operations. Figure 5 shows morphoid operations on graphs whose nodes are points.

5 Abstraction and Isomorphism

The semantic definition of isomorphism relies on an additional operation on morphoids — the operation of abstraction. As an example we consider vector spaces. In morphoid type theory an abstract vector space is one in which the vectors are points. The space $\mathbb{R}^n$ is a vector space whose vectors are n-tuples of real numbers. A tuple of real numbers is an implementation of a vector — a tuple of real numbers is not a point. However, we can define an abstraction operation such that for any morphoid value x we have that $x@\mathbf{Point}$ is a point. Details can be found in (McAllester, 2014).

There is an abstraction ordering on morphoids where $x \preceq y$ if x can be converted to y by abstracting parts of x to points. For every type there is a set of maximally abstract elements of that type. The maximally abstract graphs are the graphs whose nodes are points. The maximally abstract vector spaces are those vector spaces in which the vectors are points. The maximally abstract types are the point types. For each morphoid type σ it is possible to define an abstraction operation mapping $x \in \sigma$ to $x@\sigma$ where $x@\sigma$ is a maximally abstract member of σ. For example, for any morphoid type $\sigma \in \mathbf{type}_i$ we have that $\sigma@\mathbf{type}_i$ is the point type whose members are the points of the form $x@\mathbf{Point}$ for $x \in \sigma$. Details can be found in (McAllester, 2014).

The isomorphism relation $x =_\sigma y$ is defined to mean that $x \in \sigma$, $y \in \sigma$, and there exists $z \in \sigma$ such that $(x@\sigma) \circ z^{-1} \circ (y@\sigma)$ is defined. Condition (M) on morphoid types guarantees that this is an equivalence relation on the elements of σ.

6 The Semantic Value Function

The semantics of morphoid type theory is an extension of the sementics of predicate calculus. The semantics involves three concepts — variable interpretations, semantic entailment, and a semantic value function. These three concepts are defined by mutual recursion where the recursion reduces the size of the expressions involved. A variable interpretation assigns a value to each variable declared in a given context. More formally, for any well-formed context Σ we write $\mathcal{V}[\![\Sigma]\!]$ for the set of variable interpretations consistent with declarations and Boolean assertions in Σ. We define $\mathcal{V}[\![\Sigma]\!]$ by the following rules where this is undefined if no rule applies.

- We define $\mathcal{V}[\![\epsilon]\!]$ to be the set containing the empty variable interpretation.
- $\mathcal{V}[\![\Sigma; x:\tau]\!]$ is defined if $\mathcal{V}[\![\Sigma]\!]$ is defined, x is not declared in Σ, and $\Sigma \models \tau:\mathbf{type}_i$ in which case $\mathcal{V}[\![\Sigma; x:\tau]\!]$ is defined to be the set of variable interpretations of the form $\rho[x \leftarrow v]$ for $\rho \in \mathcal{V}[\![\Sigma]\!]$ and $v \in \mathcal{V}_\Sigma[\![\tau]\!]\rho$.
- $\mathcal{V}[\![\Sigma; \Phi]\!]$ is defined if $\mathcal{V}[\![\Sigma]\!]$ is defined and $\Sigma \models \Phi : \mathbf{Bool}$ in which case $\mathcal{V}[\![\Sigma; \Phi]\!]$ is defined to be the set of all $\rho \in \mathcal{V}[\![\Sigma]\!]$ such that $\mathcal{V}_\Sigma[\![\Phi]\!]\rho = \mathbf{True}$.

A semantic entailment is written as $\Sigma \models \Theta$ and this holds if $\mathcal{V}[\![\Sigma]\!]$ is defined and Θ holds under all variable interpretations in $\mathcal{V}[\![\Sigma]\!]$. The entailment relation $\Sigma \models \Theta$ holds if one of the following clauses applies.

- The entailment $\Sigma \models e:\tau$ holds if $\mathcal{V}[\![\Sigma]\!]$, $\mathcal{V}_\Sigma[\![e]\!]$ and $\mathcal{V}_\Sigma[\![\tau]\!]$ are all defined and for all $\rho \in \mathcal{V}[\![\Sigma]\!]$ we have that $\mathcal{V}_\Sigma[\![\tau]\!]\rho$ is a morphoid type and $\mathcal{V}_\Sigma[\![e]\!]\rho \in \mathcal{V}_\Sigma[\![\tau]\!]\rho$.
- For a Boolean expression Φ, i.e., for $\Sigma \models \Phi : \mathbf{Bool}$, we have that $\Sigma \models \Phi$ holds if for all $\rho \in \mathcal{V}[\![\Sigma]\!]$ we have $\mathcal{V}_\Sigma[\![\Phi]\!]\rho = \mathbf{True}$.
- We write $\Sigma \models e_1 \doteq e_2$ if $\mathcal{V}[\![\Sigma]\!]$, $\mathcal{V}_\Sigma[\![e_1]\!]$ and $\mathcal{V}_\Sigma[\![e_2]\!]$ are all defined and for $\rho \in \mathcal{V}[\![\Sigma]\!]$ we have that $\mathcal{V}_\Sigma[\![e_1]\!]\rho$ and $\mathcal{V}_\Sigma[\![e_2]\!]\rho$ are the same value.

For $\mathcal{V}[\![\Sigma]\!]$ defined and for an expression e that is well-formed in the context Σ, we have a semantic value function $\mathcal{V}_\Sigma[\![e]\!]$. The semantic value function $\mathcal{V}_\Sigma[\![e]\!]$ maps a variable interpretation $\rho \in \mathcal{V}[\![\Sigma]\!]$ to a morphoid value $\mathcal{V}_\Sigma[\![e]\!]\rho$. For $\mathcal{V}[\![\Sigma]\!]$ defined, the following clauses state when $\mathcal{V}_\Sigma[\![e]\!]$ is defined and, when it is defined, define the value $\mathcal{V}_\Sigma[\![e]\!]\rho$ for $\rho \in \mathcal{V}[\![\Sigma]\!]$.

• x. For x declared in Σ and for $\rho \in \mathcal{V}[\![\Sigma]\!]$ we have that $\mathcal{V}_\Sigma[\![x]\!]$ is defined with $\mathcal{V}_\Sigma[\![x]\!]\rho = \rho(x)$.

• $\mathbf{Bool}$. We have that $\mathcal{V}_\Sigma[\![\mathbf{Bool}]\!]\rho$ is the type containing the two Boolean values $\mathbf{True}$ and $\mathbf{False}$.

• $\mathbf{type}_i$. We have $\mathcal{V}_\Sigma[\![\mathbf{type}_i]\!]\rho$ is the type whose members are all morphoid types in the set-theoretic universe V_{κ_i} where κ_i is the ith inaccessible cardinal.

• $\sigma \to \tau$. If $\Sigma \models \sigma : \mathbf{type}_i$, and $\Sigma \models \tau : \mathbf{type}_i$, then $\mathcal{V}_\Sigma[\![\sigma \to \tau]\!]$ is defined with $\mathcal{V}_\Sigma[\![\sigma \to \tau]\!]\rho = (\mathcal{V}_\Sigma[\![\sigma]\!]\rho) \to (\mathcal{V}_\Sigma[\![\tau]\!]\rho)$.

• $f(e)$. If $\mathcal{V}_\Sigma[\![f]\!]$ and $\mathcal{V}_\Sigma[\![e]\!]$ are defined and for all $\rho \in \mathcal{V}[\![\Sigma]\!]$ we have that $\mathcal{V}_\Sigma[\![f]\!]\rho$ can be applied to $\mathcal{V}_\Sigma[\![e]\!]\rho$ then $\mathcal{V}_\Sigma[\![f(e)]\!]$ is defined with $\mathcal{V}_\Sigma[\![f(e)]\!]\rho = (\mathcal{V}_\Sigma[\![f]\!]\rho)(\mathcal{V}_\Sigma[\![e]\!]\rho)$.

• $\forall\, x:\tau\ \Phi[x]$. If $\Sigma;\ y:\tau \models \Phi[y] : \mathbf{Bool}$ then $\mathcal{V}_\Sigma[\![\forall\, x:\tau\ \Phi[x]]\!]$ is defined with $\mathcal{V}_\Sigma[\![\forall\, x:\tau\ \Phi[x]]\!]\rho$ being $\mathbf{True}$ if for all $v \in \mathcal{V}_\Sigma[\![\tau]\!]\rho$ we have $\mathcal{V}_{\Sigma;y:\tau}[\![\Phi[y]]\!]\rho[y \leftarrow v] = \mathbf{True}$.

• $\Phi \vee \Psi$. If $\Sigma \models \Phi : \mathbf{Bool}$ and $\Sigma \models \Psi : \mathbf{Bool}$ then $\mathcal{V}_\Sigma[\![\Phi \vee \Psi]\!]$ is defined with $\mathcal{V}_\Sigma[\![\Phi \vee \Psi]\!]\rho = \mathcal{V}_\Sigma[\![\Phi]\!]\rho \vee \mathcal{V}_\Sigma[\![\rho]\!]\rho$.

• $\neg\Phi$. If $\Sigma \models \Phi : \mathbf{Bool}$ then $\mathcal{V}_\Sigma[\![\neg\Phi]\!]$ is defined with $\mathcal{V}_\Sigma[\![\neg\Phi]\!]\rho = \neg\mathcal{V}_\Sigma[\![\Phi]\!]\rho$.

$$\mathbf{Bijection}[\sigma,\tau] \equiv \mathbf{SubType}(f\!:\!\sigma \to \tau,\ \forall y\!:\!\tau\ \exists! x\!:\!\sigma\ f(x) =_\tau y) \qquad a \leadsto_\sigma b \ \equiv\ \exists z\!:\!\mathbf{iso}(\sigma, a, b)$$

$$\frac{\begin{array}{l}\Sigma \vdash \sigma\!:\!\mathbf{type}_i \\ \Sigma \vdash a\!:\!\tau \\ \Sigma \vdash b\!:\!\eta\end{array}}{\Sigma \vdash \mathbf{iso}(\sigma, a, b)\!:\!\mathbf{type}_i} \qquad \frac{\begin{array}{l}\Sigma \vdash \sigma, \tau\!:\!\mathbf{type}_i \\ \Sigma \vdash f\!:\!\mathbf{Bijection}[\sigma,\tau]\end{array}}{\begin{array}{l}\Sigma \vdash \updownarrow(\sigma,\tau,f)\!:\!\mathbf{iso}(\mathbf{type}_i, \sigma, \tau) \\ \Sigma \vdash \forall x\!:\!\sigma\ \ x \leadsto_{\updownarrow(\sigma,\tau,f)} f(x)\end{array}} \qquad \frac{\Sigma \vdash \eta\!:\!\mathbf{iso}(\mathbf{type}_i, \sigma, \tau)}{\begin{array}{l}\Sigma \vdash \uparrow_{\eta\to\sigma}\!:\!\mathbf{Bijection}[\eta,\sigma] \\ \Sigma \vdash \downarrow_{\eta\to\tau}\!:\!\mathbf{Bijection}[\eta,\tau] \\ \Sigma \vdash \left\{\begin{array}{l}\forall x\!:\!\sigma\ \forall y\!:\!\tau \\ \quad (x \leadsto_\eta y) \Leftrightarrow \\ \quad\quad \exists z\!:\!\eta\ \uparrow_{\eta\to\sigma}(z) =_\sigma x \wedge \\ \quad\quad\quad \downarrow_{\eta\to\tau}(z) =_\tau y\end{array}\right.\end{array}}$$

$$\frac{\Sigma \vdash c\!:\!\mathbf{iso}(\sigma, a, b)}{\Sigma \vdash c\!:\!\sigma} \qquad \frac{\begin{array}{l}\Sigma \vdash a\!:\!\sigma,\ b\!:\!\sigma \\ \Sigma \vdash c\!:\!\mathbf{iso}(\sigma, a, b)\end{array}}{\begin{array}{l}\Sigma \vdash a =_\sigma c \\ \Sigma \vdash b =_\sigma c\end{array}} \qquad \frac{\Sigma \vdash a =_\sigma b}{\Sigma \vdash a \leadsto_\sigma b}$$

$$\frac{\begin{array}{l}\Sigma; x\!:\!\sigma; y\!:\!\gamma[x] \vdash e[x,y]\!:\!\tau[x,y] \\ \Sigma \vdash a_1\!:\!\sigma,\ a_2\!:\!\sigma,\ a_3\!:\!\mathbf{iso}(\sigma, a_1, a_2) \\ \Sigma \vdash b_1\!:\!\gamma[a_1],\ b_2\!:\!\gamma[a_2],\ b_3\!:\!\mathbf{iso}(\gamma[a_3], b_1, b_2)\end{array}}{\Sigma \vdash e[a_3, b_3]\!:\!\mathbf{iso}(\tau[a_3,b_3], e[a_1,b_1], e[a_2,b_2])} \qquad \frac{\begin{array}{l}\Sigma \vdash \sigma_3\!:\!\mathbf{iso}(\mathbf{type}_i, \sigma_1, \sigma_2) \\ \Sigma \vdash f_1\!:\!\sigma_1 \to \tau_1,\ f_2\!:\!\sigma_2 \to \tau_2,\ f_3\!:\!\sigma_3 \to \tau_3 \\ \Sigma;\ z\!:\!\sigma_3 \vdash f_3(z)\!:\!\mathbf{iso}(\tau_3,\ f_1(\uparrow_{\sigma_3\to\sigma_1}(z)),\ f_2(\downarrow_{\sigma_3\to\sigma_2}(z)))\end{array}}{\Sigma \vdash f_3\!:\!\mathbf{iso}(\sigma_3 \to \tau_3,\ f_1,\ f_2)}$$

$$\frac{\begin{array}{l}\Sigma \vdash c\!:\!\mathbf{iso}(\sigma, a, b) \\ \Sigma \vdash \Phi[c]\end{array}}{\Sigma \vdash c\!:\!\mathbf{iso}(\mathbf{SubType}(x\!:\!\sigma,\ \Phi[x]),\ a,\ b)} \qquad \frac{\begin{array}{l}\Sigma \vdash c\!:\!\mathbf{iso}(\tau[d], a, b) \\ \Sigma \vdash d\!:\!\sigma\end{array}}{\Sigma \vdash c\!:\!\mathbf{iso}(\exists x\!:\!\sigma\ \tau[x],\ a,\ b)}$$

Figure 6: **Internalizing Isomorphism.** Here $\mathbf{iso}(\sigma, x, y)$ is the type whose elements are the σ-isomorphisms from x to y. The fourth row gives the rules of iso-substition and iso-extensionality.

• $s =_\sigma w$. If $\Sigma \models s : \sigma$ and $\Sigma \models w : \sigma$ then $\mathcal{V}_\Sigma [\![s =_\sigma w]\!]$ is defined with $\mathcal{V}_\Sigma [\![s =_\sigma w]\!]\,\rho$ being **True** if $\mathcal{V}_\Sigma [\![s]\!]\,\rho =_{\mathcal{V}_\Sigma[\![\sigma]\!]\rho} \mathcal{V}_\Sigma [\![w]\!]\,\rho$.

• $\mathbf{The}(x\!:\!\sigma,\ \Phi[x])$. If $\Sigma \models \exists! y\!:\!\sigma\ \Phi[y]$ then

$$\mathcal{V}_\Sigma [\![\mathbf{The}(x\!:\!\sigma,\ \Phi[x])]\!]\,\rho =$$

$$\mathbf{The}(\mathcal{V}_\Sigma [\![\mathbf{SubType}(x\!:\!\sigma,\ \Phi[x])]\!]\,\rho).$$

• $\mathbf{PairOf}(x\!:\!\sigma,\ y\!:\!\tau[x])$. If $\Sigma \models \sigma : \mathbf{type}_i$ and $\Sigma; z : \sigma \models \tau[z] : \mathbf{type}_i$ then $\mathcal{V}_\Sigma [\![\mathbf{PairOf}(x\!:\!\sigma,\ y\!:\!\tau[x])]\!]$ is defined with $\mathcal{V}_\Sigma [\![\mathbf{PairOf}(x\!:\!\sigma,\ y\!:\!\tau[x])]\!]\,\rho$ being the type containing the pairs $\mathbf{Pair}(v, w)$ for $v \in \mathcal{V}_\Sigma [\![\sigma]\!]\,\rho$ and $w \in \mathcal{V}_{\Sigma;\, z:\sigma} [\![\tau[z]]\!]\,\rho[z \leftarrow v]$.

• $\mathbf{Pair}(u, w)$. If $\mathcal{V}_\Sigma [\![u]\!]$ and $\mathcal{V}_\Sigma [\![w]\!]$ are defined then $\mathcal{V}_\Sigma [\![\mathbf{Pair}(u, w)]\!]$ is defined with $\mathcal{V}_\Sigma [\![\mathbf{Pair}(u, w)]\!]\,\rho = \mathbf{Pair}(\mathcal{V}_\Sigma [\![u]\!]\,\rho,\ \mathcal{V}_\Sigma [\![w]\!]\,\rho)$.

• $\pi_i(e)$. If $\mathcal{V}_\Sigma [\![e]\!]$ is defined and for all $\rho \in \mathcal{V}[\![\Sigma]\!]$ we have that $\mathcal{V}_\Sigma [\![e]\!]\,\rho$ is a pair then $\mathcal{V}_\Sigma [\![\pi_i(e)]\!]$ is defined with $\mathcal{V}_\Sigma [\![\pi_i(e)]\!]\,\rho = \pi_i(\mathcal{V}_\Sigma [\![e]\!]\,\rho)$.

• $\mathbf{SubType}(x\!:\!\sigma,\ \Phi[x])$. If $\Sigma \models \sigma : \mathbf{type}_i$ and $\Sigma;\ y : \sigma \models \Phi[y] : \mathbf{Bool}$ then $\mathcal{V}_\Sigma [\![\mathbf{SubType}(x\!:\!\sigma,\ \Phi[x])]\!]$ is defined with $\mathcal{V}_\Sigma [\![\mathbf{SubType}(x\!:\!\sigma,\ \Phi[x])]\!]\,\rho$ being the type whose members are those values $v \in \mathcal{V}_\Sigma [\![\sigma]\!]\,\rho$ with $\mathcal{V}_{\Sigma;\, y:\sigma} [\![\Phi[y]]\!]\,\rho[y \leftarrow v] = \mathbf{True}$.

• $\exists x : \sigma\ \tau[x]$. If $\Sigma;\ y : \sigma \models \tau[y] : \mathbf{type}_i$ then $\mathcal{V}_\Sigma [\![\exists x\!:\!\sigma\ \tau[x]]\!]$ is defined with $\mathcal{V}_\Sigma [\![\exists x\!:\!\sigma\ \tau[x]]\!]\,\rho$ being the type containing those values w such that there exists $u \in \mathcal{V}_\Sigma [\![\sigma]\!]\,\rho$ with $w \in \mathcal{V}_\Sigma [\![\tau[y]]\!]\,\rho[y \leftarrow u]$.

The morphoid operations can also be defined on variable interpretations. For $\rho \in \mathcal{V}[\![\Sigma]\!]$ we have

$$\frac{\begin{array}{l}\Sigma;\alpha:\mathbf{type}_i \vdash \gamma[\alpha]:\mathbf{type}_i \\ \Sigma \vdash \sigma:\mathbf{type}_i,\ \tau:\mathbf{type}_i,\ f:\mathbf{Bijection}[\sigma,\tau] \\ \Sigma \vdash a:\gamma[\sigma],\ b:\gamma[\tau],\ a \leadsto_{\gamma[\updownarrow(\sigma,\tau,f)]} b\end{array}}{\Sigma \vdash \mathbf{Pair}(\sigma,a) =_{\mathbf{PairOf}(\alpha:\mathbf{type}_i,\ y:\gamma[\alpha])} \mathbf{Pair}(\tau,b)}$$

$$\frac{\begin{array}{l}\Sigma;\ \alpha:\mathbf{type}_i \vdash \delta[\alpha],\eta[\alpha]:\mathbf{type}_i \\ \Sigma \vdash \sigma,\tau:\mathbf{type}_i \\ \Sigma \vdash f:\mathbf{Bijection}[\sigma,\tau] \\ \Sigma \vdash a:\mathbf{PairOf}(\delta[\sigma],\ \eta[\sigma]) \\ \Sigma \vdash b:\mathbf{PairOf}(\delta[\tau],\ \eta[\tau])\end{array}}{\Sigma \vdash \left\{\begin{array}{l}(a \leadsto_{\mathbf{PairOf}(\delta[\updownarrow(\sigma,\tau,f)],\ \eta[\updownarrow(\sigma,\tau,f)])} b) \\ \quad\Leftrightarrow \\ \pi_1(a) \leadsto_{\delta[\updownarrow(\sigma,\tau,f)]} \pi_1(b) \wedge \\ \pi_2(a) \leadsto_{\eta[\updownarrow(\sigma,\tau,f)]} \pi_2(b)\end{array}\right.}$$

$$\frac{\begin{array}{l}\Sigma;\ \alpha:\mathbf{type}_i \vdash \delta[\alpha],\eta[\alpha]:\mathbf{type}_i \\ \Sigma \vdash \sigma,\tau:\mathbf{type}_i \\ \Sigma \vdash f:\mathbf{Bijection}[\sigma,\tau] \\ \Sigma \vdash g:\delta[\sigma] \to \eta[\sigma] \\ \Sigma \vdash h:\delta[\tau] \to \eta[\tau]\end{array}}{\Sigma \vdash \left\{\begin{array}{l}(g \leadsto_{\delta[\updownarrow(\sigma,\tau,f)]\to\eta[\updownarrow(\sigma,\tau,f)]} h) \\ \quad\Leftrightarrow \\ \forall x_1:\delta[\sigma] \\ \forall x_2:\delta[\tau] \\ (x_1 \leadsto_{\delta[\updownarrow(\sigma,\tau,f)]} x_2) \\ \quad \Rightarrow g(x_1) \leadsto_{\eta[\updownarrow(\sigma,\tau,f)]} h(x_2)\end{array}\right.}$$

Figure 7: **Some Derived Rules.** The rules in this figure can be derived from the rules in figure 6. The first rule constructs isomorphism relations at types of the form $\mathbf{PairOf}(\alpha:\mathbf{type}_i,\ y:\gamma[\alpha])$. The rule states that $\mathbf{Pair}(\sigma,a)$ is isomorphic to $\mathbf{Pair}(\tau,b)$ if there exists a bijection f from σ to τ that carries a to b. The rules in the second row allow one to determine whether f carries a to b in the case where $\gamma[\alpha]$ is a simple type over α (see the text). There are two base cases not listed in the figure. For $\gamma[\alpha] = \alpha$ we have that $a \leadsto_{\updownarrow(\sigma,\tau,f)} b$ if and only if $f(a) =_\tau b$. If $\gamma[\alpha]$ does not depend on α we have $a \leadsto_\gamma b$ if and only if $a =_\gamma b$.

that $\mathbf{Left}(\rho)$ is the variable interpretation that maps x to $\mathbf{Left}(\rho(x))$. $\mathbf{Right}(\rho)$ is defined similarly. For $\rho \in \mathcal{V}[\![\Sigma]\!]$ we have that ρ^{-1} maps x to $\rho(x)^{-1}$. For $\rho,\gamma \in \mathcal{V}[\![\Sigma]\!]$ we have that $\rho\circ\gamma$ is defined if and only if $\mathbf{Right}(\rho) = \mathbf{Left}(\gamma)$ in which case $\rho\circ\gamma$ is the variable interpretation mapping x to $\rho(x)\circ\gamma(x)$.

A fundamental property of Morphoid type theory is that if $\mathcal{V}[\![\Sigma]\!]$ is defined then it is closed under inverse and composition — for $\rho \in \mathcal{V}[\![\Sigma]\!]$ we have $\rho^{-1} \in \mathcal{V}[\![\Sigma]\!]$ and for $\rho,\gamma \in \mathcal{V}[\![\Sigma]\!]$ with $\rho\circ\gamma$ defined we have $(\rho\circ\gamma) \in \mathcal{V}[\![\Sigma]\!]$. Furthermore for $\mathcal{V}_\Sigma[\![e]\!]$ defined and for $\rho,\gamma \in \mathcal{V}[\![\Sigma]\!]$ we have

$$\mathcal{V}_\Sigma[\![e]\!](\rho\circ\gamma) = (\mathcal{V}_\Sigma[\![e]\!]\rho)\circ(\mathcal{V}_\Sigma[\![e]\!]\gamma)$$

and

$$\mathcal{V}_\Sigma[\![e]\!](\rho^{-1}) = (\mathcal{V}_\Sigma[\![e]\!]\rho)^{-1}.$$

Another fundamental property of the value function involves the abstraction ordering. The abstraction ordering can be extended to variable interpretations where we have $\rho \preceq \gamma$ if ρ and γ are defined on the same set of variables and for each variable x we have $\rho(x) \preceq \gamma(x)$. The value function is monotone with respect to the abstraction ordering — for $\rho \preceq \gamma$ we have $\mathcal{V}_\Sigma[\![e]\!]\rho \preceq \mathcal{V}_\Sigma[\![e]\!]\gamma$.

7 Internalizing Isomorphism

Figure 6 gives inference rules for isomorphism. Figure 7 gives rules which can be derived from the rules in figure 6. The first rule in figure 7 derives isomorphisms at types of the form PairOf$(\alpha:\mathbf{type}_i,\ \tau[\alpha])$. We note that types **Group** and **TOP** as defined in section 2 can be written as subtypes of pair types of this form. The rule states that two objects $\mathbf{Pair}(\sigma,a)$ and $\mathbf{Pair}(\tau,b)$ of type PairOf$(\alpha : \mathbf{type}_i,\ \tau[\alpha])$ are isomorphic if there exists a bijection f from σ to τ which carries a to b. The two rules in the second row of figure 7 allow one to determine whether or not f carries a to b in the case where $\tau[\alpha]$ is a "simple type" over

α. To define the simple types we first introduce the notation $\mathbf{PairOf}(\gamma, \eta)$ as an abbreviation for $\mathbf{PairOf}(x : \gamma,\ y : \eta)$ in the case where x does not occur in η. A simple type expression $\gamma[\alpha]$ over the type variable α is then defined to be either the variable α itself, a type γ not containing α, or a type of the form $\mathbf{PairOf}(\delta[\alpha],\ \eta[\alpha])$ or $\delta[\alpha] \rightarrow \eta[\alpha]$ where $\delta[\alpha]$ and $\eta[\alpha]$ are recursively simple type expressions over α. Subtypes of pair types of the form $\mathbf{PairOf}(\alpha : \mathbf{type}_i,\ \gamma[\alpha])$ where $\gamma[\alpha]$ is a simple type over α covers the types $\mathbf{Group}$ and $\mathbf{TOP}$ as well as many other mathematical concepts. We leave the derivation of the rules in the second row of figure 7 as a (tricky and tedious) exercise for the reader.

While the rules in figure 7 are adequate in many situations, they do not cover types such as

$$\mathbf{Pairof}(p{:}\mathbf{Pairof}(\mathbf{type}_i, \mathbf{type}_i),\ y{:}\gamma[p])$$

or

$$\mathbf{Pairof}(G{:}\mathbf{Group},\ H{:}\tau[G])$$

or

$$\mathbf{Pairof}(f{:}\mathbf{type}_i \rightarrow \mathbf{type}_i,\ A{:}\tau[f]).$$

For the general case we need the rules of figure 6.

8 Summary

Morphoid type theory is a type-theoretic foundation for mathematics supporting the concept of isomorphism and the substitution of isomorphics. Morphoid type theory is an extension of classical predicate calculus that avoids the use of propositions-as-types, path induction or squashing. Morphoid type theory may be more comfortable for mathemticians who take a realist or Platonic approach to the practice of mathematics.

References

Thierry Coquand and Gerard Huet. 1988. The calculus of constructions. *Information and computation*, 76(2):95–120.

Martin Hofmann and Thomas Streicher. 1994. The groupoid model refutes uniqueness of identity proofs. In *Logic in Computer Science, 1994. LICS'94. Proceedings., Symposium on*, pages 208–212. IEEE.

HoTT-Authors. 2013. Homotopy type theory, univalent foundations of mathematics. http://hottheory.files.wordpress.com/2013/03/hott-online-611-ga1a258c.pdf.

Per Martin-Löf. 1971. A theory of types.

David McAllester. 2014. Morphoid type theory. *CoRR*, abs/1407.7274.

John C. Reynolds. 1983. Types, abstraction and parametric polymorphism. In *IFIP Congress*, pages 513–523.

Giovanni Sambin and Jan M Smith. 1998. *Twenty Five Years of Constructive Type Theory*, volume 36. Oxford University Press.

General Perspective on Distributionally Learnable Classes

Ryo Yoshinaka
Kyoto University, Japan
ry@i.kyoto-u.ac.jp

Abstract

Several algorithms have been proposed to learn different subclasses of context-free grammars based on the idea generically called *distributional learning*. Those techniques have been applied to many formalisms richer than context-free grammars like multiple context-free grammars, simple context-free tree grammars and others. The learning algorithms for those different formalisms are actually quite similar to each other. We in this paper give a uniform view on those algorithms.

1 Introduction

Approaches based on the idea generically called *distributional learning* have been making great success in the algorithmic learning of various subclasses of context-free grammars (CFGs) (Clark, 2010c; Yoshinaka, 2012). Those techniques are applied to richer formalisms as well. The formalisms studied so far include multiple CFGs (Yoshinaka, 2011a), simple context-free tree grammars (CFTGs) (Kasprzik and Yoshinaka, 2011), second-order abstract categorial grammars (Yoshinaka and Kanazawa, 2011), parallel multiple CFGs (Clark and Yoshinaka, 2014), conjunctive grammars (Yoshinaka, 2015) and others. The goal of this paper is to present a uniform view on those algorithms.

Every grammar formalism for which distributional learning techniques have been proposed so far generate their languages through context-free derivation trees, whose nodes are labeled by production rules. The formalism and grammar rules determine how a context-free derivation tree τ is mapped to a derived object $\tilde{\tau} = d$. A context-free derivation tree τ can be decomposed into a subtree σ and a tree-context χ so that $\tau = \chi[\sigma]$. The subtree determines a substructure $s = \tilde{\sigma}$ of d and the tree-context determines a contextual structure $c = \tilde{\chi}$ in which the substructure is plugged to form the derived object $d = c \odot s$, where we represent the plugging operation by $\odot$. In the CFG case, c is a string pair $\langle l, r\rangle$ and s is a string u and $\langle l, r\rangle \odot u = lur$, which may correspond to a derivation $I \stackrel{*}{\Rightarrow} lXr \stackrel{*}{\Rightarrow} lur$ where I is the initial symbol and X is a nonterminal symbol. In richer formalisms those substructures and contexts may have richer structures, like tuples of strings or λ-terms. A learner does not know how a given example d is derived by a hidden grammar behind the observed examples. A learner based on distributional learning simply tries all the possible decompositions of a positive example into arbitrary two parts c' and s' such that $d = c' \odot s'$ where some grammar may derive d thorough a derivation tree $\tau' = \chi'[\sigma']$ with $\tilde{\chi}' = c'$ and $\tilde{\sigma}' = s'$. Based on observation on the relation between substructures and contexts collected from given examples, a hypothesis grammar is computed. We call properties on grammars with which distributional learning approaches work *distributional properties*.

This paper first formally defines grammar formalisms based on context-free derivation trees. We then show that grammars with different distributional properties are learnable by standard distributional learning techniques if the formalism satisfies some conditions, which include polynomial-time decomposability of objects into contexts and

Proceedings of the 14th Meeting on the Mathematics of Language (MoL 14), pages 87–98,
Chicago, USA, July 25–26, 2015.

substructures. In addition, we discuss cases where we cannot enumerate all of the possible contexts and substructures.

2 Σ-grammars

There is a number of ways to represent a *language*, a subset of an object set $\mathbb{O}_*$, whose elements are typically strings, trees but anythings encodable are eligible. Formalisms this paper discusses generate objects in $\mathbb{O}_*$ through context-free derivation trees τ, which are mapped to an element $d \in \mathbb{O}_*$ in a uniform way. The map is inductively defined and computed. Each derivation subtree τ' of τ also determines an object, which we call a *substructure* of d. Each substructure is not necessarily a member of $\mathbb{O}_*$. For example, nonterminal symbols of multiple CFGs (Seki et al., 1991) derive n-tuples of strings, where the value n is unique to each nonterminal, while the languages generated by multiple CFGs are still simply string sets. A generalization of the CFG formalism is specified by kinds of objects that each nonterminal generates and admissible operations over those objects.

Let $\mathbb{O}$ be a set of *objects*, which are identified with their codes of finite length. We have a set Ω of finite representations O which are interpreted as subsets $\mathbb{O}_O$ of $\mathbb{O}$ through an effective procedure. By a *sort* we flexibly refer to $O \in \Omega$ or $\mathbb{O}_O \subseteq \mathbb{O}$. We also have an indexed family of computable functions from tuples of objects of some sorts to objects of some sort. Let $\mathbb{F}$ be a set of function names or function indices f, which represent functions $\tilde{f}$. By $\mathbb{O}_1 \times \cdots \times \mathbb{O}_n \to \mathbb{O}_0$ we denote the set of functions whose domain is $\mathbb{O}_1 \times \cdots \times \mathbb{O}_n$ and codomain is $\mathbb{O}_0$. By $\mathbb{F}_{O_0,O_1,\ldots,O_n}$, we denote the set of function names $f \in \mathbb{F}$ with $\tilde{f} \in \mathbb{O}_{O_1} \times \cdots \times \mathbb{O}_{O_n} \to \mathbb{O}_{O_0}$. We assume that the domain sorts $O_1, \ldots, O_n$ and the codomain sort O_0 are easily computed from f. We specify a class of grammars by a triple, which we call a *signature*, $\Sigma = \langle \Omega, \mathbb{F}, O_* \rangle$ where $O_* \in \Omega$ is a special sort of objects. We write $\mathbb{O}_*$ for $\mathbb{O}_{O_*}$.

A *context-free* Σ*-grammar* (Σ*-grammar* for short) is a tuple $G = \langle N, \sigma, F, P, I \rangle$ where N is a finite set of *nonterminal symbols*, $I \subseteq N$ is a set of *initial symbols*, $\sigma \in N \to \Omega$ is a *sort assignment* on nonterminals such that $\sigma(X) = O_*$ for all $X \in I$, $F \subseteq \mathbb{F}$ is a finite set of function names, and P is a finite set of *production rules*, which are elements of $N \times F \times N^*$. Each production rule is denoted as

$$X_0 \leftarrow f\langle X_1, \ldots, X_n \rangle$$

where $X_0, \ldots, X_n \in N$ and $f \in \mathbb{F}_{O_0,O_1,\ldots,O_n}$ for $\sigma(X_i) = O_i$. For each $O \in \Omega$, $N_O = \sigma^{-1}(O) \subseteq N$ is the set of O*-nonterminals* which are assigned the sort O. By $\mathcal{G}(\Sigma)$ we denote the class of Σ-grammars.

A Σ-grammar defines its language via derivation trees, which are recursively defined as follows.

- If τ_i are X_i-derivation trees for $i = 1, \ldots, n$ and ρ is a rule of the form $X_0 \leftarrow f\langle X_1, \ldots, X_n \rangle$, then the term $\tau_0 = \rho[\tau_1, \ldots, \tau_n]$ is an X_0-derivation tree. Its *yield* $\tilde{\tau}_0$ is $\tilde{f}(\tilde{\tau}_1, \ldots, \tilde{\tau}_n) \in \mathbb{O}_{\sigma(X_0)}$ where $\tilde{\tau}_i$ is the yield of τ_i.

The case where $n = 0$ gives the base of this recursive definition. An X-derivation tree is *complete* if $X \in I$. The yield of any X-derivation tree is called an X*-substructure*. By $\mathcal{S}(G, X)$ we denote the set of X-substructures. The language of G is $\mathcal{L}(G) = \bigcup_{X \in I} \mathcal{S}(G, X)$, which we call a Σ*-language*. In other words, $\mathcal{L}(G)$ is the set of the yields of complete derivation trees. The class of Σ-languages is denoted by $\mathcal{L}(\Sigma)$.

Distributional learning is concerned with what X*-derivation contexts* represent. An X-derivation context is obtained by replacing an occurrence of an X-derivation tree in a complete derivation tree by a special symbol $\square_{\sigma(X)}$. Accordingly the yield $\tilde{\chi}$ of an X-derivation context χ should be a finite representation of a function that gives $\widetilde{\chi[\tau]}$ when applied to $\tilde{\tau}$ for any X-derivation tree τ. We assume to have a set $\mathbb{E}_O$ of representations of functions from $\mathbb{O}_O$ to $\mathbb{O}_*$ for $O \in \Omega$ to which the yields of derivation contexts belong.

- $\square_X$ is an X-derivation context for all $X \in I$ and its yield $\square_{O_*} \in \mathbb{E}_{O_*}$ represents the identity function on $\mathbb{O}_*$,
- For an X-derivation context χ_0, a rule $\rho = X \leftarrow f\langle X_1, \ldots, X_n \rangle$ and X_i-derivation trees τ_i for $i \in \{1, \ldots, n\} - \{j\}$, the term χ obtained by replacing $\square_X$ in χ_0 by $\rho[\tau_1, \ldots, \tau_{j-1}, \square_{X_j}, \tau_{j+1}, \ldots, \tau_n]$ is an X_j-derivation context. Its yield $\tilde{\chi} \in \mathbb{E}_{\sigma(X_j)}$, which

is denoted as

$$\tilde{\chi} = \tilde{\chi}_0 \odot \tilde{f}(\tilde{\tau}_1, \ldots, \tilde{\tau}_{j-1}, \square_{\sigma(X_j)}, \tilde{\tau}_{j-1}, \ldots, \tilde{\tau}_n),$$

represents the function $\phi \in \mathbb{O}_{\sigma(X_j)} \to \mathbb{O}_*$ such that for all $s \in \mathbb{O}_{\sigma(X_j)}$,

$$\phi(s) = \phi_0(\tilde{f}(\tilde{\tau}_1, \ldots, \tilde{\tau}_{j-1}, s, \tilde{\tau}_{j+1}, \ldots, \tilde{\tau}_n)),$$

where ϕ_0 is the function represented by $\tilde{\chi}_0$.

The yield of any X-derivation context is called an X-*context*. By $\mathcal{C}(G, X)$ we denote the set of X-contexts. For $c \in \mathcal{C}(G, X)$ and $s \in \mathcal{S}(G, X)$, $c \odot s$ is the result of the application of the function represented by c to s.

3 Context-substructure relation

By $\mathbb{S}$ and $\mathbb{C}$ we denote the set of substructures and contexts, respectively, which can be obtained by some grammar in $\mathcal{G}(\Sigma)$:

$$\mathbb{S} = \bigcup_{O \in \Omega} \mathbb{S}_O \text{ and } \mathbb{C} = \bigcup_{O \in \Omega} \mathbb{C}_O \text{ where}$$

$$\mathbb{S}_O = \bigcup \{\, \mathcal{S}(G, X) \mid X \text{ is an } O\text{-nonterminal of some } G \in \mathcal{G}(\Sigma) \,\}$$

$$\mathbb{C}_O = \bigcup \{\, \mathcal{C}(G, X) \mid X \text{ is an } O\text{-nonterminal of some } G \in \mathcal{G}(\Sigma) \,\}.$$

We write $\mathbb{S}_*$ for $\mathbb{S}_{O_*}$. Note that the above definition is relative to Σ. Even if $\mathbb{O}_{O_1} = \mathbb{O}_{O_2}$ for different $O_1, O_2 \in \Omega$, it can be the case that $\mathbb{S}_{O_1} \neq \mathbb{S}_{O_2}$ and $\mathbb{C}_{O_1} \neq \mathbb{C}_{O_2}$. Though usually $\mathbb{O}_O$ has a definition independent from Σ, it is possible to specify $\mathbb{O}_O$ in terms of the signature so that $\mathbb{S}_O = \mathbb{O}_O$. Clearly if $s \in \mathbb{S}_O$ and $c \in \mathbb{C}_O$, then there is a grammar $G \in \mathcal{G}(\Sigma)$ generating $c \odot s$ using a nonterminal of sort O. Therefore, $c \odot s$ is well defined for any $s \in \mathbb{S}_O$ and $c \in \mathbb{C}_O$ without specifying a particular Σ-grammar. Similarly $c \odot \tilde{f}(s_1, \ldots, s_{j-1}, \square_{O_j}, s_{j+1}, s_n)$ is well defined for any $c \in \mathbb{C}_{O_0}$, $f \in \mathbb{F}_{O_0, \ldots, O_n}$ and $s_i \in \mathbb{S}_{O_i}$. This operation is generalized to sets $S \subseteq \mathbb{S}_O$ and $C \subseteq \mathbb{C}_O$ in the straightforward way, like $C \odot S = \{\, c \odot s \mid c \in C \text{ and } s \in S \,\}$.

Hereafter, whenever we write $c \odot s$ and $\tilde{f}(s_1, \ldots, s_n)$, we assume they are well-formed. That is, the domains of the functions represented by c and f match the sorts to which s and $s_1, \ldots, s_n$ belong, respectively. Accordingly we drop the subscript O from $\square_O$ and write $\tilde{f}(s_1, \ldots, s_{j-1}, \square, s_{j+1}, \ldots, s_n)$. When we have a substructure set S, we assume $S \subseteq \mathbb{S}_O$ for some $O \in \Omega$. We often identify s with $\{s\}$ unless confusion arises. Also we assume $\mathbb{S}_O \neq \varnothing$ for all $O \in \Omega$. The same assumptions apply to contexts.

We are interested in whether the composition $c \odot s$ belongs to a concerned language $L \in \mathcal{L}(\Sigma)$. Clark (2010b) has introduced *syntactic concept lattices* to analyze the context-substring relation on string languages and particularly to design a distributional learning algorithm for CFLs. Generalizing his discussion, we define an O-concept lattice $\mathfrak{B}_O(L)$ of a language $L \subseteq O_*$ for respective sorts $O \in \Omega$. Assuming L and O understood from the context, let us write

$$S^{\ddagger} = \{\, c \in \mathbb{C}_O \mid c \odot S \subseteq L \,\},$$
$$C^{\dagger} = \{\, s \in \mathbb{S}_O \mid C \odot s \subseteq L \,\}$$

for $S \subseteq \mathbb{S}_O$ and $C \subseteq \mathbb{C}_O$. We write $S^{\dagger}$ for $S^{\ddagger\dagger}$ and $C^{\ddagger}$ for $C^{\dagger\ddagger}$.

We call a pair $\langle S, C \rangle \subseteq \mathbb{S}_O \times \mathbb{C}_O$ a *concept* iff $S^{\dagger} = C$ and $C^{\ddagger} = S$. For any $S \subseteq \mathbb{S}_O$ and $C \subseteq \mathbb{C}_O$, $\langle S^{\dagger}, S^{\ddagger} \rangle$ and $\langle C^{\dagger}, C^{\ddagger} \rangle$ are concepts. We call them the concepts *induced by* S and C, respectively. For two concepts $\langle S_1, C_1 \rangle$ and $\langle S_2, C_2 \rangle$ in $\mathfrak{B}_O(L)$, we write $\langle S_1, C_1 \rangle \leq_L^O \langle S_2, C_2 \rangle$ if $S_1 \subseteq S_2$, which is equivalent to $C_2 \subseteq C_1$. With this partial order, $\mathfrak{B}_O(L)$ is a complete lattice.

We can introduce a partial order to substructure sets based on the concepts that they induce. Let us write $S_1 \leq_L^O S_2$ if $S_2^{\ddagger} \subseteq S_1^{\ddagger}$. The relation represents the substitutability of S_1 for S_2.

Lemma 1. *The following three are equivalent for* $S, T \subseteq \mathbb{S}_O$*:*

- $S \leq_L^O T$,
- $c \odot T \subseteq L$ *implies* $c \odot S \subseteq L$ *for all* $c \in \mathbb{C}_O$,
- $T^{\ddagger} \odot S \subseteq L$.

If $S_i \leq_L^{O_i} T_i$ *for* $i = 1, \ldots, n$, *then for any* $f \in \mathbb{F}_{O_0, O_1, \ldots, O_n}$, *we have* $f(S_1, \ldots, S_n) \leq_L^{O_0} f(T_1, \ldots, T_n)$.

If $S_1 \leq_L S_2$ and $S_2 \leq_L S_1$, we write $S_1 \equiv_L S_2$.

4 Conditions to be distributionally learnable

Distributional learning algorithms decompose examples $d \in \mathbb{S}_*$ into contexts $c \in \mathbb{C}_O$ and substructures $s \in \mathbb{S}_O$ so that $c \odot s = d$. Then a *primal* approach uses substructures or sets of substructures as nonterminals of a conjecture grammar. We want each nonterminal $[\![S]\!]$ indexed by $S \subseteq \mathbb{S}_O$ to satisfy $\mathcal{S}(G, [\![S]\!]) = S^\dagger$. On the other hand, a *dual* approach uses contexts or sets of contexts as nonterminals where the semantics of the nonterminal is $\mathcal{S}(G, [\![C]\!]) = C^\dagger$. For an object $d \in \mathbb{S}_*$, $O \in \Omega$ and $\vec{O} = (O_0, \dots, O_n)$ with $O_i \in \Omega$, we define

$$\mathbb{S}_{O|d} = \{\, s \in \mathbb{S}_O \mid c \odot s = d \text{ for some } c \in \mathbb{C}_O \,\}$$
$$\mathbb{C}_{O|d} = \{\, c \in \mathbb{C}_O \mid c \odot s = d \text{ for some } s \in \mathbb{S}_O \,\}$$
$$\mathbb{F}_{\vec{O}|d} = \{\, f \in \mathbb{F}_{\vec{O}} \mid c \odot f(s_1, \dots, s_m) = d \text{ for some } c \in \mathbb{C}_{O_0|d} \text{ and } s_i \in \mathbb{S}_{O_i|d} \,\},$$

$\mathbb{S}_{|D} = \bigcup_{O\in\Omega}^{d\in D} \mathbb{S}_{O|d}$, $\mathbb{C}_{O|D} = \bigcup_{O\in\Omega}^{d\in D} \mathbb{C}_{O|d}$ and $\mathbb{F}_{|D} = \bigcup_{\vec{O}\in\Omega^*}^{d\in D} \mathbb{F}_{\vec{O}|d}$ for $D \subseteq O_*$. Let $\Omega_{|D} = \bigcup_{d\in D} \Omega_{|d}$ for $\Omega_{|d} = \{\, O \mid \mathbb{S}_{O|d} \neq \varnothing \,\}$.

We require $\mathcal{G}(\Sigma)$ to be a tractable formalism such that composition and decomposition can be done efficiently.

Assumption 1. *There are polynomial-time algorithms which*

- *decide whether $s \in \mathbb{S}_O$ from $s \in \mathbb{S}$ and $O \in \Omega$,*
- *compute $\tilde{f}(s_1, \dots, s_n)$ from $s_i \in \mathbb{S}_{O_i}$ and $f \in \mathbb{F}_{O_0, O_1, \dots, O_n}$,*
- *decide whether $c \in \mathbb{C}_O$ from $c \in \mathbb{C}$ and $O \in \Omega$,*
- *compute $c \odot s$ from $c \in \mathbb{C}_O$ and $s \in \mathbb{S}_O$ for any $O \in \Omega$.*
- *decide whether $s \in \mathcal{L}(G)$ from $s \in \mathbb{S}_*$ and $G \in \mathcal{G}(\Sigma)$,*

Assumption 2. *There is $p \in \mathbb{N}$ such that the arity of every $f \in \mathbb{F}$ is at most p.*

Assumption 3. *There are polynomial-time algorithms that compute* $\text{SUB}(d)$, $\text{CON}(d)$ *and* $\text{FUN}(d)$ *from $d \in \mathbb{S}_*$ such that $\mathbb{S}_{|d} \subseteq \text{SUB}(d) \subseteq \mathbb{S}$, $\mathbb{C}_{|d} \subseteq \text{CON}(d) \subseteq \mathbb{C}$ and $\mathbb{F}_{|d} \subseteq \text{FUN}(d) \subseteq \mathbb{F}$.*

Actually by Assumptions 2 and 3, one can derive the polynomial-time uniform membership decidability. Moreover, it is easy to filter out nonmembers of $\mathbb{S}_{|d}$, $\mathbb{C}_{|d}$ and $\mathbb{F}_{|d}$ from $\text{SUB}(d)$, $\text{CON}(d)$ and $\text{FUN}(d)$, respectively, but it is not necessary. Assumption 3 implies $|\Omega_{|d}|$ is polynomially bounded, since $O \in \Omega_{|d}$ iff $\mathbb{F}_{O, O_1, \dots, O_n} \neq \varnothing$ for some $O_1, \dots, O_n$.

We write $\text{SUB}_O(D) = \text{SUB}(D) \cap \mathbb{S}_O$, $\text{CON}_O(D) = \text{CON}(D) \cap \mathbb{C}_O$ and $\text{FUN}_{\vec{O}}(D) = \text{FUN}(D) \cap \mathbb{F}_{\vec{O}}$.

It is often the case that elements of Ω represents pairwise disjoint sets. Actually for any signature Σ, one can find $\Sigma' = \langle \Omega', \mathbb{F}', O_* \rangle$ that satisfies this condition such that $\mathcal{L}(\Sigma) = \mathcal{L}(\Sigma')$. Let $\Omega' = \{\, O' \mid O \in \Omega \,\} \cup \{O_*\}$ and $\mathbb{O}_{O'} = \mathbb{O}_O \times \{O\}$ for each $O \in \Omega - \{O_*\}$. For $f \in \mathbb{F}_{O_0, \dots, O_n}$ with $\tilde{f}(s_1, \dots, s_n) = s_0$, we have $f' \in \mathbb{F}'_{O'_0, \dots, O'_n}$ with $\tilde{f}'(s'_1, \dots, s'_n) = s'_0$ where $s'_i = (s_i, O_i)$ if $O_i \in \Omega - \{O_*\}$ and $s'_i = s_i$ if $O_i = O_*$. Clearly every Σ-grammar has an equivalent Σ'-grammar. Moreover, this makes it clear that from $s \in \mathbb{S}$ one can immediately specify the unique sort $O' \in \Omega'$ such that $s \in \mathbb{O}_{O'}$. Similarly we may assume that each $c \in \mathbb{C}$ has unique $O \in \Omega$ such that $c \in \mathbb{C}_O$ and finding that O is a trivial task. Hereafter we work under this assumption. By O and f we mean $\mathbb{O}_O$ and $\tilde{f}$ for notational convenience.

Example 1. A right regular grammar over an alphabet Δ is a Σ_{reg}-grammar for $\Sigma_{\text{reg}} = \langle \{\Delta^*\}, \mathbb{F}, \Delta^* \rangle$. $\mathbb{F}$ has nullary functions which are members of $\Delta \cup \{\varepsilon\}$ and unary functions f_a for $f_a(w) = aw$ for all $w \in \Sigma^*$ for some $a \in \Delta$. Clearly the class of right regular grammars satisfies Assumptions 1, 2 and 3.

Example 2. A CFG is a Σ_{cfg}-grammar for $\Sigma_{\text{cfg}} = \langle \{\Delta^*\}, \mathbb{F}, \Delta^* \rangle$ where each $f \in \mathbb{F}$ is represented as an $(n+1)$-dimension vector of strings $\langle u_0, \dots, u_n \rangle$ such that $f(v_1, \dots, v_n) = u_0 v_1 u_1 \dots v_n u_n$ for all $v_i \in \Delta^*$. The class of CFGs itself satisfies Assumption 1 but not Assumptions 2 and 3, since we have no limit on n. But several normal forms fulfill Assumptions 2 and 3.

Example 3. Let $\Omega = \{O_1, O_2, \dots\}$ where O_m denotes the set of m-tuples of strings. Linear context-free rewriting systems, equivalent to non-deleting multiple CFGs, are Σ_{mcfg}-grammars where $O_* = O_1 = \Delta^*$ and every $f \in \mathbb{F}_{O_{m_0}, O_{m_1}, \dots, O_{m_n}}$ concatenates strings $u_{i,j}$ occurring in an input $\langle \langle u_{1,1}, \dots, u_{1,m_1} \rangle, \dots, \langle u_{n,1}, \dots, u_{n,m_n} \rangle \rangle$ in some way to form an m_0-tuple of strings. The uniform membership problem of this class is PSPACE-complete (Kaji et al., 1992). There are infinitely many ways to decompose a string d into substructures and contexts as $O_m \in \Omega_{|d}$ for all m. Assumptions 1 and 3 will be fulfilled when we restrict admissible functions so that $\mathbb{F}_{O_{m_0},\dots,O_{m_n}} \neq \varnothing$ only if $n \leq p$ and $m_i \leq q$ for all i.

As is the case for multiple CFGs, Assumption 2 is often needed to make the uniform membership problem solvable in polynomial-time (Assumption 1).

5 Learning models

Learning algorithms in this paper work under three different learning models.

A *positive presentation (text)* of a language $L_* \subseteq \mathbb{O}_*$ is an infinite sequence $d_1, d_2, \dots \in \mathbb{O}_*$ such that $L_* = \{ d_i \mid i \geq 1 \}$. In the framework of *identification in the limit from positive data*, a learner is given a positive presentation of the language $L_* = \mathcal{L}(G_*)$ of the target grammar G_* and each time a new example d_i is given, it outputs a grammar G_i computed from $d_1, \dots, d_i$. We say that a learning algorithm $\mathcal{A}$ *identifies G_* in the limit from positive data* if for any positive presentation $d_1, d_2, \dots$ of $\mathcal{L}(G_*)$, there is an integer n such that $G_n = G_m$ for all $m \geq n$ and $\mathcal{L}(G_n) = \mathcal{L}(G_*)$. We say that $\mathcal{A}$ *identifies a class $\mathbb{G}$ of grammars in the limit from positive data* iff $\mathcal{A}$ identifies all $G \in \mathbb{G}$ in the limit from positive data.

We say that $\mathcal{A}$ *identifies a class $\mathbb{G}$ of grammars in the limit from positive data and membership queries* when we allow $\mathcal{A}$ to ask *membership queries* (MQs) to an oracle when it computes a hypothesis grammar. An instance of an MQ is an object $d \in \mathbb{O}_*$ and the oracle answers whether $d \in L_*$ in constant time.

The third model is the learning with a *minimally adequate teacher* (MAT). A learner is not given a positive presentation but it may ask *equivalence queries* (EQs) to an oracle in addition to MQs. An instance of an EQ is a grammar G. If $\mathcal{L}(G) = L_*$, the oracle answers "*Congratulations!*" and the learning process ends. Otherwise, the oracle returns a counerexample $d \in (L_* - \mathcal{L}(G)) \cup (\mathcal{L}(G) - L_*)$, which is called *positive* if $d \in L_* - \mathcal{L}(G)$ and *negative* if $d \in \mathcal{L}(G) - L_*$.

When we have an oracle, the learning task itself is trivial unless we show some favorable property on the learning efficiency.

6 Learnable subclasses

This section presents how Σ-grammars with distributional properties can be learned. Note that all of those properties are relative to Σ. We assume Σ-grammars $G_* = \langle N_*, \sigma_*, F_*, P_*, I_* \rangle$ in this section have no useless nonterminals or functions. That is, $\mathcal{S}(G_*, X) \neq \varnothing$, $\mathcal{C}(G_*, X) \neq \varnothing$ for all $X \in N_*$ and every $f \in F_*$ appears in some rule in P_*.

6.1 Substitutable Languages

Definition 1 (Clark and Eyraud (2007)). A language $L \in \mathcal{L}(\Sigma)$ is said to be *substitutable* if for any $O \in \Omega$, $c_1, c_2 \in \mathbb{C}_O$ and $s_1, s_2 \in \mathbb{S}_O$,

$$c_1 \odot s_1,\ c_1 \odot s_2,\ c_2 \odot s_1 \in L \text{ implies } c_2 \odot s_2 \in L\,.$$

The definition can be rephrased as follows:

$$s_1^{\ddagger} \cap s_2^{\ddagger} \neq \varnothing \text{ implies } s_1 \equiv_L s_2\,.$$

Example 4. Yoshinaka (2008) has proposed a learning algorithm for k, l-substitutable CFLs, which satisfy the following property:

$$x_1uy_1vz_1, x_1uy_2vz_1, x_2uy_1vz_2 \in L \implies x_2uy_2vz_2 \in L$$

for any $x_i, y_i, z_i \in \Delta^*$, $u \in \Delta^k$ and $v \in \Delta^l$. We define a signature $\Sigma_{k,l} = \langle \Omega_{k,l}, \mathbb{F}_{k,l}, O_* \rangle$ as follows. Let $\Omega_{k,l} = \{O_*\} \cup \{ O_{u,v} \mid u \in \Delta^k \text{ and } v \in \Delta^l \} \cup \{ O_u \mid u \in \Delta^{<k+l} \}$, where $O_* = \Delta^*$, $O_{u,v} = \{ \overline{uwv} \mid w \in \Delta^* \}$ and $O_u = \{\overline{u}\}$. Here we put overlines to make elements of Ω pairwise disjoint. Let $\mathbb{F}_{k,l} = \{ +_{\alpha,\beta} \mid \alpha, \beta \in \Omega - \{\Delta^*\} \} \cup \{ \square_\alpha^{O_*} \mid \alpha \in \Omega - \{O_*\} \} \cup \Delta$. The binary function $+_{\alpha,\beta}$ concatenates two strings from sorts α and β and gives the right sort in $\Omega - \{O_*\}$. For example, $+_{\alpha,\beta} \in \mathbb{F}$ with $\alpha = O_{u,v}$, $\beta = O_w$ has codomain $O_{u,x}$ where x is the suffix of vw of length l. The unary operation $\square_\alpha^{O_*} \in \mathbb{F}_{O_*,\alpha}$ simply removes the overline and "promotes" $\bar{u} \in \alpha$ to $u \in O_*$. Δ consists of the nullary functions giving a single letter from $\overline{\Delta}$. It is not hard to see that every CFG has an equivalent $\Sigma_{k,l}$-grammar. Note that O_*-nonterminals never occur on the right hand side of a rule in a $\Sigma_{k,l}$-grammar. Hence $\mathbb{C}_{O_*}$ is just the singleton $\{\square_{O_*}\}$ such that $\square_{O_*} \odot u = u$ for all $u \in \Delta^*$, whereas $\mathbb{C}_\alpha \neq \mathbb{C}_{O_*}$ contains arbitrary pairs of strings $\langle l, r \rangle \in \Delta^* \times \Delta^*$ such that $\langle l, r \rangle \odot \bar{u} = lur$ for any $\bar{u} \in \alpha$. The $\Sigma_{k,l}$-substitutability is exactly the k, l-substitutability.

Theorem 1. *The class of substitutable Σ-languages is identifiable in the limit from positive data.*

The theorem follows Lemmas 2 and 3 below.

From a finite set D of positive examples, Algorithm 1 computes the grammar $\textsc{SubstP}(D) = \langle N, \sigma, F, P, I\rangle$ defined as follows:

- $N_O = \{ [\![s]\!] \mid s \in \textsc{Sub}_O(D) \}$ for $O \in \Omega_{|D}$,
- $I = \{ [\![s]\!] \mid s \in D \}$,
- $F = \textsc{Fun}(D)$,
- P consists of the rules of the form

  $$[\![s_0]\!] \leftarrow f\langle [\![s_1]\!], \dots, [\![s_n]\!] \rangle$$

  where $f \in \textsc{Fun}_{O_0,\dots,O_n}(D)$ for $[\![s_i]\!] \in N_{O_i}$ if there is $c \in \textsc{Con}_O(D)$ such that

  $$c \odot s_0,\ c \odot f(s_1, \dots, s_n) \in D\,.$$

Since we assume elements of Ω are pairwise disjoint, each $[\![s]\!]$ belongs to a unique sort. Otherwise, each nonterminal should be tagged with a sort like $[\![s, O]\!]$.

Algorithm 1 Learning substitutable Σ-grammars

Data: A positive presentation $d_1, d_2, \dots$
Result: A sequence of grammars $G_1, G_2, \dots$
let $\hat{G}$ be a grammar such that $\mathcal{L}(\hat{G}) = \varnothing$;
for $n = 1, 2, \dots$ **do**
  let $D = \{d_1, \dots, d_n\}$;
  if $D \not\subseteq \mathcal{L}(\hat{G})$ **then**
    let $\hat{G} = \textsc{SubstP}(D)$;
  end if
  output $\hat{G}$ as G_n;
end for

An alternative way to construct a grammar is to use contexts rather than substructures for nonterminals. One can replace $\textsc{SubstP}(D)$ in the algorithm by $\textsc{SubstD}(D)$ which is defined as follows.

- $N_O = \{ [\![c]\!] \mid c \in \textsc{Con}_O(D) \}$ for $O \in \Omega_{|D}$,
- $I = \{ [\![\Box_{O_*}]\!] \}$,
- $F = \textsc{Fun}(D)$,
- P consists of the rules of the form

  $$[\![c_0]\!] \leftarrow f\langle [\![c_1]\!], \dots, [\![c_n]\!] \rangle$$

  where $f \in \textsc{Fun}_{O_0,\dots,O_n}$ for $[\![c_i]\!] \in N_{O_i}$ if there are $s_i \in \textsc{Sub}_{O_i}(D)$ such that

  $$c_i \odot s_i \in D \text{ for all } i \text{ and } c_0 \odot f(s_1, \dots, s_n) \in D\,.$$

The existing algorithms for different classes of substitutable languages (Clark and Eyraud, 2007; Yoshinaka, 2008; Yoshinaka, 2011a) are based on slight variants of SubstP. This paper shows the correctness of the algorithm using SubstD.

Lemma 2. *Let D be a finite subset of a Σ-substitutable language L_* and G the grammar output by $\textsc{SubstD}(D)$. Then $\mathcal{L}(G) \subseteq L_*$.*

Proof. One can show by induction on the derivation that if $s \in \mathcal{S}(G, [\![c]\!])$ then $c \odot s \in L_*$. Suppose that G has a rule $[\![c]\!] \leftarrow f\langle [\![c_1]\!], \dots, [\![c_n]\!] \rangle$, $s_i \in \mathcal{S}(G, [\![c_i]\!])$ and $s = f(s_1, \dots, s_n)$. The induction hypothesis says $c_i \odot s_i \in L_*$ for all i. By the rule construction, there are t_i for $i = 1, \dots, n$ such that $c_i \odot t_i \in D \subseteq L_*$ and $c \odot f(t_1, \dots, t_n) \in D \subseteq L_*$. We have $s_i \equiv_{L_*} t_i$ since they occur in the same context c_i. By Lemma 1, $c \odot f(s_1, \dots, s_n) \in L_*$. □

Let $G_* = \langle N_*, \sigma_*, F_*, P_*, I_* \rangle$ be a Σ-grammar generating L_*. Fix $s_X \in \mathcal{S}(G_*, X)$ and $c_X \in \mathcal{C}(G_*, X)$ where $c_X = \Box_{O_*}$ for $X \in I_*$. Define D_* by

$$D_* = \{ c_X \odot s_X \mid X \in N_* \} \cup \{ c_{X_0} \odot f(s_{X_1}, \dots, s_{X_n}) \mid X_0 \leftarrow f\langle X_1, \dots, X_n \rangle \in P_* \}\,.$$

Lemma 3. *If $D_* \subseteq D$, then $\mathcal{S}(G_*, X) \subseteq \mathcal{S}(\textsc{SubstD}(D), [\![s_X]\!])$ for all X.*

Proof. Let $G = \textsc{SubstD}(D)$. If G_* has a rule $X_0 \leftarrow f\langle X_1, \dots, X_n \rangle$ then G has the corresponding rule $[\![c_{X_0}]\!] \leftarrow f\langle [\![c_{X_1}]\!], \dots, [\![c_{X_n}]\!] \rangle$, since

$$c_{X_0} \odot f(s_{X_1}, \dots, s_{X_n}), c_{X_i} \odot s_{X_i} \in D\,.$$

In particular since c_X for $X \in I$ is the identity function $\Box_{O_*}$, the corresponding nonterminal $[\![c_X]\!] = [\![\Box_{O_*}]\!]$ is the initial symbol of G, too. □

This shows that we do not need too many data to achieve a right grammar, since $|D_*| \le |P_*| + |N_*|$, where $|\cdot|$ denotes the cardinality of a set. Moreover, it is easy to see Algorithm 1 updates its conjecture in polynomial time in the total size of D by Assumptions 1, 2 and 3.

6.2 Finite kernel property

Definition 2 (Clark et al. (2009), Yoshinaka (2011b)). A nonempty finite set $S \subseteq \mathbb{S}_{\sigma(X)}$ is called a *k-kernel* of a nonterminal X if $|S| \le k$ and

$$\mathcal{S}(G, X) \equiv_{\mathcal{L}(G)} S\,.$$

A Σ-grammar G is said to have the *k-finite kernel property* (k-FKP) if every nonterminal X has a k-kernel S_X.

Theorem 2. *Under Assumptions 1, 2 and 3, Algorithm 2 identifies Σ-grammars with the k-FKP in the limit from positive data and membership queries.*

Algorithm 2 Learning Σ-grammars with k-FKP

Data: A positive presentation $d_1, d_2, \dots$ of L_*;
Result: A sequence of Σ-grammars $G_1, G_2, \dots$;
let $D := K := F := J := \varnothing$;
let $\hat{G} := \text{PRIMAL}_k(K, F, J)$;
for $n = 1, 2, \dots$ **do**
  let $D := D \cup \{d_n\}$; $J := \text{CON}(D)$;
  if $D \not\subseteq \mathcal{L}(\hat{G})$ **then**
    let $K := \text{SUB}(D)$ and $F := \text{FUN}(D)$;
  end if
  output $\hat{G} = \text{PRIMAL}_k(K, F, J)$ as G_n;
end for

The conjecture grammar $\text{PRIMAL}_k(K, F, J) = \langle N, \sigma, F, P, I\rangle$ of Algorithm 2 is defined from finite sets of substructures $K \subseteq \mathbb{S}$, functions $F \subseteq \mathbb{F}$ and contexts $J \subseteq \mathbb{C}$. The subsets of those sets corresponding to respective sorts are denoted as $K_O = K \cap \mathbb{S}_O$, $J_O = J \cap \mathbb{C}_O$ and $F_{\vec{O}} = F \cap \mathbb{F}_{\vec{O}}$.

- $N_O = \{\, [\![S]\!] \mid S \subseteq K_O \text{ with } 1 \le |S| \le k \,\}$ for each $O \in \Omega_{|D}$,
- $I = \{\, [\![S]\!] \in N_{O_*} \mid S \subseteq L \,\}$,
- P consists of the rules of the form

$$[\![S_0]\!] \leftarrow f\langle [\![S_1]\!], \dots, [\![S_n]\!]\rangle$$

where $f \in F_{O_0,O_1,\dots,O_n}$ for $[\![S_i]\!] \in N_{O_i}$ if

$$({S_0}^{\ddagger} \cap J_{O_0}) \odot f(S_1, \dots, S_n) \subseteq L_*\,. \quad (1)$$

The grammar is constructed by the aid of finitely many MQs. $\text{PRIMAL}_k(K, F, J)$ can be computed in polynomial time by Assumptions 1, 2 and 3. A rule $[\![S_0]\!] \leftarrow f\langle [\![S_1]\!], \dots, [\![S_n]\!]\rangle$ is compatible with the semantics of the nonterminals if $S_0^{\dagger} \supseteq f(S_1^{\dagger}, \dots, S_n^{\dagger})$, which is equivalent to

$${S_0}^{\ddagger} \odot f(S_1, \dots, S_n) \subseteq L_* \quad (2)$$

by Lemma 1. However, this condition (2) cannot be checked by finitely many MQs. The condition (1) can be seen as an approximation of (2), which is decidable by finitely many MQs. Clearly (2) implies (1) but not vice versa. If a rule satisfies (1) but not (2), we call the rule *incorrect*. If a rule is incorrect, there is a *witness* $c \in \mathbb{C}_{O_0} - J_{O_0}$ such that $c \odot S_0 \in L_*$ and $c \odot f(S_1, \dots, S_n) \notin L_*$.

Lemma 4. *For every finite $K \subseteq \mathbb{S}$ and $F \subseteq \mathbb{F}$ there is $J \subseteq \mathbb{C}$ such that $\hat{G} = \text{PRIMAL}(K, F, J)$ has no incorrect rules and $|J| \le |F||K|^{k(p+1)}$, in which case $\mathcal{L}(\hat{G}) \subseteq L_*$.*

Let S_X be a k-kernel of each nonterminal X of a grammar $G_* = \langle N_*, \sigma_*, F_*, P_*, I_*\rangle$ generating L_*.

Lemma 5. *There is a finite subset $D \subseteq L_*$ such that $S_X \subseteq \mathbb{S}_{|D}$ for all $X \in N_*$, $F_* \subseteq \mathbb{F}_{|D}$ and $|D| \le k|N_*| + |P_*|$. Moreover, if $S_X \subseteq K$ for all $X \in N_*$ and $F_* \subseteq F$, then $L_* \subseteq \mathcal{L}(\hat{G})$.*

We prove Theorem 2 discussing the efficiency.

Proof of Theorem 2. Clearly Algorithm 2 updates its conjecture in polynomial time in the data size. Polynomially (in the size of G_*) many positive examples will stabilize K and F by Lemma 5. After K and F stabilized, all the incorrect rules will be removed with at most polynomially (in $|K||F|$) many examples by Lemma 4. After that point Algorithm 2 never changes the conjecture, which generates the target language L_*. □

6.3 Congruential grammars

Definition 3 (Clark (2010a)). A Σ-grammar G is said to be *congruential* if every $s \in \mathcal{S}(G, X)$ is a 1-kernel of every $X \in N$.

Congruential Σ-grammars have the 1-FKP. Under the following additional assumption, this special case will be polynomial-time learnable with a minimally adequate teacher.

Assumption 4. *For any derivation tree τ, the size of its yield $\tilde{\tau}$ is polynomially bounded by that of τ.*

Theorem 3. *Under Assumptions 1, 2, 3 and 4, Algorithm 3 learns any language L_* generated by a* *congruential Σ-grammar G_* with a minimally adequate teacher in time polynomial in $|N_*|, |F_*|, \ell$ where ℓ is the total size of counterexamples given to the learner.*

Algorithm 3 Learning congruential Σ-grammars

let $K := F := J := \varnothing$;
let $\hat{G} :=$ PRIMAL$_1(K, F, J)$;
for $n = 1, 2, \ldots$ **do**
  if $\mathcal{L}(\hat{G}) = L_*$ (equivalence query) **then**
    output $\hat{G}$ and halt;
  else if the given counterexample d is positive ($d \in L_* - \mathcal{L}(\hat{G})$) **then**
    let $K := K \cup$ SUB(d) and $F := F \cup$ FUN(d);
  else
    let $J := J \cup$ WITNESSP$(\tau_d, \square_{O_*})$ where τ_d is an (implicit) parse tree of d by $\hat{G}$
  end if
  let $\hat{G} =$ PRIMAL$_1(K, F, J)$;
end for

Algorithm 3 uses the same grammar construction PRIMAL as Algorithm 2 where the parameters K and F are calculated from positive counterexamples given by the oracle. On the other hand, J is computed in a different way. By Lemma 4, when the oracle answers a negative counterexample d towards an EQ, our conjecture $\hat{G}$ must use an incorrect rule to derive d. To find and remove such an incorrect rule, Algorithm 3 calls a subroutine WITNESSP with input $(\tau_d, \square)$, where τ_d is a derivation tree of $\hat{G}$ whose yield is d. To be precise, τ_d does not have to be a derivation tree. Rather what we require is that for each $s \in \mathbb{S}_{|d}$, one can compute at least one tuple of $s_1, \ldots, s_n \in \mathbb{S}_{|d}$ and $f \in \mathbb{F}_{|d}$ such that $s = f(s_1, \ldots, s_n)$ and the height of the lowest derivation tree of each s_i is strictly lower than that of s. Indeed one can do this in polynomial time by a dynamic programming method from SUB(d) and FUN(d). Yet for explanatory easiness, we treat such information as an (implicit) derivation tree τ_d. The procedure WITNESSP returns a context that witnesses an incorrect rule that contributes to generating d by searching τ_d recursively calling itself. The procedure WITNESSP in general takes a pair (τ, c) such that τ is an $[\![s]\!]$-derivation tree of $\hat{G}$ and $c \in s^\ddagger - \tilde{\tau}^\ddagger$. Let $\tau = \rho(\tau_1, \ldots, \tau_n)$ where $\rho = [\![s]\!] \leftarrow f\langle [\![s_1]\!], \ldots, [\![s_n]\!]\rangle$. If $c \odot f(s_1, \ldots, s_n) \notin L_*$ then the rule ρ is incorrect. So WITNESSP returns c which witnesses the incorrectness of the rule. Otherwise, we have

$$c \odot f(s_1, \ldots, s_n) \in L_*$$
$$c \odot f(\tilde{\tau}_1, \ldots, \tilde{\tau}_n) \notin L_*$$

for the yields $\tilde{\tau}_i$ of τ_i. One can find i such that

$$c \odot f(s_1, \ldots, s_{i-1}, s_i, \tilde{\tau}_{i+1}, \ldots, \tilde{\tau}_n) \in L_*$$
$$c \odot f(s_1, \ldots, s_{i-1}, \tilde{\tau}_i, \tilde{\tau}_{i+1}, \ldots, \tilde{\tau}_n) \notin L_* .$$

This means an incorrect rule is in τ_i. We call WITNESSP$(\tau_i, c \odot f(s_1, \ldots, s_{i-1}, \square, \tilde{\tau}_{i+1}, \ldots, \tilde{\tau}_n))$.

Lemma 6. *The procedure* WITNESSP$(\tau_d, \square)$ *runs in polynomial time in ℓ and $|d|$.*

Proof. The number of recursive calls of WITNESSP is no more than the height of τ_d, which is at most $|\mathbb{S}_{|d}|$. Let the instance of the j-th recursive call be (τ_j, c_j) and χ_j the derivation context for $c = \tilde{\chi}_j$. χ_{j+1} is obtained from χ_j by replacing at most p subtrees by a derivation tree whose yield is an element of K. By Assumption 4, the size of c_j and thus the size of an instance of an MQ is polynomially bounded by $|d|\ell$. WITNESSP runs in polynomial time. □

Lemma 7. *Each time Algorithm 3 receives a negative counterexample, at least one incorrect rule is removed.*

Lemma 8. *Let $G_* = \langle N_*, \sigma, F_*, P_*, I_* \rangle$ be a congruential grammar generating L_*. Each time Algorithm 3 receives a positive counterexample, the cardinality of the set $\{ X \in N_* \mid K \cap \mathcal{L}(G_*, X) = \varnothing \} \cup (F_* - F)$ decreases strictly.*

Proof of Theorem 3. Time between an EQ and another is polynomially bounded by Lemma 6. By Lemmas 5 and 8, Algorithm 3 gets at most $|N_*| + |F_*|$ positive counterexamples. The grammar $\hat{G} =$ PRIMAL(K, F, J) is constructed from those positive counterexamples, so it has polynomially many rules. Therefore, by Lemma 7, after getting polynomially many negative counterexamples, which suppress all the incorrect rules, Algorithm 3 gets a right grammar representing L_*. □

6.4 Finite context property

Definition 4 (Clark (2010b), Yoshinaka (2011b)[1]). A nonempty finite set $C \subseteq \mathbb{C}$ is called a *k-context* of a nonterminal X if $|C| \leq k$ and

$$\mathcal{S}(G, X) \equiv_{\mathcal{L}(G)} C^{\dagger}.$$

A Σ-grammar G is said to have the *k-(weak) finite context property* (k-FCP) if every nonterminal X has a k-context C_X.

Theorem 4. *Under Assumptions 1, 2 and 3, Algorithm 4 identifies Σ-grammars with the k-FCP in the limit from positive data and membership queries.*

The theorem can be shown by an argument similar to the proof of Theorem 2 based on Lemmas 9 and 10 below. The discussion on the learning efficiency of Algorithm 2 is applied to Algorithm 4 as well.

Algorithm 4 Learning Σ-grammars with k-FCP

Data: A positive presentation $d_1, d_2, \ldots$ of L_*;
Result: A sequence of Σ-grammars $G_1, G_2, \ldots$;
let $D := J := F := K := \varnothing$;
let $\hat{G} := \mathrm{DUAL}_k(J, F, K)$;
for $n = 1, 2, \ldots$ **do**
  let $D := D \cup \{d_n\}$; $K := \mathrm{SUB}(D)$;
  if $D \not\subseteq \mathcal{L}(\hat{G})$ **then**
    let $J := \mathrm{CON}(D)$ and $F := \mathrm{FUN}(D)$;
  end if
  output $\hat{G} = \mathrm{DUAL}_k(J, F, K)$ as G_n;
end for

The conjecture grammar $\mathrm{DUAL}_k(J, F, K) = \langle N, \sigma, F, P, I \rangle$ of Algorithm 4 is defined from finite sets of contexts $J \subseteq \mathbb{C}$, functions $F \subseteq \mathbb{F}$ and substructures $K \subseteq \mathbb{S}$. For each $C \subseteq J_O$, we write $C^{(K)}$ to mean $C^{\dagger} \cap K_O$. This set can be seen as a finite approximation of $C^{\dagger}$, which is computable with MQs.

- $N_O = \{ [\![C]\!] \mid C \subseteq J_O \text{ with } 1 \leq |C| \leq k \}$ for $O \in \Omega_{|D}$,
- $I = \{ [\![\{\square_*\}]\!] \}$,
- P consists of the rules of the form

$$[\![C_0]\!] \leftarrow f\langle [\![C_1]\!], \ldots, [\![C_n]\!] \rangle$$

where $f \in F_{O_0, \ldots, O_n}$ for $[\![C_i]\!] \in N_{O_i}$ if $C_0 \odot f(C_1^{(K)}, \ldots, C_m^{(K)}) \subseteq L_*$.

[1]We adopt the definition by Yoshinaka, which is slightly weaker than Clark's.

We say that a rule $[\![C_0]\!] \leftarrow f([\![C_1]\!], \ldots, [\![C_n]\!])$ is *incorrect* if $C_0 \odot f(C_1^{\dagger}, \ldots, C_n^{\dagger}) \not\subseteq L_*$. In that case, there are $s_i \in C_i^{\dagger}$ such that $C_0 \odot f(s_1, \ldots, s_n) \not\subseteq L_*$.

Lemma 9. *For every finite $J \subseteq \mathbb{C}$ and $F \subseteq \mathbb{F}$ there is $K \subseteq \mathbb{S}$ such that $\hat{G} = \mathrm{DUAL}(J, F, K)$ has no incorrect rules and $|K| \leq p|F||J|^{k(p+1)}$, in which case $\mathcal{L}(\hat{G}) \subseteq L_*$.*

Let $G_* = \langle N_*, \sigma_*, F_*, P_*, I_* \rangle$ generate L_* and C_X a k-context of each nonterminal $X \in N_*$.

Lemma 10. *There is a finite subset $D \subseteq L_*$ such that $C_{|D} \supseteq C_X$ for all $X \in N_*$, $F_{|D} \supseteq F_*$ and $|D| \leq k|N_*| + |P_*|$. Moreover, if $J \supseteq C_X$ for all $X \in N_*$ and $F \supseteq F_*$, then $L_* \subseteq \mathcal{L}(\hat{G})$.*

6.5 Context-deterministic grammars

Definition 5 (Shirakawa and Yokomori (1993), Yoshinaka (2012)[2]). A Σ-grammar G is said to be *(weakly) context-deterministic* if every $c \in \mathcal{C}(G, X)$ is a 1-context of every $X \in N_O$.

Differently from Theorem 3, we do not need Assumption 4 for learning context-deterministic grammars with a minimally adequate teacher.

Theorem 5. *Under Assumptions 1, 2 and 3, Algorithm 3 learns any language L_* generated by a context-deterministic Σ-grammar G_* with a minimally adequate teacher in time polynomial in $|N_*|, |F_*|, \ell$ where ℓ is the total size of counterexamples given to the learner.*

Proof. By Lemmas 11, 12 and 13 below. □

Algorithm 5 uses the same grammar construction DUAL as Algorithm 4. By Lemma 9, when the oracle answers a negative counterexample d towards an EQ, our conjecture $\hat{G}$ must use an incorrect rule to derive d. To find and remove such an incorrect rule, Algorithm 5 calls a subroutine WITNESSD with a derivation tree τ_d of $\hat{G}$ whose yield is d. The procedure WITNESSD returns a finite set of substructures that witnesses an incorrect rule that contributes to generating d. An input given to WITNESSD is in general a $[\![c]\!]$-derivation tree τ such that $c \odot \tilde{\tau} \notin L_*$. Let $\tau = \rho[\tau_1, \ldots, \tau_n]$ where $\rho = [\![c]\!] \leftarrow f\langle [\![c_1]\!], \ldots, [\![c_n]\!] \rangle$. If there is i such that $c_i \odot \tilde{\tau}_i \notin L_*$, we recursively call WITNESSD(τ_i).

[2]We adopt the definition by Yoshinaka, which is slightly weaker than Shirakawa and Yokomori's.

Algorithm 5 Learning context-deterministic Σ-grammars

let $J := F := K := \varnothing$;
let $\hat{G} := \text{DUAL}_1(J, F, K)$;
for $n = 1, 2, \dots$ **do**
  if $\mathcal{L}(\hat{G}) = L_*$ (equivalence query) **then**
    output $\hat{G}$ and halt;
  else if the given counterexample d is positive ($d \in L_* - \mathcal{L}(\hat{G})$) **then**
    let $J := J \cup \text{CON}(d)$ and $F := F \cup \text{FUN}(d)$;
  else
    let $K := K \cup \text{WITNESSD}(\tau)$ where τ is an (implicit) parse tree of d by $\hat{G}$
  end if
  let $\hat{G} = \text{DUAL}_1(J, F, K)$;
end for

Otherwise, $\tilde{\tau}_i \in c_i^{\dagger}$ for all i, which means the rule ρ is incorrect. $\text{WITNESSD}(\tau)$ returns the set $\{\tilde{\tau}_1, \dots, \tilde{\tau}_n\}$. Differently from the case of WITNESSP, an instance of a recursive call is always an (implicit) derivation tree of some $s \in \mathbb{S}_{|d}$. This explains why we do not need Assumption 4 in this case.

Lemma 11. *Time between an* EQ *and another is polynomially bounded.*

Lemma 12. *Each time Algorithm 5 receives a negative counterexample, at least one incorrect rule is removed.*

Lemma 13. *Let* $G_* = \langle N_*, \sigma, F_*, P_*, I_* \rangle$ *be a context-deterministic grammar for* L_*. *Each time Algorithm 5 receives a positive counterexample, the set* $\{ X \in N_* \mid J \cap \mathcal{C}(G_*, X) = \varnothing \} \cup (F_* - F)$ *gets shrunk.*

6.6 Combined approaches

By combining primal and dual approaches, one can obtain stronger approaches (Yoshinaka, 2012). The class of Σ-grammars whose nonterminals admit either a k-kernel or l-context can be learned by combining the techniques presented in Sections 6.2 and 6.4 under Assumptions 1, 2 and 3. Also Σ-grammars whose nonterminals satisfy either the requirement to be congruential or to be context-deterministic can be learned with a minimally adequate teacher under Assumptions 1, 2, 3 and 4 (Sections 6.3 and 6.5).

7 Restricted cases

In some grammar classes, it may be the case that only (supersets of) $\mathbb{C}_{|d}$ and $\mathbb{F}_{|d}$ are computable in polynomial-time but $\mathbb{S}_{|d}$ is not, or the other way around: $\mathbb{S}_{|d}$ and $\mathbb{F}_{|d}$ are efficiently computable but $\mathbb{C}_{|d}$ is not. For example, in non-permuting parallel multiple CFGs (Seki et al., 1991), elements of $\mathbb{S}_{|d}$ for a string d are tuples of strings of the form $\langle v_1, \dots, v_m \rangle$ for $d = u_0 v_1 u_1 \dots v_m u_m$ and such substrings are polynomially many if m is fixed. However, $\mathbb{C}_{|d}$ contains exponentially many contexts. Clark and Yoshinaka (2014) showed that still a dual approach works for parallel multiple CFGs if nonterminals are known to have k-contexts belonging to a certain subset $\mathbf{C} \subseteq \mathbb{C}$ such that $\mathbf{C}_{|d} = \mathbb{C}_{|d} \cap \mathbf{C}$ is polynomial-time computable. A symmetric result of a primal approach has also been obtained by Kanazawa and Yoshinaka (2015) targeting a certain kind of tree grammars. This section does not postulate Assumption 3.

Definition 6. A Σ-grammar G is said to have the $(k, \mathbf{S})$-FKP if every nonterminal admits a k-kernel which is a subset of $\mathbf{S}$.

Assumption 5. *There are polynomial-time algorithms that compute* $\text{SUB}(d)$, $\text{CON}(d)$ *and* $\text{FUN}(d)$ *such that* $\mathbf{S}_{|d} \subseteq \text{SUB}(d) \subseteq \mathbb{S}$, $\mathbb{C}_{|d} \subseteq \text{CON}(d) \subseteq \mathbb{C}$ *and* $\mathbb{F}_{|d} \subseteq \text{FUN}(d) \subseteq \mathbb{F}$, *where* $\mathbf{S}_{|d} = \mathbf{S} \cap \mathbb{S}_{|d}$.

It is not hard to see that Algorithm 2 works for learning Σ-grammars with $(k, \mathbf{S})$-FKP under Assumptions 1, 2 and 5. All discussions in Section 6.2 hold for this restricted case.

The symmetric definition and assumption are as follows.

Definition 7. A Σ-grammar G is said to have the $(k, \mathbf{C})$-FCP if every nonterminal admits a k-context which is a subset of $\mathbf{C}$.

Assumption 6. *There are polynomial-time algorithms that compute* $\text{SUB}(d)$, $\text{CON}(d)$ *and* $\text{FUN}(d)$ *such that* $\mathbb{S}_{|d} \subseteq \text{SUB}(d) \subseteq \mathbb{S}$, $\mathbf{C}_{|d} \subseteq \text{CON}(d) \subseteq \mathbb{C}$ *and* $\mathbb{F}_{|d} \subseteq \text{FUN}(d) \subseteq \mathbb{F}$.

It is not hard to see that under Assumptions 1, 2 and 6, Algorithms 4 work for learning Σ-grammars with $(k, \mathbf{C})$-FCP Σ-grammars. All discussions in Section 6.4 hold for this restricted case.

When learning substitutable languages, even a weaker assumption suffices.

Assumption 7. *There are sets* $\mathbf{S} \subseteq \mathbb{S}$ *and* $\mathbf{C} \subseteq \mathbb{C}$ *such that for every nonterminal* X *of* $G \in \mathcal{G}(\Sigma)$, *we have* $\mathcal{S}(G,X) \cap \mathbf{S} \neq \varnothing$ *and* $\mathcal{C}(G,X) \cap \mathbf{C} \neq \varnothing$. *Moreover, there are polynomial-time algorithms that compute* $\text{SUB}(d)$, $\text{CON}(d)$ *and* $\text{FUN}(d)$ *such that* $\mathbf{S}_{|d} \subseteq \text{SUB}(d) \subseteq \mathbb{S}$, $\mathbf{C}_{|d} \subseteq \text{CON}(d) \subseteq \mathbb{C}$ *and* $\mathbb{F}_{|d} \subseteq \text{FUN}(d) \subseteq \mathbb{F}$.

Under Assumptions 1, 2 and 7, Algorithm 1 works using either SUBSTP or SUBSTD.

On the other hand, the results on the polynomial-time MAT learnability of congruential and context-deterministic Σ-grammars do not hold anymore under any of Assumptions 5, 6 and 7.

8 Extending learnable classes

This section compares learnable classes of Σ-languages for different Σ with the same special sort O_*. For Σ_1 and Σ_2 with $\Sigma_i = \langle \Omega^i, \mathbb{F}^i, O_* \rangle$, if $\Omega^1 \subseteq \Omega^2$ and $\mathbb{F}^1 \subseteq \mathbb{F}^2$, every Σ_1-grammar is a Σ_2-grammar, so $\mathcal{L}(\Sigma_1) \subseteq \mathcal{L}(\Sigma_2)$. However, since the distributional properties defined so far are relative to a signature, a Σ_1-grammar with a distributional property under Σ_1 does not necessarily have the corresponding property under Σ_2. Yet if $\mathbb{S}_O$ and $\mathbb{C}_O$ are preserved by moving from Σ_1 to Σ_2, the distributional properties other than the substitutability are preserved.

Let us define the direct union $\Sigma_0 = \langle \Omega^0, \mathbb{F}^0, O_* \rangle$ of arbitrary signatures Σ_1 and Σ_2 by $\Omega^0 = \{O_*\} \cup \{\,(O,i) \mid O \in \Omega^i \text{ with } i \in \{1,2\}\,\}$ where $\mathbb{O}_{(O,i)} = \{\,(s,i) \mid s \in O\,\}$ and $\mathbb{F}^0 = \mathbb{G}^1 \cup \mathbb{G}^2 \cup \{\square_1, \square_2\}$, where $\mathbb{G}^i$ is a trivial variant of $\mathbb{F}^i$ working on the new domain and codomain of the form (O,i) and $\square_i(s,i) = s$ for all $s \in O_*$. Then every Σ_i-grammar G can be seen as a special type of Σ_0-grammar by adding a new initial symbol Z and rules of the form $Z \leftarrow \square_{\Sigma_i}\langle X \rangle$ for all initial symbols X of G. We have $\mathcal{L}(\Sigma_1) \cup \mathcal{L}(\Sigma_2) \subseteq \mathcal{L}(\Sigma_0)$. Every Σ_i-grammar that is congruential, context-deterministic, with the k-FKP or with the k-FCP for $i = 1, 2$ can be seen as a Σ_0-grammar with those properties. Note that $\mathbb{C}_{O_*}$ is the singleton of the identity function in Σ_0, which means any element of $\mathcal{L}(G)$ is a 1-kernel of the new initial symbol Z. In this way, from two signatures, one can obtain a richer learnable class of languages.

The above argument on signature generalization does not hold for substitutable case. Rather the opposite holds. If $\Omega^1 \subseteq \Omega^2$ and $\mathbb{F}^1 \subseteq \mathbb{F}^2$, then a language substitutable under Σ_2 is substitutable under Σ_1 but not vice versa.

Let us say that Σ_2 is *finer* than Σ_1 if every sort of Ω^1 is partitioned into finite number of sorts in Ω^2 and every function of $\mathbb{F}^2$ is a subfunction of some function in $\mathbb{F}^1$ which accords with the partition. That is, every sort O of Ω^1 has a finite set $\Omega^2_O \subseteq \Omega^2$ such that $O = \bigcup \Omega^2_O$ and $\mathbb{F}^1_{O_0,\dots,O_n} = \bigcup \{\, \mathbb{F}^2_{O'_0,\dots,O'_n} \mid O'_i \in \Omega^2_{O_i} \,\}$. For instance, $\Sigma_{k,l}$ is finer than $\Sigma_{k',l'}$ for $k' \le k$ and $l' \le l$ in Example 4. If Σ_2 is finer than Σ_1, $\mathcal{L}(\Sigma_1) = \mathcal{L}(\Sigma_2)$ holds. Every language substitutable under Σ_1 is substitutable under Σ_2 but not vice versa. Moreover, every congruential (resp. context-deterministic) Σ_1-grammar has an equivalent congruential (resp. context-deterministic) Σ_2-grammar but not vice versa.

9 Grammars with partial functions

Yoshinaka (2015) showed that a dual approach can be applied to the learning of *conjunctive grammars*. Conjunctive grammars (Okhotin, 2001) are CFGs extended with the conjunctive operation $\&$ so that one can extract the intersection of the languages of nonterminals. For example, a conjunctive rule $A_0 \to A_1 \& A_2$ means that if *both* A_1 and A_2 generate the same string u then so does A_0. Conjunctive grammars cannot be seen as Σ-grammars, since the conjunctive operation $\&$ is a *partial* function whose domain is not represented as the direct product of two sorts, which is not legitimate in the general framework of Σ-grammars.

A *partial signature* is a triple $\Pi = \langle \Omega, \mathbb{F}, O_* \rangle$ which is defined in the way similar to a (total) signature but $\mathbb{F}$ may have partial functions. Accordingly contexts in $\mathbb{C}$ will be partial functions. We do not have $\mathcal{C}(G,X) \odot \mathcal{S}(G,X) \subseteq \mathcal{L}(G)$ any more, since $c \odot s$ may not be defined for some elements $c \in \mathcal{C}(G,X)$ and $s \in \mathcal{S}(G,X)$. The correspondence between O-concept lattices and Σ-grammars collapses. This prevents the application of the theory of distributional learning developed in this paper to Π-grammars. Still we can generalize the discussion on the learning of conjunctive grammars.

Definition 8. A Π-grammar G is said to have the *strong* k-FCP if for any $X \in N_O$, there is a finite set

$C_X \subseteq \mathbb{C}_O$ with $|C_X| \leq k$ such that

$$\mathcal{S}(G, X) = \{\, s \mid c \odot s \in L \text{ for all } c \in C_X \,\}\,.$$

Definition 8 requires every $c \in C_X$ to be total on $\mathcal{S}(G, X)$. One can learn Π-grammars with the strong k-FCP under Assumptions 1, 2 and 6, where **C** consists of total functions only. The grammar construction DUAL_k should be modified so that we have a rule $[\![C_0]\!] \leftarrow f\langle [\![C_1]\!], \ldots, [\![C_n]\!] \rangle$ if $c \odot f(s_1, \ldots, s_n) \in L_*$ for any $c \in C_0$ and $s_i \in C_i^{(K)}$ such that $f(s_1, \ldots, s_n)$ is defined. One might think that one can naturally define context-deterministic grammars accordingly: Every $c \in \mathcal{C}(G, X)$ should be a 1-context of X. However, this means that functions in such a Π-grammar are essentially total.

Acknowledgments

The view presented in this paper has been sharpened through the interactions and discussions with several researchers with whom I worked on the distributional learning of generalized CFGs. I would like to show my deepest gratitude to Alexander Clark, Makoto Kanazawa, Anna Kasprzik and Gregory Kobele. Without those people this work would have been hard to accomplish. Any insufficiency or errors in this paper are of course of my own.

References

Alexander Clark and Rémi Eyraud. 2007. Polynomial identification in the limit of substitutable context-free languages. *Journal of Machine Learning Research*, 8:1725–1745.

Alexander Clark and Ryo Yoshinaka. 2014. Distributional learning of parallel multiple context-free grammars. *Machine Learning*, 96(1-2):5–31.

Alexander Clark, Rémi Eyraud, and Amaury Habrard. 2009. A note on contextual binary feature grammars. In *EACL 2009 workshop on Computational Linguistic Aspects of Grammatical Inference*, pp. 33–40.

Alexander Clark. 2010a. Distributional learning of some context-free languages with a minimally adequate teacher. In J. Sempere and P. García, editors, *ICGI*, LNCS 6339, pp. 24–37. Springer.

Alexander Clark. 2010b. Learning context free grammars with the syntactic concept lattice. In J. Sempere and P. García, editors, *ICGI*, LNCS 6339, pp. 38–51. Springer.

Alexander Clark. 2010c. Towards general algorithms for grammatical inference. In M. Hutter, F. Stephan, V. Vovk, and T. Zeugmann, editors, *ALT*, LNCS 6331, pp. 11–30. Springer.

Yuichi Kaji, Ryuichi Nakanishi, Hiroyuki Seki, and Tadao Kasami. 1992. The universal recognition problems for parallel multiple context-free grammars and for their subclasses. *IEICE Transaction on Information and Systems*, E75-D(7):499–508.

Makoto Kanazawa and Ryo Yoshinaka. 2015. Distributional learning and context/substructure enumerability in non-linear tree grammars. In *Formal Grammar - 20th International Conference, FG 2015, Barcelona, Spain, August 8-9, 2015. Proceedings*. to appear.

Anna Kasprzik and Ryo Yoshinaka. 2011. Distributional learning of simple context-free tree grammars. In J. Kivinen, C. Szepesvári, E. Ukkonen, and T. Zeugmann, editors, *ALT*, LNCS 6925, pp. 398–412. Springer.

Alexander Okhotin. 2001. Conjunctive grammars. *Journal of Automata, Languages and Combinatorics*, 6(4):519–535.

Hiroyuki Seki, Takashi Matsumura, Mamoru Fujii, and Tadao Kasami. 1991. On multiple context-free grammars. *Theoretical Computer Science*, 88(2):191–229.

Hiromi Shirakawa and Takashi Yokomori. 1993. Polynomial-time MAT learning of c-deterministic context-free grammars. *Transaction of Information Processing Society of Japan*, 34:380–390.

Ryo Yoshinaka and Makoto Kanazawa. 2011. Distributional learning of abstract categorial grammars. In S. Pogodalla and J.-P. Prost, editors, *LACL*, LNCS 6736, pp. 251–266. Springer.

Ryo Yoshinaka. 2008. Identification in the limit of k, l-substitutable context-free languages. In A. Clark, F. Coste, and L. Miclet, editors, *ICGI*, LNCS 5278, pp. 266–279. Springer.

Ryo Yoshinaka. 2011a. Efficient learning of multiple context-free languages with multidimensional substitutability from positive data. *Theoretical Computer Science*, 412(19):1821–1831.

Ryo Yoshinaka. 2011b. Towards dual approaches for learning context-free grammars based on syntactic concept lattices. In G. Mauri and A. Leporati, editors, *DLT*, LNCS 6795, pp. 429–440. Springer.

Ryo Yoshinaka. 2012. Integration of the dual approaches in the distributional learning of context-free grammars. In A. H. Dediu and C. Martín-Vide, editors, *LATA*, LNCS 7183, pp. 538–550. Springer.

Ryo Yoshinaka. 2015. Learning conjunctive grammars and contextual binary feature grammars. In A. H. Dediu, E. Formenti, C. Martín-Vide, and B. Truthe, editors, *LATA*, LNCS 8977, pp. 623–635. Springer.

Canonical Context-Free Grammars and Strong Learning: Two Approaches

Alexander Clark
Department of Philosophy,
King's College London
alexander.clark@kcl.ac.uk

Abstract

Strong learning of context-free grammars is the problem of learning a grammar which is not just weakly equivalent to a target grammar but isomorphic or structurally equivalent to it. This is closely related to the problem of defining a canonical grammar for the language. The current proposal for strong learning of a small class of CFGs uses grammars whose nonterminals correspond to congruence classes of the language, in particular to a subset of those that satisfy a *primality* condition. Here we extend this approach to larger classes of CFGs where the nonterminals correspond instead to closed sets of strings; to elements of the syntactic concept lattice. We present two different classes of canonical context-free grammars. One is based on all of the primes in the lattice: the other, more suitable for strong learning algorithms is based on a subset of primes that are irreducible in a certain sense.

1 Introduction

This paper is concerned with the problem of strong learning of context-free grammars in the distributional framework. One approach, initiated in (Clark, 2014) is to develop strong learning algorithms by defining canonical grammars based on properties of algebraic structures associated with the language: specifically the syntactic monoid of the language. In that paper a strong learning result was presented for a subclass of the class of substitutable languages, languages which have a simple language theoretic closure property.

In this paper we will extend the canonical grammar ideas to a larger class of grammars, while not presenting a full strong learning result, for reasons of space and some technical details not yet resolved. Rather than using the syntactic monoid, we use the syntactic concept lattice (SCL), (Clark, 2013), a richer structure that is suitable for modeling all context-free grammars. In the case of substitutable languages the syntactic monoid is almost identical to the syntactic concept lattice.

We want these canonical grammars to be as unambiguous as possible, and to use as few nonterminals as possible. These two obvious principles pull in the same direction: a grammar with extra nonterminals will typically have extra derivations and thus a higher degree of ambiguity. Finding some global minimum leads in general to intractable computational problems – the set covering problem, a classic NP-hard problem – and the answer may be indeterminate (in that there may be two structurally distinct minima). So rather we stipulate some technical notion which is more determinate, and can be efficiently identified (though we do not talk about the algorithmic issues here). In particular, we want the grammars defined to be compatible with efficient learning algorithms for context-free grammars (Yoshinaka, 2012a; Leiß, 2014).

In the case of the monoid, we only have one operation, concatenation, and given a derivation tree with unlabeled interior nodes, each node in the tree can only be legally labeled with the unique congruence class of the yield of the subtree. Thus given the unlabeled trees, the labeling is determined.

In the case of the SCL, we have a lattice struc-

Proceedings of the 14th Meeting on the Mathematics of Language (MoL 14), pages 99–111,
Chicago, USA, July 25–26, 2015. ©2015 Association for Computational Linguistics

ture, and so there are many different possible ways of modeling the structure, and many different ways of labeling a given tree from the very specific to the very general. We previously argued in (Clark, 2011) that the most general labelings would be optimal; that view now seems simplistic.

We argue that we should only model those unpredictable parts of the structure, that is to say those places where the structure differs from the free structure $\mathcal{P}(\Sigma^*)$. The grammar does not need to state that $\{u\} \circ \{v\} = \{uv\}$ or that $\{u\} \cup \{v\} = \{u, v\}$: these are true in the free structure. It is only when these are not equal that we need to represent the difference.

We will give definitions of the basic mathematical concepts we use in Section 2, including a brief introduction to the syntactic concept lattice in Section 2.4 to make the paper self-contained.

In Section 3 we explain the relation between strong learning and canonical grammars.

Then in Section 4 we extend the definition of primes from congruence classes to closed sets of strings. Section 5 presents our first family of canonical grammars that are based directly on all of the primes in the language.

In the case of concepts the lattice structure means that there may be many different concepts that contain a given string, and so in Section 6 we discuss how to exploit the lattice structure to select a smaller set of categories that are irreducible in some sense; and then in Section 7 we present a second family of canonical grammars based on this restricted subset of the primes. We finish with a worked example to illustrate the abstract mathematical discussion and some discussion.

2 Preliminaries

2.1 Strings, Languages and Contexts

We assume a fixed alphabet Σ and write Σ^* for the set of strings. A language is a subset of Σ^*, we write concatenation of languages L, M as $L \cdot M$, or sometimes just LM. The empty string is λ. We take a symbol $\Box \notin \Sigma$, and using this we define a context as an element of $\Sigma^*\Box\Sigma^*$, written $l\Box r$. We define $\odot$ as $l\Box r \odot w = lwr$, and extend these to sets of strings and contexts in the usual way. The empty context $\lambda\Box\lambda = \Box$ is particularly important: of course $\Box \odot w = w$.

2.2 Grammars

We define CFGs standardly as a tuple $\langle \Sigma, V, S, P \rangle$ where $S \in V$ is a single start symbol, V is the set of nonterminals and P is a finite subset of $V \times (\Sigma \cup V)^*$, written as $N \to \alpha$. The derivation process is denoted by $\Rightarrow$ and $\Rightarrow^*$. We define

$$\mathcal{L}(G, N) = \{w \in \Sigma^* \mid N \Rightarrow_G^* w\}$$

and define $\mathcal{L}(G) = \mathcal{L}(G, S)$. We also define the set of derivation contexts:

$$\mathcal{C}(G, N) = \{l\Box r \in \Sigma^*\Box\Sigma^* \mid S \Rightarrow_G^* lNr\}$$

The following property corresponds to the *context-free* property of the derivation process:

$$\mathcal{C}(G, N) \odot \mathcal{L}(G, N) \subseteq \mathcal{L}(G).$$

Two grammars $G_1 = \langle \Sigma, V_1, S_1, P_1 \rangle$, $G_2 = \langle \Sigma, V_2, S_2, P_2 \rangle$ are weakly equivalent if $\mathcal{L}(G_1) = \mathcal{L}(G_2)$. They are isomorphic if there is a bijection $\phi : V_1 \to V_2$, such that $\phi(S_1) = S_2$ and $\phi(P_1) = P_2$, extending ϕ to productions and sets of productions in the natural way. Isomorphic grammars are identical except for a relabeling of the nonterminal symbols. Clearly isomorphism implies weak equivalence.

2.3 Lattices

We assume some familiarity with lattices: see (Davey and Priestley, 2002) for basic definitions. We write $\top, \bot, \vee, \wedge$ as standard. An element x is join-irreducible iff $x = a \vee b$ implies $a = x$ or $b = x$; dually it is meet irreducible iff $x = a \wedge b$ implies $a = x$ or $b = x$. For some lattices, the set of join irreducible elements and the set of meet irreducible elements can form a "basis" for the lattice, in that every element can be represented as a finite join of join-irreducible elements and/or a finite meet of meet-irreducible elements. In the lattice of all subsets of Σ^*, $\mathcal{P}(\Sigma^*)$ the join irreducible sets are the singleton sets $\{w\}$ for any string, and the meet irreducible sets are $\Sigma^* \setminus \{w\}$.

A descending chain is a strictly descending sequence of elements of a lattice $X_0 \supset X_2 \supset \ldots X_n$. A lattice satisfies the descending chain condition (DCC) if there are no infinite descending chains. If a lattice satisfies the DCC, then every nonempty subset has at least one minimal element. We define the ascending chain condition (ACC) dually.

2.4 The Syntactic Concept Lattice

We now describe the syntactic concept lattice briefly; for fuller descriptions see e.g. (Clark, 2013; Leiß, 2014; Wurm, 2012). Given a fixed language L, we have a Galois connection between sets of strings and sets of contexts defined, where S is a set of strings and C is a set of contexts:

$$S^{\triangleright} = \{l\Box r \mid l\Box r \odot S \subseteq L\}$$

and

$$C^{\triangleleft} = \{w \in \Sigma^* \mid C \odot w \subseteq L\}$$

.

A closed set of strings is a set of strings S such that $S = S^{\triangleright\triangleleft}$, a closed set of contexts is one such that $C = C^{\triangleleft\triangleright}$. A concept is an ordered pair $\langle S, C\rangle$ such that $S = C^{\triangleleft}$ and $C = S^{\triangleright}$. In this case both S and C are closed. We will therefore often refer to a concept through the corresponding closed set of strings. Note that for any such concept, the following property holds, which corresponds to the context-free property of the CFG derivations:

$$C \odot S \subseteq L$$

Clearly $w \in L$ iff $\Box \in \{w\}^{\triangleright}$, and so L is closed.

The Syntactic Concept Lattice of L, written $\mathfrak{B}(L)$ is the collection of concepts, with the following constants, relations and operations, which we define in terms of the closed sets of strings alone: $S \vee T = (S \cup T)^{\triangleright\triangleleft}, S \wedge T = S \cap T$, $S \circ T = (S \cdot T)^{\triangleright\triangleleft}$, $S \leq T$ iff $S \subseteq T$, $\top = \Sigma^*$, $\bot = \emptyset^{\triangleright\triangleleft}$. With these operations $\mathfrak{B}(L)$ is a complete idempotent semiring, and furthermore a complete residuated lattice.

This lattice forms a hierarchy of all distributionally definable sets of strings in the language. There will be a finite number of elements iff the language is regular. Minimal grammars will have nonterminals that correspond to elements of the syntactic concept lattice as shown by (Clark, 2013). Given a context-free grammar G such that $\mathcal{L}(G) = L$, we define a universal morphism $h_L : V \to \mathfrak{B}(L)$ given by $h_L(N) = \mathcal{L}(G, N)^{\triangleright\triangleleft}$. We extend this to a CFG-morphism in the obvious way. (Clark, 2013) proved that for all CFGs $\mathcal{L}(h_L(G)) = L$. Therefore any CFG for L can be mapped to a possibly smaller grammar whose nonterminals are elements of $\mathfrak{B}(L)$. We can therefore assume that the nonterminals of the grammar are elements of $\mathfrak{B}(L)$.

3 Weak and Strong Learning

We will not present any learning algorithms here, but the work is motivated by learning considerations and so we need to make the background assumptions clear. In standard models of learning, there is a target grammar G_* and the learner, using information only about $\mathcal{L}(G_*)$, must eventually return a grammar $\hat{G}$ such that $\mathcal{L}(\hat{G}) = \mathcal{L}(G_*)$. In strong learning (Clark, 2014) in contrast, given the same information source, the learner must pick a $\hat{G}$ such that $\hat{G}$ is isomorphic to G_*.

(Clark, 2014) observes that the existence of a canonical grammar is a necessary condition for a strong learning algorithm. Any strong learning algorithm will implicitly define a canonical grammar for any language in the class of languages that it learns. Much of that paper is in fact concerned with precisely that definition. Accordingly in this paper we focus on defining a canonical grammar rather than directly presenting a learning algorithm.

The universal property of the syntactic concept lattice is an important tool. This means that rather than dealing directly with CFGs which are arbitrary and intractable we can deal with the lattices $\mathfrak{B}(L)$ which have nice mathematical properties. We can assume without loss of generality that the nonterminals of the grammar will correspond to concepts or closed sets of strings: to elements of the lattice. Given this, there is a natural notion of a production being *correct*: $X \to w$ is correct if $w \in X$, $X \to Y_1 \ldots Y_n$ is correct if $X \supseteq Y_1 \cdot \cdots \cdot Y_n$.

Given that a language that is not regular will have an infinite number of concepts, we need a principled way of selecting a finite number of these in an appropriate way so that we have a finite grammar. The general approach we take is to identify some elements that are irreducible in some sense with respect to the algebraic structure of the residuated lattice.

In the case of substitutable grammars, the closed sets of strings are almost exactly the congruence classes—except for $\top, \bot$ and $\{\lambda\}^{\triangleright\triangleleft}$, every closed set of strings is either equal to a congruence class or a congruence class together with λ. There seems to be only one plausible way to define a grammar, given that the mathematical structure of the congruence classes is just a monoid. Since this structure is so simple, there is only one reasonable irreducibility property that we can use to select from the congruence classes: primality, which we define later. In the case of general CFLs things are unsurprisingly much more complicated. There seem to be two different factors to be considered. One factor concerns, as in the case of the monoid, the concatenation structure of the strings—the monoid structure of $\mathfrak{B}(L)$—and the other concerns the partial order: the lattice structure of $\mathfrak{B}(L)$.

We start by discussing the concatenation structure in Section 4 and discuss the lattice structure later in Section 6.

4 Primes

Since a monoid is a very simple algebraic structure, with a single associative binary operation, there is only one reasonable technique to define a subset of elements of the syntactic monoid in such a way that the grammar based on those elements is well behaved. (Clark, 2014) argues that we should represent only those elements where concatenation differs from the free operation of concatenation: in other words where $[uv] \supset [u][v]$. Language theoretically these represent places where the monoid has some nontrivial structure and grammatically they provide evidence for a nontrivial nonterminal: a nonterminal which occurs on the left hand side of more than one production. Congruence classes which have this desirable property are called *primes*.

For a congruence class the definition of a prime is straightforward. If a nonzero nonunit[1] congruence class X has a nontrivial decomposition into two congruence classes Y, Z such that $X = Y \cdot Z$ then it is composite. The trivial decompositions are $X = X \cdot [\lambda] = [\lambda] \cdot X$, where $[\lambda] = \{w \in \Sigma^* \mid \{w\}^{\triangleright} = \{\lambda\}^{\triangleright}\}$.

In the case of a CFL which is not substitutable, we need a different criterion, since we may use concepts that are not congruence classes but unions of congruence classes. This is complicated by the fact that the empty string may occur in many different closed sets of strings.

Definition 1. *$X \in \mathfrak{B}(L_*)$ is composite if there are $Y, Z \in \mathfrak{B}(L)$ such that $X \neq Y$ and $X \neq Z$ and $X = Y \cdot Z$. An element of $\mathfrak{B}(L_*)$ is prime if it is not composite. We write $\mathfrak{P}(L_*)$ for the set of primes of a language L_*. The unit prime is $\{\lambda\}^{\triangleright\triangleleft}$.*

We define $\mathfrak{P}(L_*)_\lambda = \mathfrak{P}(L_*) \setminus \{\{\lambda\}^{\triangleright\triangleleft}\}$, the set of nonunit primes.

Note that here we do not exclude $\{\lambda\}^{\triangleright\triangleleft}$. Clearly for any closed set of strings $X = X \cdot \{\lambda\}^{\triangleright\triangleleft} = \{\lambda\}^{\triangleright\triangleleft} \cdot X$. For the condition to be nontrivial, we clearly need to exclude such cases, but we also want to exclude cases such as $a^* = a^* \cdot a^*$ and $a^*b^* = (a^*b^*) \cdot b^*$.

Lemma 1. *For every $a \in \Sigma$, $\{a\}^{\triangleright\triangleleft}$ is prime.*

Proof. Supppose $\{a\}^{\triangleright\triangleleft} = B \cdot C$. Since $a \in \{a\}^{\triangleright\triangleleft}$ either $a \in B$ and $\lambda \in C$ or vice versa. Assume the former. Since $a \in B$, this means that $B \supseteq \{a\}^{\triangleright\triangleleft}$ and since $\lambda \in C$, this means that $B \subseteq \{a\}^{\triangleright\triangleleft}$. Therefore $B = \{a\}^{\triangleright\triangleleft}$, or, by a similar argument $C = \{a\}^{\triangleright\triangleleft}$. Therefore it is prime. □

Lemma 2. *$\{\lambda\}^{\triangleright\triangleleft}$ is prime.*

Proof. Suppose $\{\lambda\}^{\triangleright\triangleleft} = X \cdot Y$. Clearly $\lambda \in X, \lambda \in Y$. Therefore $X \subseteq \{\lambda\}^{\triangleright\triangleleft}$ and $Y \subseteq \{\lambda\}^{\triangleright\triangleleft}$. But if X, Y are in $\mathfrak{B}(L_*)$, they must both be equal to $\{\lambda\}^{\triangleright\triangleleft}$ since that is the smallest concept that contains λ. □

Definition 2. *If $X \in \mathfrak{B}(L)$ and $\alpha \in \mathfrak{P}(L)^+$, a nonempty string of primes, we write $\bar{\alpha}$ for the concatenation of the primes in α. So if $\alpha = \langle A_1, \ldots, A_n \rangle$ then $\bar{\alpha} = A_1 \cdot \cdots \cdot A_n$. We say that α is a prime decomposition of X iff $\bar{\alpha} = X$, and none of the elements of α are unit. In the special case where $X = \{\lambda\}^{\triangleright\triangleleft}$ we consider $\langle \{\lambda\}^{\triangleright\triangleleft} \rangle$ to be a prime decomposition.*

Example 1. *If $L = \{a^n b^n \mid n > 0\}$, then the primes are $\{a\}, \{b\}$ and $L' = L \cup \{\lambda\}$, together with $\top, \bot$ and $\{\lambda\}^{\triangleright\triangleleft}$. L has the prime decomposition $\langle \{a\}, L', \{b\} \rangle$.*

We need to consider two cases: one where a prime contains λ and one where it does not. If a closed set of strings contains the empty string, that means that it represents an optional category; it can be replaced by the empty string. If X is a closed set of strings that contains λ and $X = Y \cdot Z$, then clearly $\lambda \in Y \cap Z$ and $Y \subseteq X$ and $Z \subseteq X$. Therefore a decomposition of a concept that contains λ will be into

[1] The zero congruence class, if it exists, is $\emptyset^{\triangleleft}$ and the unit is $[\lambda]$.

proper subsets of that concept. Decompositions using concepts with λ may not terminate, if the lattice has infinite descending chains.

Example 2. *Let $L_1 = (ba)^*$, and $L_n = L_{n-1} \cdot (ba^n)^*$. Consider the language $L = \bigcup_n L_n$. This is a closed set of strings with an infinite descending chain $L \supset L \setminus L_1 \supset L \setminus (L_1 \cup L_2) \cdots$. For each n, $(ba^n)^*$ is closed and prime, and L has no finite prime decomposition.*

Lemma 3. *If $\mathfrak{B}(L_*)$ satisfies DCC then every element has a prime decomposition.*

Proof. If X is prime, then it has a length one decomposition, $\langle X \rangle$. Define the width of a non empty set of strings to be the minimum length of a string in the set. If it contains the empty string, then the width is zero.

Let M be the set of all non-zero non-unit concepts without prime decompositions. Suppose it is nonempty; then it has at least one minimal element, by the DCC. Take a minimal element of minimal width, X. Suppose X has width n. It is not prime by assumption and so $X = Y \cdot Z$. Case 1: width of X is zero and so both Y and Z contain the empty string: therefore Y, Z are both proper subsets of X. Therefore they are not in M (since X is minimal). Moreover they are not zero or unit, therefore they have prime decompositions such that $Y = \bar{\alpha}$ and $Z = \bar{\beta}$ and therefore $\alpha\beta$ is a decomposition of X. Case 2: the width of X is greater than zero; and the width of Y and Z are both less than width of X. Then Y and Z are both not in M and therefore both have prime decompositions and therefore so does X. Case 3: the width of X is greater than zero, and one of Y or Z had width zero. Assume that $\lambda \in Y$ (the other case is identical), then Z is a proper subset of X and therefore has a prime decomposition, and Y has width less than X and is therefore also has a prime decomposition. □

These decompositions aren't necessarily unique, but this lemma shows that the set of primes is sufficiently large to express any concept we want through finite concatenations.

It is not the case that every closed set of strings that is composite has a unique prime decomposition; even if we restrict ourselves to maximal decompositions: decompositions where no element can be replaced with a larger one. Clearly we can decompose, for example a^* into $\{\lambda, a^k\} \cdot (a^* \setminus \{a^k\})$ for any k, and for a suitable language these can be prime.

Example 3. *Consider the language $L = \{a, aa\}$ this has closed sets of strings $A = \{a\}$ and $B = \{\lambda, a\}$. $L = A \cdot B = B \cdot A$. So L is composite and it has two distinct prime factorisations, which are clearly both maximal.*

It simplifies the analysis here if we assume that all the concepts in a language have unique prime decompositions; accordingly we will restrict ourselves to that case for the moment, though we will remove this requirement in Section 7.

Definition 3. *A language has the unique factorisation property (UFP) if every closed set of strings with a nonempty distribution has a unique maximal prime factorisation; as before we stipulate that $\{\lambda\}^{\triangleright\triangleleft}$ has such a unique factorisation.*

If a language has the UFP, and P is a closed set of strings, then we can write $\Phi(P)$ for the unique prime factorisation, which is a string of primes, in the case of P, is the length 1 string. If $\alpha = \Phi(P)$ then $\bar{\alpha} = P$.

Now if we have languages which are UFP, DCC and a finite number of primes, then we can define a unique grammar.

Definition 4. *We define the class of languages $\mathfrak{L}_{\mathrm{P}}$ to be the set of languages which satisfy all of the following three conditions:*

1. *the unique prime decomposition property,*
2. *have no infinite ascending or descending chains,*
3. *and have a finite number of primes.*

All substitutable context free languages with a finite number of primes satisfy these conditions and so this is an extension of the approach in (Clark, 2014). All regular languages have a finite number of primes and therefore satisfy the chain conditions, but may not be uniquely decomposable.

Suppose we have a language $L \in \mathfrak{L}_{\mathrm{P}}$ and a prime N, we can construct a set of productions with N on the left hand side as follows.

Definition 5. *Let $\Gamma \subseteq \mathfrak{B}(L)^+$. We define a preorder on these strings by $\alpha \precsim \beta$ iff $\bar{\alpha} \subseteq \bar{\beta}$.*

The maximal elements of Γ wrt this pre-order are the elements

$$\max \Gamma = \{\alpha \in \Gamma \mid \forall \beta \in \Gamma, \bar{\beta} \supseteq \bar{\alpha} \Rightarrow \bar{\alpha} = \bar{\beta}\}.$$

Definition 6. *Define* $\Gamma(N) = \{\alpha \in \mathfrak{P}(L)^+_\lambda \mid N \supset \bar{\alpha}\}$. *We can take the maximal elements of this set:* $\max(\Gamma(N)) = \{\alpha \in \Gamma(N) \mid \forall \alpha' \in \Gamma(N), \bar{\alpha'} \supseteq \bar{\alpha} \Rightarrow \alpha' = \alpha\}$

The elements of $\max(\Gamma(N))$ will be the right hand sides of productions with N on the left hand side.

Lemma 4. *If* $\alpha \in \Gamma(N)$ *then there is an* $\alpha' \in \max(\Gamma(N))$ *such that* $\bar{\alpha} \subseteq \bar{\alpha'}$.

Proof. Consider the set $\{\gamma \in \Gamma(N) \mid \bar{\gamma} \supseteq \bar{\alpha}\}$. If $\mathfrak{B}(L)$ has no infinite ascending chains, then every non empty set has a maximal element; so let α' be a maximal element of this set which clearly satisfies the condition required. □

Note that $\langle N \rangle$ is not in $\Gamma(N)$, since we require a strict superset relation; this saves the definition from vacuity. If we have infinite ascending chains then we may not have a maximal element. This motivates the use of an ascending chain condition.

5 Canonical Grammar based on primes

We are now in a position to define a set of canonical grammars for $\mathfrak{L}_P$. Given the nature of the problem it is inevitable that we will have to have a large number of restrictions on the sorts of grammars that we can learn. There are two ways of framing these restrictions either as restrictions on the grammars that generate the language or as language theoretic restrictions themselves. Here we stick to the latter approach; we take a language, and define criteria that define a class of languages. As the reader will see though, we can express the constraints we need in purely language theoretic terms, though those constraints will correspond naturally to finiteness constraints on the various sets of nonterminals and productions.

Definition 7. *For each language* $L \in \mathfrak{L}_P$ *we define the grammar* $\mathcal{G}_P(L)$ *as the tuple* $\langle \Sigma, V, S, P \rangle$ *where*

- *S is a distinguished symbol,*
- $V = \{S\} \cup \mathfrak{P}(L)_\lambda$. *If* $\top^\triangleright = \emptyset$ *then we remove* $\top$ *from* V*: if* $\bot = \emptyset$ *then we remove* $\bot$,
- *P is the union of the following sets of productions, for all* $N \in \mathfrak{P}(L)_\lambda$

  $N \to a$ *if* $a \in N$, *and* $a \in \Sigma$

  $N \to \lambda$ *if* $\lambda \in N$

  $N \to \alpha$ *for all* $\alpha \in \max(\Gamma(N))$

  $S \to \Phi(L)$.

Lemma 5. *For every prime* N, $\mathcal{L}(G, N) \supseteq N$.

Proof. Induction on length of w. Base case; $|w| \leq 1$. in which case if $w \in N$ then there is a rule $N \to w$. Otherwise if $w \in N$ and $w = a_1 \dots a_n$ where $a_i \in \Sigma$, then since $N \supseteq a_1^{\triangleright\triangleleft} \dots a_n^{\triangleright\triangleleft}$, and each $a_i^{\triangleright\triangleleft}$ is prime, we know that we have a correct production $\langle a_1^{\triangleright\triangleleft} \dots a_n^{\triangleright\triangleleft} \rangle \in \beta(N)$. Therefore there is some $\alpha \in \gamma(N)$ such that $w \in \bar{\alpha}$, and a production $N \to \alpha$. Since $w \in \bar{\alpha}$ we must have strings $w_1 \dots w_k = w$ such that $k = |\alpha|$, and for each $1 \leq i \leq k$ $w_i \in N_i$ and $\alpha = N_1 \dots N_k$.

By induction otherwise each w_i has length less than n and thus $N_i \stackrel{*}{\Rightarrow} w_i$, and therefore $N \stackrel{*}{\Rightarrow} w$. We should also consider the case where $k = 1$, in which case we have a unary rule, and the case where all but one of the w_i have length 0. In both cases we have one prime Q strictly less than N, and since we have a finite number of primes, and no unary cycles a simple induction on the derivation height will suffice. □

Lemma 6. *The canonical grammar for G is finite and generates L.*

Proof. Note that all rules are correct and thus by induction we can show that $\mathcal{L}(G, N) \subseteq N$, which combined with the previous lemma tells us that $\mathcal{L}(G, P) = P$ for all primes.

It is finite since we cannot have two productions whose right hand sides start with the same prime. Note that $\mathfrak{B}(L)$ is a residuated lattice and that for any two closed sets of strings X, Y, $X \backslash Y = \{w \mid X \cdot \{w\} \subseteq Y\}$ is closed. If $P \to N\alpha$ and $P \to N\beta$ are both productions with $|\alpha| > 0$, $|\beta| > 0$ then since $\alpha \leq N \backslash P$ we have that $\alpha = \beta$ since they are both equal to the unique prime factorisation of $N \backslash P$. Therefore if there are only n primes, we can have at most $2n + |\Sigma| + 1$ productions with P on the left hand side.

Finally we verify that S generates the right strings, which is immediate. Either L is prime, in which case it is trivial since we have a production $S \rightarrow L$, or L has prime decomposition $N_1 \ldots N_k$, in which case we can see that the production $S \rightarrow N_1 \ldots N_k$ combined with the fact that $\mathcal{L}(N_i) = N_i$, gives the fact that $\mathcal{L}(S) = L$. □

This gives us a canonical grammar class for $\mathfrak{L}_P$ but the grammars are still very redundant and ambiguous. In the next section we consider how to select a smaller set of nonterminals.

Lemma 7. *Every language with a finite number of primes and no infinite chains is a context free language.*

Proof. Clearly $\mathfrak{L}_P$ is a proper subset of the context free languages, but even if we have no unique prime decomposition, we can still get a CFG by picking some shortest prime decomposition nondeterministically. □

We can weaken these conditions as we shall see as they are not necessary conditions; here is an example of a CFL with infinite chains that still receives an adequate canonical grammar.

Example 4. *Let* $L = \{w \in \{a,b\}^+ : |w|_a > |w|_b\}$ *The congruence classes are obviously indexed by* $|w|_a - |w|_b$. *Call these* $E_n = \{w \mid |w|_a - |w|_b = n\}$. *The closed sets of strings are* $C_n = \{w \mid |w|_a - |w|_b \geq n\}$. $L = C_1$ *is prime,* C_{-1} *is prime,* C_0 *is prime.* C_n *for* $n > 1$ *are composite.* $C_n \circ C_m = C_{n+m}$.

$C_2 = C_1 \circ C_1 = b^{\triangleright\triangleleft} \circ b^{\triangleright\triangleleft}$ *and it is composite. Since* $C_1 = E_1 \cup C_0$, $C_1 \cdot C_1 = (E_1 \cup C_0) \cdot (E_1 \cup C_0) = E_1 \cdot E_1 \cup (E_1 \cdot C_0) \cup (C_0 \cdot C_0) \cup (C_0 \cdot E_1)$. *Since* $\lambda \in C_0$ *this means that* $C_1 \cdot C_1 \supseteq C_0 \cup E_1 \cup E_2 = C_2$. *Ergo this is composite. Therefore there are exactly three primes. However this still has a simple canonical grammar.*

6 Lattice structure

For general context-free languages there may be very many prime concepts, and as a result grammars based on all primes may be excessively ambiguous, and unsuitable for the description of natural languages.[2] Indeed there are finite languages where the number of primes is exponential in the number of strings in the language.

Example 5. *For some large* $\Sigma = \{a_1, \ldots, a_n\}$ *define* $L = \{a_i a_j \mid i \neq j\}$. *Clearly* $|L| = n(n-1)$. *Every nonempty proper subset* X *of* Σ *is closed; defined by the set of contexts* $\{\Box a_i \mid a_i \notin X\}$. $\mathfrak{B}(L)$ *therefore has* $2^n + 1$ *concepts, none of which are composite.*

In as case such as this the grammar defined by $\mathcal{G}_P(L)$ will be exponentially larger than $|L|$; which is clearly undesirable.

Moreover, the previous approach relies on the number of primes in the language being finite. Many simple languages have an infinite number of primes though, and so it is natural to try to extend this approach by considering some additional properties that might serve to pick out a finite subset of these primes. We need some additional constraints to get smaller and less ambiguous grammars. While it is natural to look to the meet and join irreducible elements of the syntactic concept lattice, it seems better to use a slightly larger set; the images of the irreducible elements of the free lattice. We call these semi-irreducible; we would like to use the terms join-prime and meet-prime, but these already have different meanings in lattice theory.

Definition 8. *Suppose* $X \in \mathfrak{B}(L)$; *we say that* X *is join-semi-irreducible* (JSI) *if* $X = \{w\}^{\triangleright\triangleleft}$ *for some* $w \in \Sigma^*$. *we say that* X *is meet-semi-irreducible* (MSI) *if* $X = \{l \Box r\}^{\triangleleft}$ *for some* $l, r \in \Sigma^*$.

Observe that if X is join-semi-irreducible then it contains some strings that are not in any lower concepts. Similarly if X is meet-semi-irreducible then it contains some contexts that do not occur in any higher concepts. Note that L is always MSI.

We will illustrate this with a simple example, using a finite language.

Example 6. *Consider the language generated by the* CFG, $S \rightarrow AX, S \rightarrow BY, A \rightarrow a, A \rightarrow c, A \rightarrow m, B \rightarrow b, B \rightarrow c, X \rightarrow x, X \rightarrow z, X \rightarrow m, Y \rightarrow y, Y \rightarrow z$. *This is a finite language which consists of 11 strings, all of length 2. The string* cz *receives two parses under this grammar:*

[2] We do not discuss here the difficult question of whether natural language grammars exhibit spurious ambiguity—syntactic ambiguity that does not relate to semantic ambiguity.

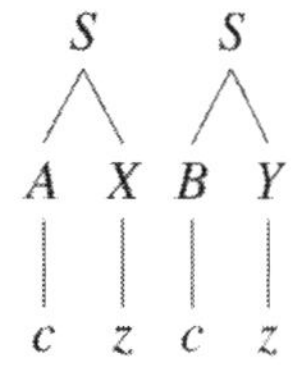

Figure 1 shows the lattice for this language. Note that the ambiguous letters/words $\{c, m, z\}$ are at the bottom of the diagram. For example $\{c\}^{\triangleright} = \mathcal{C}(A) \cup \mathcal{C}(B)$; this is clearly JSI, *since it is defined by c, but not* MSI. *At the top of the diagram are concepts that are* MSI, *but not* JSI. *In the middle, marked with boxes we have the concepts that are* MSI-JSI.

We can represent every element either as a meet of MSI-concepts or a join of JSI-concepts.

Lemma 8. *For any $X \in \mathfrak{B}(L)$, $X = \bigcup\{\{w\}^{\triangleright\triangleleft} \mid w \in X\}$ and $X^{\triangleright} = \bigcup\{\{l\Box r\}^{\triangleleft\triangleright} \mid l\Box r \in X^{\triangleright}\}$.*

We can now discuss the role of these concepts. Suppose we have some derivation in a grammar $S \stackrel{*}{\Rightarrow} lNr \stackrel{*}{\Rightarrow} lwr$, where N is the nonterminal corresponds to some concept.

This places two constraints on N. On the one hand $l\Box r \in N^{\triangleright}$, in other words $N \leq \{l\Box r\}^{\triangleleft}$, which is an MSI-concept. On the other hand $w \in N$ in other words $N \geq w^{\triangleright\triangleleft}$, a JSI-concept. Clearly $\{l\Box r\}^{\triangleleft} \geq w^{\triangleright\triangleleft}$ since $lwr \in L$. In the special case where $\{l\Box r\}^{\triangleleft} = w^{\triangleright\triangleleft}$, we know then that this must be the value of N. Therefore the elements that are MSI, JSI and primes are very special.

Definition 9. *A closed set of strings is an* MSI-JSI-*prime, if it is* MSI, JSI *and prime.*

7 Grammar based on MSI-JSI-primes

We now consider how to define a class of grammars where the nonterminals consist only of the set of MSI-JSI-primes. Rather than requiring that the lattice contains no infinite chains, and using the UFP, we define a weaker property, which more directly determines the finiteness of the relevant sets of productions. We define the non-standard term *finitely Noetherian*.

Definition 10. *We say that a set $\Gamma \subseteq \mathfrak{B}(L)^{+}$, with the pre-order $\precsim$, is finitely Noetherian if $|\max \Gamma|$ is finite and every element of Γ is less than some maximal element.*

A set can fail to be finitely Noetherian either because it has infinite ascending chains with no maximal elements, or because it has an infinite number of maximal elements.

Definition 11. *$\mathfrak{L}_{\mathrm{MJ}}$ is the set of all languages L such that*

1. *L is nonempty and does not contain λ.*
2. *There are a finite number of* MSI-JSI-*primes; we let V denote the set of* MSI-JSI-*primes, except for $\{\lambda\}^{\triangleright\triangleleft}$, $\top$ if $\top_{\triangleright} = \emptyset$, and $\bot$ if $\bot = \emptyset$.*
3. *For each $a \in \Sigma$ and each $l\Box r \in a^{\triangleright}$ there is some $X \in V$ such that $a \in X$ and $l\Box r \in X^{\triangleright}$.*
4. *For every $X \in V$, $\{\alpha \in V^{+} \mid \bar{\alpha} \subset X\}$ is finitely Noetherian.*
5. *$\{\alpha \in V^{+} \mid \bar{\alpha} \subseteq L\}$ is finitely Noetherian.*

$\mathfrak{L}_{\mathrm{MJ}}$ is incomparable with $\mathfrak{L}_{\mathrm{P}}$, as the following two examples show.

Example 7. *$L = \{ax, bx, ay, cy, bz, cz\}$. This language in is $\mathfrak{L}_{\mathrm{P}}$ but not $\mathfrak{L}_{\mathrm{MJ}}$ as there are no* MSI-JSI-*primes that contain for example a. $\{a, b\}$ is* MSI *as it is defined by $\Box x$, but not* JSI. *$\{a\}$ on the other hand is* JSI *but not* MSI.

Example 8. *$L = \{a^n x b^n \mid x \geq 0\} \cup \{a^n x b^n \mid x \geq 0\}$. This is in $\mathfrak{L}_{\mathrm{MJ}}$ but not $\mathfrak{L}_{\mathrm{P}}$. Note that for all n, $\{b^n, c^n\}$ is closed and* MSI *as it is defined by the single context $a^n x\Box$, and is clearly prime, but not* JSI. *Therefore there are infinitely many primes. The* MSI-JSI-*primes are only $\{a^n x b^n \mid x \geq 0\}$, $\{a^n x c^n \mid x \geq 0\}$, $\{a\}$, $\{b\}$, $\{c\}$ and $\{x\}$.*

Definition 12. *If $L \in \mathfrak{L}_{\mathrm{MJ}}$ and V is the set of* MSI-JSI-*primes, then for a set of strings X we write $\Gamma(X) = \{\alpha \in V^{+} \mid \bar{\alpha} \subset X\}$, and $\Delta(X) = \{\alpha \in V^{+} \mid \bar{\alpha} \subseteq X\}$.*

Definition 13. *For every language L in $\mathfrak{L}_{\mathrm{MJ}}$, we define a grammar*

$$\mathcal{G}_{\mathrm{MJ}}(L) = \langle V \cup \{S\}, S, P_L \cup P_B \cup P_S \rangle$$

where

- *V is the set of* MSI-JSI-*primes of L.*
- *S is a distinguished symbol.*

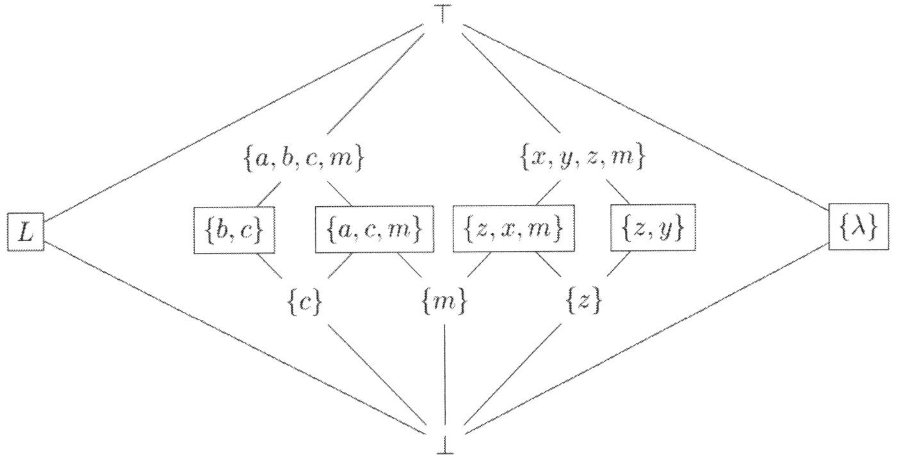

Figure 1: Hasse diagram of the syntactic concept lattice of language in Example 6. All of these elements are prime. The boxed elements are those which are both MSI and JSI, which, apart from the empty string concept, correspond to the nonterminals of the original grammar. $\{a, b, c, m\}$ is MSI, since it is equal to $\{\Box z\}^{\triangleleft}$, but it is not JSI.

- *P_L is the set of all productions $X \rightarrow a$ such that $X \in V$, $a \in \Sigma \cup \lambda$ and $a \in X$, and there is no $Y \in V$ such that $a \in Y$ and $Y < X$.*
- *P_B is the set of productions $X \rightarrow \alpha$ for every $X \in V$ and for every $\alpha \in \max \Gamma(X)$*
- *P_S is the set of productions $S \rightarrow \alpha$ for every $\alpha \in \max \Delta(L)$.*

Note that by Definition 11, this is finite and a CFG.

We now show that this grammar will generate all of the strings in the language. We do this by a joint induction on the length of the strings and the height of the nonterminals in the lattice; it seems easier to write these proofs as *reductios*.

Lemma 9. *For any $X \in V$, If $\lambda \in X$ then*

$$X \overset{*}{\Rightarrow}_{G_*} \lambda.$$

Proof. Suppose this is false. Take a minimal $X \in V$ such that $\lambda \in X$ but it is not the case that $X \overset{*}{\Rightarrow}_{G_*} \lambda$. If X is a minimal element of the set of nonterminals that contain λ then there would be a production $X \rightarrow \lambda$ in G_*; therefore X is not minimal. Let Y be some nonterminal less than X such that $\lambda \in Y$. Since X is minimal we have $Y \overset{*}{\Rightarrow} \lambda$. Now $\langle Y \rangle \in \Gamma(X)$ so there is some production $X \rightarrow \alpha$ such that $\bar{\alpha} \supseteq Y$. Now if $\alpha = Z_1 \ldots Z_k$ then each of the Z_i must be a proper subset of X that contains λ, (since $\lambda \in Y \subseteq \bar{\alpha}$). Since they are proper subsets we have $Z_i \overset{*}{\Rightarrow} \lambda$ and thus $X \overset{*}{\Rightarrow} \lambda$ which is a contradiction. □

Lemma 10. *For any $a \in \Sigma$ and $X \in V$, if $a \in X$ then*

$$X \overset{*}{\Rightarrow}_{G_*} a.$$

Proof. We use just the same argument as the previous proof, except that when we consider $\alpha = Z_1 \ldots Z_k$, there must be at least one i such that $a \in Z_i$ and $\lambda \in Z_j$ for all $j \neq i$. By Lemma 9, $Z_j \overset{*}{\Rightarrow} \lambda$, by minimality of the counterexample $Z_i \overset{*}{\Rightarrow} a$, and therefore $X \overset{*}{\Rightarrow} a$. □

Lemma 11. *For any $w = a_1 \ldots a_n \in \{l\Box r\}^{\triangleleft}$ for some $l\Box r$, there are $A_1, \ldots, A_n \in V$ such that $a_i \in A_i$, and*

$$\overline{\langle A_1 \ldots A_n \rangle} \subseteq \{l\Box r\}^{\triangleleft}.$$

Proof. By induction on n. Base case, $n = 1$ is trivial by Part 3 of Definition 11.

Clearly $a_1 \in \{l\Box a_2 \ldots a_n r\}^{\triangleleft}$. Pick some $A_1 \in V$ such that $a_1 \in A_1$ and $l\Box a_2 \ldots a_n r \in A_1^{\triangleright}$. Since $A_1 \in V$ there is some v_1 such that $A_1 = v_1^{\triangleright\triangleleft}$. Since $a_2 \ldots a_n \in \{lv_1\Box r\}^{\triangleleft}$ then by the inductive hypothesis there are $A_2, \ldots, A_n$ such that $\overline{\langle A_2 \ldots A_n \rangle} \subseteq \{lv_1\Box r\}^{\triangleleft}$, and therefore

$$\overline{\langle A_1 \ldots A_n \rangle} \subseteq \{l\Box r\}^{\triangleleft}.$$

The result then follows by induction. □

Lemma 12. *For any $X \in V$, $w \in X$ if $|w| > 1$ then*

$$X \stackrel{*}{\Rightarrow} w.$$

Proof. Suppose this is false for some w and X; pick a shortest w and a minimal X.

By Lemma 10, $|w| > 1$. Let $w = a_1 \ldots a_n$. Let $l\Box r$ be some context such that $X = \{l\Box r\}^{\triangleleft}$. By Lemma 11, we know that we have some $A_1 \ldots A_n$ such that $\overline{\langle A_1 \ldots A_n \rangle} \subseteq X$, $a_i \in A_i$, for $1 \leq i \leq n$. Since X is a prime, we know that $\overline{\langle A_1 \ldots A_n \rangle} \subset X$. Therefore there is some $X \rightarrow \beta$ such that $\overline{\langle A_1 \ldots A_n \rangle} \subseteq \overline{\beta}$. Let $\beta = B_1 \ldots B_k$ for some $k \geq 1$. Now $w \in \overline{\beta}$. Therefore there are strings $v_1 \ldots v_k = w$ such that $v_j \in B_j$, $1 \leq j \leq k$. Now for each v_j, B_j either $|v_j| < |w|$ or $B_j \subset X$, and so by the assumption at the beginning of the proof, $B_j \stackrel{*}{\Rightarrow} v_j$. Therefore $X \Rightarrow B_1 \ldots B_k \Rightarrow v_1 \ldots v_k = w$ which is a contradiction. □

Theorem 1. *For every $L_* \in \mathfrak{L}_{\text{MJ}}$, $\mathcal{L}(\mathcal{G}_{\text{MJ}}(L_*)) = L_*$.*

Proof. Write G_* for $\mathcal{G}_{\text{MJ}}(L_*)$. It is easy to see that each production $X \rightarrow \alpha$ is correct. A simple induction establishes that $\mathcal{L}(G_*) \subseteq L_*$. The nontrivial part of the proof is to show that every string in L_* is generated from S. Given the previous lemmas, this is straightforward; the proofs are a bit repetitive, following the earlier lemmas with minor variations. Suppose $w \in L_*$.

- If $w = a$ for some $a \in \Sigma$ then pick some $X \in V$ such that $\Box \in X^{\triangleright}$ and $a \in X$. By Lemma 10, $X \stackrel{*}{\Rightarrow} a$ and $X \subseteq L$; If $X = L$ then there is a unary rule $S \rightarrow X$. Otherwise $\langle X \rangle \in \Delta(L)$ and so there is some production $S \rightarrow B_1 \ldots B_k$ such that $a \in \overline{\langle B_1 \ldots B_k \rangle}$. So there must be some B_i such that $a \in B_i$ and $\lambda \in B_j$ for $j \neq i$, and the result follows by Lemmas 10 and 9.

- If $|w| > 1$, then there are two cases. First it might be that $L \in V$, in which case there is a single rule $S \rightarrow L$, and it is immediate by Lemma 12. If not, then we can use the same argument as in Lemma 12, which we take rapidly here. If $w = a_1 \ldots a_n \in L$, then we have some $A_1, \ldots, A_n$ such that $\langle A_1 \ldots A_n \rangle \in \Delta(L)$ and $A_i \stackrel{*}{\Rightarrow} a_i$ for $1 \leq i \leq n$. Therefore we have some rule $S \rightarrow B_1 \ldots B_k$ where $w \in \overline{\langle B_1, \ldots B_k \rangle}$ and therefore strings $v_1, \ldots v_k$ such that $w = v_1 \ldots v_k$ and $v_i \in B_i$; therefore $S \Rightarrow B_1 \ldots B_k \stackrel{*}{\Rightarrow} v_1 \ldots v_k = w$.

Therefore $\mathcal{L}(G_*) \supseteq L_*$ which establishes the theorem. □

Label	Strings	Context
OAUX	can, λ	eagles □ eat.
AUX	can	□ eagles eat?
N	eagles	□ that eat fly.
NP	eagles that eat	□ can eat.
RC	that can eat, λ, ...	eagles □ can fly.
VI	eat, fly	eagles that eat can □.
VP	can eat, died	eagles that □ can fly.
C	that	eagles □ can fly fly.
STOP	.	eagles fly □
Q	?	can eagles fly □

Table 1: Prime concepts of the example; note that they are all MSI-JSI-primes except for OAUX.

8 Example

We will illustrate some properties of our proposed solution with a simple toy example, based approximately on an example in (Berwick et al., 2011). We use a finite language as this shows most sharply the distinction between weak learning, which is trivial, and strong learning, which even in the case of acyclic CFGs is either impossible or intractable in the general case, depending on how it is formalised. The language is generated by the following grammar; optional elements are in brackets.

S → NP VP ., S → CAN NP VI ?
NP → EAGLES (RC), NP → THEY
RC → THAT VP
VP → (CAN) VI
VP → DIED
VI → EAT, VI → FLY

It contains examples like "can eagles that fly eat?" and "eagles that can fly can eat." "can eagles that died eat?". The language contains some optional elements and as a result is not substitutable. This is not the minimal example as for technical reasons the example needs to be sufficiently complex. Very simple examples may not contain enough informa-

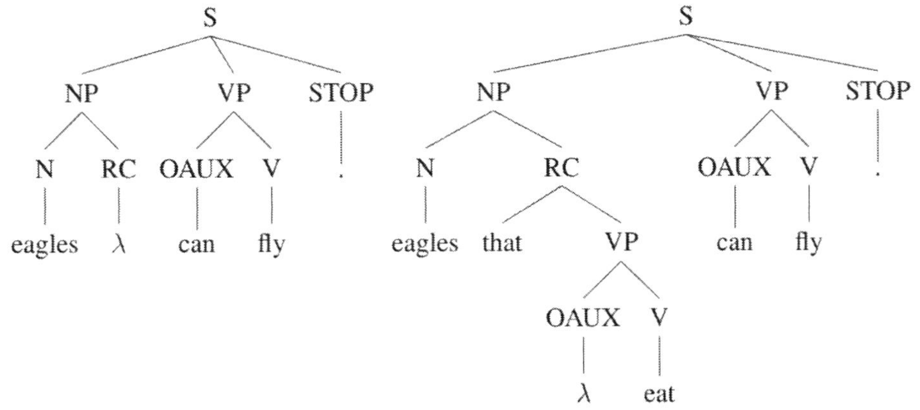

Figure 2: Example parse trees using $\mathcal{G}_P(\cdot)$. Note that where optional elements do not appear, we have an empty (phonologically null) constituent. This is a notational variant of having a unary rule.

tion to indicate the structure unequivocally. So for example, we added a word "died". This alters the structure of the lattice and means that the class VP is then a prime. Without this addition, the concept corresponding to VP would be decomposable as $\{\text{CAN}, \lambda\} \cdot \{\text{EAT}, \text{FLY}\}$. Similarly we added the word THEY to the class of NP. Without these additions, the language would not contain enough information to distinguish whether, for example, the relative clause attaches to the preceding noun or the following verb. Consider the language $L = \{ab, acb\}$. There is no motive, based on the strings, for the claim that c attaches to the left or the right. But if we enlarge it slightly to $\{ab, acb, ad\}$ it is more natural to attach the c optionally to the b.

The lattice contains 43 elements of which 10 are prime, which are listed in Table 1; we label each prime with a mnemonic label, reusing the nonterminal symbols from the target grammar for ease of reading. The composite elements contain for example the set of four strings {EAT EAT, EAT FLY, FLY EAT, FLY FLY } which is clearly not prime.

Figure 2 and 3 show some trees of some of the sentences that illustrate the structure of the derived grammars.

9 Discussion and Conclusion

The two methods we present here do not exhaust the possibilities: rather they present two extremes. One method uses all primes, the other uses what

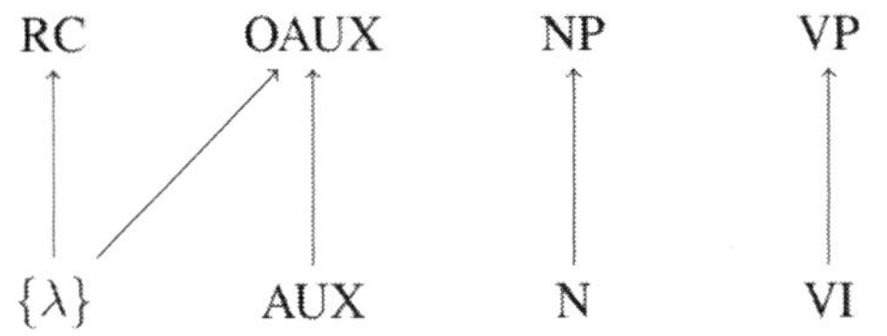

Figure 4: Lattice fragments showing some of the primes. The word CAN, forms a concept on its own, AUX, since when it occurs at the beginning of a sentence, it is obligatory. The concept RC which contains the sequences THAT EAT, THAT CAN FLY and so on, is always optional, and so there is no corresponding concept that does not contain λ.

seems to be the smallest possible set of primes that we can define using these techniques. Our reliance on individual contexts and substrings in the definition of MSI-JSI-primes is both natural but inadequate. In terms of earlier traditions of analytical linguistics, we are following Sestier-Kunze domination rather than Dobrushin-domination (Marcus, 1994). (Kunze, 1967 1968) argues that for the adequate description of German lexical categories simple Dobrushin domination is inadequate. Note also the similar notions in (Adriaans, 1999) of context-separability and expression-separability. The set of MSI-JSI-primes may well be too restricted; it seems necessary to have nonterminals that correspond to cases where $\{l\Box r\}^{\triangleleft} \supsetneq \{w\}^{\triangleright\triangleleft}$. We leave this problem for future work.

It also seems quite natural to define a dual of primality. For a closed set of strings X, we can de-

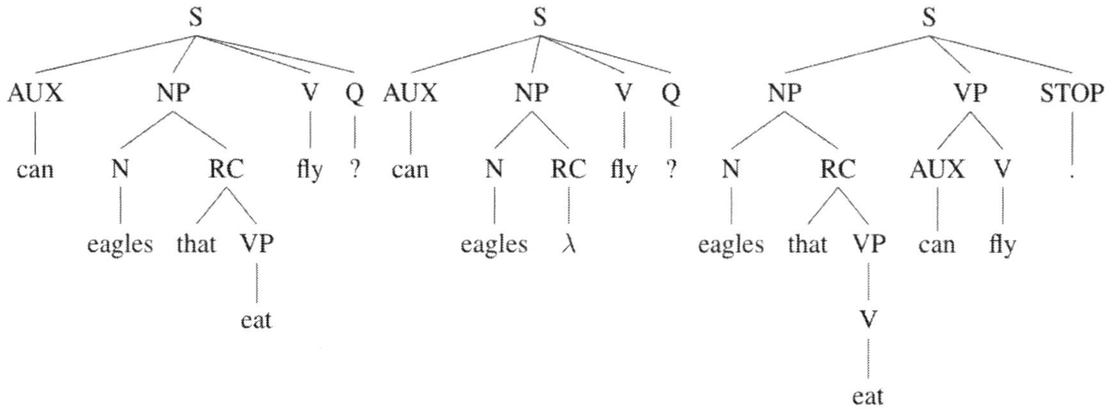

Figure 3: Example parse trees using the second approach, $\mathcal{G}_{\text{MJ}}(\cdot)$. Here we do not have OAUX, and so we have unary rules VP $\rightarrow$ V.

fine conditions $X^{\triangleright} = Y \square Z$, for some closed sets of strings Y, Z. There are also weaker variants of this. These correspond to a condition that the non-terminal occur on the left hand side of more than one production.

From a linguistic point of view, it is interesting that these approaches give local trees of potentially unbounded ranks as well as empty constituents and unary rules. This therefore means that the claim that natural languages only use a binary syntactic operation (MERGE) becomes a contentful empirical claim.

There is a close relation between the models used here and the primal and dual weak learning algorithms presented in (Yoshinaka, 2012b); in particular, the categories that are MSI, have the 1-finite-context-property (FCP) and the JSI-concepts have the 1-finite-kernel-property (FKP). The languages in $\mathfrak{L}_{\text{MJ}}$ therefore will be weakly learnable under certain paradigms. In order to turn the results here into a full strong learning result requires then the efficient computation of the canonical grammar given a sufficiently large weakly correct grammar. There do appear to be some technical problems to overcome: for example showing that the number of errors made in selecting the MSI-JSI-primes will only ever be finite.

Given the extension of distributional learning to multiple context-free grammars (MCFGs) (Seki et al., 1991) by (Yoshinaka, 2011), and the extension of the syntactic concept lattice in (Clark and Yoshinaka, 2014), it seems possible to straightforwardly extend these methods to at least some MCFGs. In particular the notion of a closed set of strings being composite is naturally generalised by replacing the single concatenation operation $\cdot$, with the family of all non-deleting non-permuting linear regular functions of appropriate arities.

The existence of these canonical grammars seems to be related in an interesting way to algebraic properties of the syntactic concept lattice. Indeed the finite cardinality of the lattice is exactly equivalent to the regularity of the language. It seems that other finiteness properties of the lattice—for example, compactness, the chain conditions etc.—may be crucial. More generally, the results presented here show that it may be possible to have strong learning algorithms for some quite large classes of languages. This suggests that the orthodox view that semantic information is required to learn syntactic structure may be mistaken; the set of strings of the language may define an intrinsic structure that can be learned purely distributionally. If the structures so defined can support compositional interpretation of the semantics, then this would provide strong empirical support for this approach.

Acknowledgments

I am grateful to Ryo Yoshinaka, Makoto Kanazawa and Greg Kobele, for technical comments; and to Bob Berwick, Paul Pietroski and George Walkden for discussions that have helped to motivate this approach.

References

Pieter Adriaans. 1999. Learning shallow context-free languages under simple distributions. Technical Report ILLC Report PP-1999-13, Institute for Logic, Language and Computation, Amsterdam.

R.C. Berwick, P. Pietroski, B. Yankama, and N. Chomsky. 2011. Poverty of the stimulus revisited. *Cognitive Science*, 35:1207–1242.

Alexander Clark and Ryo Yoshinaka. 2014. An algebraic approach to multiple context-free grammars. In Sergei Soloviev and Nicholas Asher, editors, *Logical Aspects of Computational Linguistics*. Springer.

Alexander Clark. 2011. A language theoretic approach to syntactic structure. In Makoto Kanazawa, András Kornai, Marcus Kracht, and Hiroyuki Seki, editors, *The Mathematics of Language*, pages 39–56. Springer Berlin Heidelberg.

Alexander Clark. 2013. The syntactic concept lattice: Another algebraic theory of the context-free languages? *Journal of Logic and Computation*.

Alexander Clark. 2014. Learning trees from strings: A strong learning algorithm for some context-free grammars. *Journal of Machine Learning Research*, 14:3537–3559.

B. A. Davey and H. A. Priestley. 2002. *Introduction to Lattices and Order*. Cambridge University Press.

J. Kunze. 1967–1968. Versuch eines objektivierten Grammatikmodells I, II. Z. *Zeitschriff Phonetik Sprachwiss. Kommunikat*, 20-21.

Hans Leiß. 2014. Learning context free grammars with the finite context property: A correction of A.Clark's algorithm. In Glyn Morrill, Reinhard Muskens, Rainer Osswald, and Frank Richter, editors, *Formal Grammar*, volume 8612 of *Lecture Notes in Computer Science*, pages 121–137. Springer Berlin Heidelberg.

S Marcus. 1994. The status of research in the field of analytical algebraic models of language. In Carlos Martín-Vide, editor, *Current Issues in Mathematical Linguistics*, pages 3–23. Elsevier.

H. Seki, T. Matsumura, M. Fujii, and T. Kasami. 1991. On multiple context-free grammars. *Theoretical Computer Science*, 88(2):229.

Christian Wurm. 2012. Completeness of full Lambek calculus for syntactic concept lattices. In *Proceedings of the 17th conference on Formal Grammar 2012 (FG)*.

R. Yoshinaka. 2011. Efficient learning of multiple context-free languages with multidimensional substitutability from positive data. *Theoretical Computer Science*, 412(19):1821 - 1831.

R. Yoshinaka. 2012a. Integration of the dual approaches in the distributional learning of context-free grammars. In Adrian-Horia Dediu and Carlos Martín-Vide, editors, *Language and Automata Theory and Applications*, volume 7183 of *Lecture Notes in Computer Science*, pages 538–550. Springer Berlin Heidelberg.

Ryo Yoshinaka. 2012b. Integration of the dual approaches in the distributional learning of context-free grammars. In Adrian-Horia Dediu and Carlos Martín-Vide, editors, *Language and Automata Theory and Applications*, volume 7183 of *Lecture Notes in Computer Science*, pages 538–550. Springer Berlin Heidelberg.

Output Strictly Local Functions

Jane Chandlee
University of Delaware
janemc@udel.edu

Rémi Eyraud
QARMA Team
LIF Marseille
remi.eyraud@
lif.univ-mrs.fr

Jeffrey Heinz
University of Delaware
heinz@udel.edu

Abstract

This paper characterizes a subclass of subsequential string-to-string functions called Output Strictly Local (OSL) and presents a learning algorithm which provably learns any OSL function in polynomial time and data. This algorithm is more efficient than other existing ones capable of learning this class. The OSL class is motivated by the study of the nature of string-to-string transformations, a cornerstone of modern phonological grammars.

1 Introduction

Motivated by questions in phonology, this paper studies the Output Strictly Local (OSL) functions originally defined by Chandlee (2014) and Chandlee et al. (2014). The OSL class is one way Strictly Local (SL) stringsets can be generalized to string-to-string maps. Their definition is a functional version of a defining characteristic of SL stringsets called Suffix Substitution Closure (Rogers and Pullum, 2011). Similar to SL stringsets, the OSL functions contain nested subclasses parameterized by a value k, which is the length of the suffix of *output strings* that matters for computing the function.

As Chandlee (2014) argues, almost all local phonological processes can be modeled with *Input* Strictly Local (ISL) functions. Yet there is one notable class of exceptions: so-called spreading processes, in which a feature like nasality iteratively assimilates over a contiguous span of segments. As we show, the OSL functions are needed to describe this sort of phenomenon.

Here we provide a slight, but important, revision to the original definition of OSL functions in Chandlee (2014) and Chandlee et al. (2014), which allows two important theoretical contributions while preserving the previous results The first is a finite-state transducer (FST) characterization of OSL functions, which leads to the second result, the OSLFIA (OSL Function Inference Algorithm) and a proof that it efficiently identifies the k-OSL functions from positive examples. We compare this algorithm to OSTIA (Onward Subsequential Transducer Inference Algorithm, Oncina et al. (1993)) which identifies total subsequential functions in cubic time, its modifications OSTIA-D and OSTIA-R, which can learn particular subclasses of subsequential functions using domain and range information, respectively, in at least cubic time (Oncina and Varò, 1996; Castellanos et al., 1998), and SOSFIA (Structured Onward Subsequential Inference Algorithm, Jardine et al. (2014)), which can learn particular subclasses of subsequential functions in linear time and data. We show these algorithms either cannot learn the OSL functions or do so less efficiently than the OSLFIA. These contributions were missing from the initial research on OSL functions (except for a preliminary FST characterization in Chandlee (2014)). Finally, we explain how a unified theory of local phonology will have to draw insights from both the ISL and OSL classes and offer an idea of how this might work. Thus, this paper is a crucial and necessary intermediate step towards an empirically adequate but restrictive characterization of phonological locality.

The remainder of the paper is organized as follows. Motivation and related work are given in sec-

Proceedings of the 14th Meeting on the Mathematics of Language (MoL 14), pages 112–125,
Chicago, USA, July 25–26, 2015.

tion 2, including an example of the spreading processes that cannot be modeled with ISL functions. Notations and background concepts are presented in section 3. In section 4 we define OSL functions, and the theoretical characterization and learning results are given in sections 5 and 6. In section 7, we explain how OSL functions model spreading processes. In section 8 we elaborate on a few important areas for future work, and in section 9 we conclude.

2 Background and related work

A foundational principle of modern generative phonology is that systematic variation in morpheme pronunciation is best explained with a single underlying representation of the morpheme that is transformed into various surface representations based on context (Kenstowicz and Kisseberth, 1979; Odden, 2014). Thus, much of generative phonology is concerned with the nature of these transformations.

One way to better understand the nature of linguistic phenomena is to develop strong computational characterizations of them. Discussing SPE-style phonological rewrite rules (Chomsky and Halle, 1968), Johnson (1972, p. 43) expresses the reasoning behind this approach:

> It is a well-established principle that any mapping whatever that can be computed by a finitely statable, well-defined procedure can be effected by a rewriting system (in particular, by a Turing machine, which is a special kind of rewriting system). Hence any theory which allows phonological rules to simulate arbitrary rewriting systems is seriously defective, for it asserts next to nothing about the sorts of mappings these rules can perform.

This leads to the important question of what kinds of transformations ought a theory of phonology allow?

Earlier work suggests that phonological theories ought to exclude nonregular relations (Johnson, 1972; Kaplan and Kay, 1994; Frank and Satta, 1998; Graf, 2010). More recently, it has been hypothesized that phonological theory ought to only allow certain subclasses of the regular relations (Gainor et al., 2012; Chandlee et al., 2012; Chandlee and Heinz, 2012; Payne, 2013; Luo, 2014; Heinz and Lai, 2013). This research places particular emphasis on *subsequential* functions, which can informally be characterized as functions definable with a weighted, deterministic finite-state acceptor where the weights are strings and multiplication is concatenation. The aforementioned work suggests that this hypothesis enjoys strong support in segmental phonology, with interesting and important exceptions in the domain of tone (Jardine, 2014).

Recent research has also showed an increased awareness and understanding of subregular classes of *stringsets* (formal languages) and their importance for theories of *phonotactics* (Heinz, 2007; Heinz, 2009; Heinz, 2010; Rogers et al., 2010; Rogers and Pullum, 2011; Rogers et al., 2013). While many of these classes and their properties were studied much earlier (McNaughton and Papert, 1971; Thomas, 1997), little to no attention has been paid to similar classes properly contained within the subsequential functions. Thus, at least within the domain of segmental phonology, there is an important question of whether stronger computational characterizations of phonological *transformations* are possible, as seems to be the case for phonotactics.

As mentioned above, Chandlee (2014) shows that many phonological processes belong to a subclass of subsequential functions, the Input Strictly Local (ISL) functions. Informally, a function is k-ISL if the output of every input string $a_0a_1\cdots a_n$ is $u_0u_1\cdots u_n$ where u_i is a string which only depends on a_i and the $k-1$ input symbols before a_i (so $a_{i-k+1}a_{i-k+2}\cdots a_{i-1}$). (A formal definition is given in section 4). ISL functions can model a range of processes including local substitution, epenthesis, deletion, and metathesis. For more details on the exact range of ISL processes, see Chandlee (2014) and Chandlee and Heinz (to appear).

Processes that aren't ISL include long-distance processes as well as local iterative spreading processes. As an example of the latter, consider nasal spreading in Johore Malay (Onn, 1980). As shown in (1), contiguous sequences of vowels and glides are nasalized following a nasal:

(1) /pəŋawasan/ $\mapsto$ [pəŋãw̃ãsan], 'supervision'

This process is not ISL, because the initial trigger of the spreading (the nasal) can be arbitrarily far from a target (as suggested by the nasalization of the glide and the second [ã]) when the distance is measured on the *input* side. However, on the *output* side, the triggering context is local; the second [ã] is nasalized because the preceding glide on the *output* side is nasalized. Every segment between the trigger and target is affected; nasalization applies to a contiguous, but arbitrarily long, substring. It is this type of process that we will show requires the notion of Output Strict Locality.

Processes in which a potentially unbounded number of *unaffected* segments can intervene between the trigger and target - such as long-distance consonant agreement (Hansson, 2010; Rose and Walker, 2004), vowel harmony (Nevins, 2010; Walker, 2011), and consonant dissimilation (Suzuki, 1998; Bennett, 2013) - are neither ISL nor OSL. More will be said about such long-distance processes in §7.

3 Preliminaries

The set of all possible finite strings of symbols from a finite alphabet Σ and the set of strings of length $\leq n$ are Σ^* and $\Sigma^{\leq n}$, respectively. The cardinality of a set S is denoted $\texttt{card}(S)$. The unique empty string is represented with λ. The length of a string w is $|w|$, so $|\lambda| = 0$. If w_1 and w_2 are strings then w_1w_2 is their concatenation. The prefixes of w, $\texttt{Pref}(w)$, is $\{p \in \Sigma^* \mid (\exists s \in \Sigma^*)[w = ps]\}$, and the suffixes of w, $\texttt{Suff}(w)$, is $\{s \in \Sigma^* \mid (\exists p \in \Sigma^*)[w = ps]\}$. For all $w \in \Sigma^*$ and $n \in \mathbb{N}$, $\texttt{Suff}^n(w)$ is the single suffix of w of length n if $|w| \geq n$; otherwise $\texttt{Suff}^n(w) = w$. The following reduction will prove useful later.

Remark 1. *For all* $w, v \in \Sigma^*, n \in \mathbb{N}$, $\texttt{Suff}^n(\texttt{Suff}^n(w)v) = \texttt{Suff}^n(wv)$.

If $w = uv$ is a string then let $v = u^{-1} \cdot w$ and $u = w \cdot v^{-1}$. Trivially, $\lambda^{-1} \cdot w = w = w \cdot \lambda^{-1}$, $uu^{-1} \cdot w = w$, and $w \cdot v^{-1}v = w$.

We assume a fixed but arbitrary total order $<$ on the letters of Σ. As usual, we extend $<$ to Σ^* by defining the *hierarchical order* (Oncina et al., 1993), denoted $\triangleleft$, as follows: $\forall w_1, w_2 \in \Sigma^*, w_1 \triangleleft w_2$ iff

$$\begin{cases} |w_1| < |w_2| \text{ or} \\ |w_1| = |w_2| \text{ and } \exists u, v_1, v_2 \in \Sigma^*, \exists a_1, a_2 \in \Sigma \\ \textit{s.t. } w_1 = ua_1v_1, w_2 = ua_2v_2 \text{ and } a_1 < a_2. \end{cases}$$

$\triangleleft$ is a total strict order over Σ^*, and if $\Sigma = \{a, b\}$ and $a < b$, then $\lambda \triangleleft a \triangleleft b \triangleleft aa \triangleleft ab \triangleleft ba \triangleleft bb \triangleleft aaa \triangleleft \ldots$

The *longest common prefix* of a set of strings S, $\texttt{lcp}(S)$, is $p \in \cap_{w \in S}\texttt{Pref}(w)$ such that $\forall p' \in \cap_{w \in S}\texttt{Pref}(w), |p'| < |p|$. Let $f : A \to B$ be a function f with domain A and co-domain B. When A and B are stringsets, the input and output languages of f are $\texttt{pre\_image}(f) = \{x \mid (\exists y)[x \mapsto_f y]\}$ and $\texttt{image}(f) = \{y \mid (\exists x)[x \mapsto_f y]\}$, respectively.

Jardine et al. (2014) introduce delimited subsequential FSTs (DSFSTs). The class of functions describable with DSFSTs is exactly the class representable by traditional subsequential FSTs (Oncina and Garcia, 1991; Oncina et al., 1993; Mohri, 1997), but DSFSTs make explicit use of symbols marking *both* the beginnings and ends of input strings.

Definition 1. *A* delimited subsequential FST (DSFST) *is a 6-tuple* $\langle Q, q_0, q_f, \Sigma, \Delta, \delta \rangle$ *where* Q *is a finite set of states,* $q_0 \in Q$ *is the initial state,* $q_f \in Q$ *is the final state,* Σ *and* Δ *are finite alphabets of symbols,* $\delta \subseteq Q \times (\Sigma \cup \{\rtimes, \ltimes\}) \times \Delta^* \times Q$ *is the transition function (where* $\rtimes \notin \Sigma$ *indicates the 'start of the input' and* $\ltimes \notin \Sigma$ *indicates the 'end of the input'), and the following hold:*

1. *if* $(q, \sigma, u, q') \in \delta$ *then* $q \neq q_f$ *and* $q' \neq q_0$,
2. *if* $(q, \sigma, u, q_f) \in \delta$ *then* $\sigma = \ltimes$ *and* $q \neq q_0$,
3. *if* $(q_0, \sigma, u, q') \in \delta$ *then* $\sigma = \rtimes$ *and if* $(q, \rtimes, u, q') \in \delta$ *then* $q = q_0$,
4. *if* $(q, \sigma, w, r), (q, \sigma, v, s) \in \delta$ *then* $(r = s) \wedge (w = v)$.

In words, in DSFST, initial states have no incoming transitions (1) and exactly one outgoing transition for input $\rtimes$ (3) which leads to a nonfinal state (2), and final states have no outgoing transitions (1) and every incoming transition comes from a non-initial state and has input $\ltimes$ (2). DSFSTs are also deterministic on the input (4). In addition, the transition function may be partial. We extend the transition function to δ^* recursively in the usual way: δ^* is the smallest set containing δ and which is closed under the following condition: if $(q, w, u, q') \in \delta^*$ and $(q', \sigma, v, q'') \in \delta$ then $(q, w\sigma, uv, q'') \in \delta^*$. Note no elements of the form $(q, \lambda, \lambda, q')$ are elements of δ^*.

The size of a DSFST $\mathcal{T} = \langle Q, q_0, q_f, \Sigma, \Delta, \delta \rangle$ is $|\mathcal{T}| = \texttt{card}(Q) + \texttt{card}(\delta) + \sum_{(q,a,u,q') \in \delta} |u|$.

A DSFST $\mathcal{T}$ defines the following relation:

$$R(\mathcal{T}) = \Big\{ (x, y) \in \Sigma^* \times \Delta^* \mid \big[(q_0, \rtimes x \ltimes, y, q_f) \in \delta^*\big] \Big\}$$

Since DSFSTs are deterministic, the relations they recognize are (possibly partial) functions. *Sequential functions* are defined as those representable with DSFSTs for which for all $(q, \ltimes, u, q_f) \in \delta$, $u = \lambda$.[1]

For any function $f : \Sigma^* \to \Delta^*$ and $x \in \Sigma^*$, let the *tails* of x with respect to f be defined as

$$\texttt{tails}_f(x) = \{(y, v) \mid f(xy) = uv \land u = \texttt{lcp}(f(x\Sigma^*))\} .$$

If $x_1, x_2 \in \Sigma^*$ have the same set of tails with respect to f, they are *tail-equivalent* with respect to f, written $x_1 \sim_f x_2$. Clearly, $\sim_f$ is an equivalence relation which partitions Σ^*.

Theorem 1 (Oncina and Garcia, 1991). *A function f is* subsequential *iff $\sim_f$ partitions Σ^* into* finitely *many blocks.*

The above theorem can be seen as the functional analogue to the Myhill-Nerode theorem for regular languages. Recall that for any stringset L, the tails of a word w w.r.t. L is defined as $\texttt{tails}_L(w) = \{u \mid wu \in L\}$. These tails can be used to partition Σ^* into a finite set of equivalence classes iff L is regular. Furthermore, these equivalence classes are the basis for constructing the (unique up to isomorphism) smallest deterministic acceptor for a regular language. Likewise, Oncina and Garcia's proof of Theorem 1 shows how to construct the (unique up to isomorphism) smallest subsequential transducer for a subsequential function f. With little modification to their proof, the smallest DSFST for f can also be constructed. We refer to this DSFST as the *canonical* DSFST for f and denote it $\mathcal{T}_C(f)$. (If f is understood from context, we may write $\mathcal{T}_C$.) States of $\mathcal{T}_C(f)$ which are neither initial nor final are in one-to-one correspondence with $\texttt{tails}_f(x)$ for all $x \in \Sigma^*$ (Oncina and Garcia, 1991). To construct $\mathcal{T}_C(f)$ we first let, for all $x \in \Sigma^*$ and $a \in \Sigma$, the *contribution* of a w.r.t. x be $\texttt{cont}_f(a, x) = \texttt{lcp}(f(x\Sigma^*)^{-1} \cdot \texttt{lcp}(f(xa\Sigma^*))$. Then,

- $Q = \{\texttt{tails}_f(x) \mid x \in \Sigma^*\} \cup \{q_0, q_f\}$,
- $(q_0, \rtimes, \texttt{lcp}(f(\Sigma^*)), \texttt{tails}_f(\lambda)) \in \delta$
- For all $x \in \Sigma^*$, $(\texttt{tails}_f(x), \ltimes, \texttt{lcp}(f(x\Sigma^*))^{-1} \cdot f(x), q_f) \in \delta$ iff $x \in \texttt{pre\_image}(f)$
- For all $x \in \Sigma^*, a \in \Sigma$, if $\exists y \in \Sigma^*$ with $xay \in \texttt{pre\_image}(f)$ then $(\texttt{tails}_f(x), a, \texttt{cont}_f(a, x), \texttt{tails}_f(xa)) \in \delta$.
- Nothing else is in δ.

Observe that unlike the traditional construction, the initial state q_0 is not $\texttt{tails}_f(\lambda)$. The single outgoing transition from q_0, however, goes to this state with the input $\rtimes$. Canonical DSFSTs have an important property called *onwardness*.

Definition 2 (onwardness). *A DSFST $\mathcal{T}$ is* onward *if for every $w \in \Sigma^*$, $u \in \Delta^*$, $(q_0, \rtimes w, u, q) \in \delta^* \iff u = \texttt{lcp}(\{f(w\Sigma^*)\})$.*

Informally, this means that the writing of output is never delayed. For all $q \in Q$ let the *outputs* of the edges out of q be $\texttt{outputs}(q) = \{u \mid (\exists \sigma \in \Sigma \cup \{\rtimes, \ltimes\})(\exists q' \in Q)[(q, \sigma, u, q') \in \delta]\}$.

Lemma 1. *If DSFST $\mathcal{T}$ recognizes f and is onward then $\forall q \neq q_0$ $\texttt{lcp}(\texttt{outputs}(q)) = \lambda$ and $\texttt{lcp}(\texttt{outputs}(q_0)) = \texttt{lcp}(f(\Sigma^*))$.*

Proof. By construction of a DSFST, only one transition leaves q_0: $(q_0, \rtimes, u, q)$. This implies $(q_0, \rtimes\lambda, u, q) \in \delta^*$ and as the transducer is onward we have $\texttt{lcp}(\texttt{outputs}(q_0)) = \texttt{lcp}(u) = u = \texttt{lcp}(f(\lambda\Sigma^*)) = \texttt{lcp}(f(\Sigma^*))$. Now take $q \neq q_0$ and $w \in \Sigma^*$ such that $(q_0, \rtimes w, u, q) \in \delta^*$. Suppose $\texttt{lcp}(\texttt{outputs}(q)) = v \neq \lambda$. Then v is a prefix of $\texttt{lcp}(\{f(w\sigma x) \mid \sigma \in \Sigma \cup \{\ltimes\}, x \in \Sigma^*\})$ which implies uv is a prefix of $\texttt{lcp}(f(w\Sigma^*))$. But $v \neq \lambda$, contradicting the fact that $\mathcal{T}$ is onward. □

Readers are referred to Oncina and Garcia (1991), Oncina et al. (1993), and Mohri (1997) for more on subsequential transducers, and Eisner (2003) for generalizations regarding onwardness.

4 Output Strictly Local functions

Here we define Output Strictly Local (OSL) functions, which were originally introduced by Chandlee (2014) and Chandlee et al. (2014) along with the Input Strictly Local (ISL) functions. Both classes generalize SL stringsets to functions based on a defining property of SL languages, the Suffix Substitution Closure (Rogers and Pullum, 2011).

Theorem 2 (Suffix Substitution Closure). *L is Strictly Local iff for all strings u_1, v_1, u_2, v_2, there*

[1] Sakarovitch (2009) inverts these terms.

exists $k \in \mathbb{N}$ *such that for any string* x *of length* $k-1$, *if* $u_1xv_1, u_2xv_2 \in L$, *then* $u_1xv_2 \in L$.

An important corollary of this theorem follows.

Corollary 1 (Suffix-defined Residuals). *L is Strictly Local iff* $\forall w_1, w_2 \in \Sigma^*$, *there exists* $k \in \mathbb{N}$ *such that if* $\mathtt{Suff}^{k-1}(w_1) = \mathtt{Suff}^{k-1}(w_2)$ *then the* residuals *(the tails) of* w_1, w_2 *with respect to L are the same; formally,* $\{v \mid w_1v \in L\} = \{v \mid w_2v \in L\}$.

Input and Output Strictly Local functions were defined by Chandlee (2014) and Chandlee et al. (2014) in the manner suggested by the corollary.

Definition 3 (Input Strictly Local Functions). *A function* $f : \Sigma^* \to \Delta^*$ *is ISL if there is a k such that for all* $u_1, u_2 \in \Sigma^*$, *if* $\mathtt{Suff}^{k-1}(u_1) = \mathtt{Suff}^{k-1}(u_2)$ *then* $\mathtt{tails}_f(u_1) = \mathtt{tails}_f(u_2)$.

Definition 4 (Output Strictly Local Functions (original)). *A function* $f : \Sigma^* \to \Delta^*$ *is OSL if there is a k such that for all* $u_1, u_2 \in \Sigma^*$, *if* $\mathtt{Suff}^{k-1}(f(u_1)) = \mathtt{Suff}^{k-1}(f(u_2))$ *then* $\mathtt{tails}_f(u_1) = \mathtt{tails}_f(u_2)$.

While Definition 3 lead to an automata-theoretic characterization and learning results for ISL (Chandlee et al., 2014), such results do not appear possible with the original definition of OSL. The trouble is with subsequential functions that are not sequential. The value returned by the function includes the writing that occurs when the input string has been fully read (i.e., the output of transitions going to the final state in a corresponding DSFST). This creates a problem because it does not allow for separation of what happens during the computation from what happens at its end.

Figure 1 illustrates the distinction Definition 4 is unable to make.[2] Function f is sequential, but g is not. Otherwise, they are identical. While $f(bab) = bba$, $g(bab) = bbaa$. With the original OSL definition, there is no way to refer to the output for input *bab before* the final output string has been appended.

To deal with this problem we first define the prefix function associated to a subsequential function.

Definition 5 (Prefix function). *Let* $f : \Sigma^* \to \Delta^*$ *be a subsequential function. We define the prefix function* $f^p : \Sigma^* \to \Delta^*$ *associated to f such that* $f^p(w) = \mathtt{lcp}(\{f(w\Sigma^*)\})$.

[2]Here and in Figure 3, state q_f is not shown. Non-initial states are labeled $q : u$ with q being the state's name and $(q, \ltimes, u, q_f) \in \delta$.

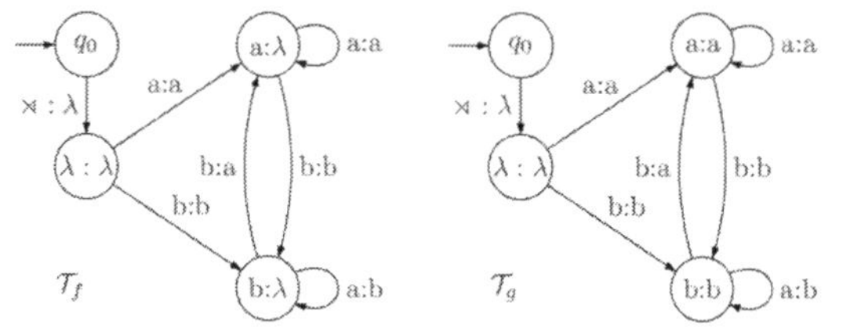

Figure 1: Two DSFST recognizing functions f and g. Except for their final transitions, $\mathcal{T}_f$ and $\mathcal{T}_g$ are identical.

Remark 2. *If* $\mathcal{T}$ *is an onward DSFST for f, then* $\forall w \in \Sigma^*, f^p(w) = u \iff \exists q, (q_0, \rtimes w, u, q) \in \delta^*$.

Remark 3. *If f is sequential then* $f = f^p$.

We can now revise the definition of OSL functions.

Definition 6 (Output Strictly Local Function (revised)). *We say that a subsequential function f is k-OSL if for all* w_1, w_2 *in* Σ^*, $\mathtt{Suff}^{k-1}(f^p(w_1)) = \mathtt{Suff}^{k-1}(f^p(w_2)) \Rightarrow \mathtt{tails}_f(w_1) = \mathtt{tails}_f(w_2)$.

Chandlee et al. (2014) provide several theorems which relate ISL functions, OSL functions (defined as in Definition 4), and SL stringsets. Here we explain why those results still hold with Definition 6. The proofs for those results depend on the six functions ($f_i, 1 \leq i \leq 6$) reproduced here in Figure 2. The transducers shown there are not DSFSTs but traditional subsequential transducers; readers are referred to Chandlee et al. (2014) for formal definitions. With the exception of f_5, these functions are clearly sequential since each state outputs λ on $\ltimes$ (shown as # in Figure 2). The transducer for f_5 is not onward, but an onward, sequential version of this transducer recognizing exactly the same function is obtained by suffixing a (which is the lcp of the outputs of state 1) onto the output of state 1's incoming transition. Thus, f_5 is also sequential. By Remark 3 then, Theorems 4, 5, 6, and 7 of that paper still hold under Definition 6.

5 Automata characterization

First we show, for any non-initial state of any canonical transducer recognizing an OSL function, that if reading a letter a implies writing λ, then this corresponds to a self-loop. So writing the empty string never causes a change of state (except from q_0).

Lemma 2. *For any OSL function f whose canonical DSFST is* $\mathcal{T}_C$, *if* $\exists q \neq q_0$, $a \in \Sigma$, *and* $q' \in Q$ *such*

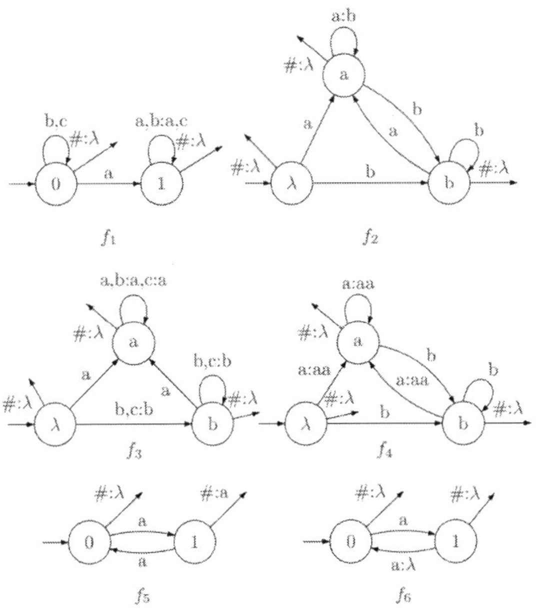

Figure 2: Examples used in proofs of Theorems 4 to 7 of Chandlee et al. (2014, see Figure 2).

that $(q, a, \lambda, q') \in \delta_C$ *then* $q' = q$.

Proof. Consider w and u such that $(q_0, \rtimes w, u, q) \in \delta_C^*$ and suppose $(q, a, \lambda, q') \in \delta_C$. Then $f^p(w) = f^p(wa)$ which implies $\texttt{Suff}^{k-1}(f^p(w)) = \texttt{Suff}^{k-1}(f^p(wa))$. As f is k-OSL, $\texttt{tails}_f(w) = \texttt{tails}_f(wa)$. As $\mathcal{T}_C$ is canonical the non-initial and non-final states correspond to unique tail-equivalence classes, and two distinct states correspond to two different classes. Therefore $q' = q$. □

Next we define k-OSL transducers.

Definition 7 (k-OSL transducer). *An onward DSFST* $\mathcal{T} = \langle Q, q_0, q_f, \Sigma, \Delta, \delta\rangle$ *is k-OSL if*

1. $Q = S \cup \{q_0, q_f\}$ *with* $S \subseteq \Delta^{\leq k-1}$
2. $(\forall u \in \Delta^*)\big[(q_0, \rtimes, u, q') \in \delta \Longrightarrow q' = \texttt{Suff}^{k-1}(u)\big]$
3. $(\forall q \in Q\backslash\{q_0\}, \forall a \in \Sigma, \forall u \in \Delta^*)$ $\big[(q, a, u, q') \in \delta \Longrightarrow q' = \texttt{Suff}^{k-1}(qu)\big]$.

Next we show that k-OSL functions and functions represented by k-OSL DSFSTs exactly correspond.

Lemma 3 (extended transition function). *Let* $\mathcal{T} = \langle Q, q_0, q_f, \Sigma, \Delta, \delta\rangle$ *be a k-OSL DSFST. We have*

$$(q_0, \rtimes w, u, q) \in \delta^* \Longrightarrow q = \texttt{Suff}^{k-1}(u)$$

Proof. By recursion on the size of $w \in \Sigma^*$. The initial case is valid for $|w| = 0$ since if $(q_0, \rtimes, u, q) \in \delta^*$ then $(q_0, \rtimes, u, q) \in \delta$. By Definition 7, $q = \texttt{Suff}^{k-1}(u)$. Suppose now that the lemma holds for inputs of size $n \neq 0$. Let w be of size n such that $(q_0, \rtimes w, u, q) \in \delta^*$ and suppose $(q, a, v, q') \in \delta$ (i.e., $(q_0, \rtimes wa, uv, q') \in \delta^*$). By recursion, we know that $q = \texttt{Suff}^{k-1}(u)$. By Definition 7, $q' = \texttt{Suff}^{k-1}(qv) = \texttt{Suff}^{k-1}(\texttt{Suff}^{k-1}(u)v) = \texttt{Suff}^{k-1}(uv)$ (by Remark 1). □

Lemma 4. *Any k-OSL DSFST corresponds to a k-OSL function.*

Proof. Let $\mathcal{T} = \langle Q, q_0, q_f, \Sigma, \Delta, \delta\rangle$ be a k-OSL DSFST computing f and let $w_1, w_2 \in \Sigma^*$ such that $\texttt{Suff}^{k-1}(f^p(w_1)) = \texttt{Suff}^{k-1}(f^p(w_2))$. Since $\mathcal{T}$ is onward, by Remark 2 there exists $q, q' \in Q$ such that $(q_0, \rtimes w_1, f^p(w_1), q) \in \delta^*$ and $(q_0, \rtimes w_2, f^p(w_2), q') \in \delta^*$. By Lemma 3, $q = \texttt{Suff}^{k-1}(f^p(w_1)) = \texttt{Suff}^{k-1}(f^p(w_2)) = q'$ which implies $\texttt{tails}_f(w_1) = \texttt{tails}_f(w_2)$. Therefore f is a k-OSL function. □

We now need to show that every k-OSL function can be represented by a k-OSL DSFST. An issue here is that one cannot work from $\mathcal{T}_C$ since its states are defined in terms of its tails, which themselves are defined in terms of *input* strings, not output strings. Hence, the proof below is constructive.

Theorem 3. *Let f be a k-OSL function. The DSFST* $\mathcal{T}$ *defined as followed computes* f:

- $Q = S \cup \{q_0, q_f\}$ *with* $S \subseteq \Delta^{\leq k-1}$
- $(q_0, \rtimes, u, \texttt{Suff}^{k-1}(u)) \in \delta \Longleftrightarrow u = f^p(\lambda)$
- $a \in \Sigma$, $(q, a, u, \texttt{Suff}^{k-1}(qu)) \in \delta$, $\Longleftrightarrow$ $(\exists w)\big[\texttt{Suff}^{k-1}(f^p(w)) = q \wedge f^p(wa) = vqu$ *with* $v = f^p(w) \cdot q^{-1}\big]$,
- $(q, \ltimes, u, q_f) \in \delta \Longleftrightarrow u = f^p(w_q)^{-1} \cdot f(w_q)$, *where* $w_q = \min_{\triangleleft}\{w \mid \exists u, (q_0, \rtimes w, u, q) \in \delta^*\}$.

The diagram below helps express pictorially how the transitions are organized per the second and third bullets above. The input is written above the arrows, and the output written below.

$$q_0 \xrightarrow[f^p(w)=vq]{\rtimes w} q \xrightarrow[u]{a} q'$$

Note that $\mathcal{T}$ is a k-OSL SFST. To prove this result, we first show the following lemma:

Lemma 5. *Let $\mathcal{T}$ be the transducer defined in Theorem 3. We have:*

$$(q_0, \rtimes w, u, q) \in \delta^* \iff f^p(w) = u$$

Proof. ($\Rightarrow$) By recursion on the length of w. Suppose $(q_0, \rtimes w, u, q) \in \delta^*$ and $|w| = 0$. Then $(q_0, \rtimes, u, q) \in \delta$; by construction, $q = \mathtt{Suff}^{k-1}(u)$ and $f^p(\lambda) = u$ which validates the initial case.

Suppose the result holds for w of size n and pick such a w. Suppose then that $(q_0, \rtimes wa, u, q) \in \delta^*$. By definition of δ^*, there exists u_1, u_2, q' such that $u = u_1u_2$, $(q_0, \rtimes w, u_1, q') \in \delta^*$ and $(q', a, u_2, q) \in \delta$. We have $f^p(w) = u_1$ (by recursion) and thus $q' = \mathtt{Suff}^{k-1}(f^p(w))$ (by Lemma 3).

By construction of $\mathcal{T}$, $q = \mathtt{Suff}^{k-1}(q'u_2)$ and thus $f^p(wa) = vq'u_2$ with $v = f^p(w) \cdot \mathtt{Suff}^{k-1}(f^p(w))^{-1}$. Therefore $f^p(wa) = vq'u_2 = f^p(w) \cdot \mathtt{Suff}^{k-1}(f^p(w))^{-1}q'u_2 = f^p(w) \cdot \mathtt{Suff}^{k-1}(f^p(w))^{-1}\mathtt{Suff}^{k-1}(f^p(w))u_2 = f^p(w)u_2 = u_1u_2 = u$.

($\Leftarrow$) By recursion on the length of w. If $|w| = 0$, then $f^p(\lambda) = u$. By construction of $\mathcal{T}$, $(q_0, \rtimes, u, \mathtt{Suff}^{k-1}(u)) \in \delta$, which validates the base case.

Now fix $n > 0$ and suppose the result holds for all w of size n. Pick such a w and let $f^p(wa) = u$. As f is subsequential, there exists u_1 such that $f^p(w) = u_1$. By recursion, there exists q such that $(q_0, \rtimes w, u_1, q) \in \delta^*$. By Lemma 3, $q = \mathtt{Suff}^{k-1}(u_1) = \mathtt{Suff}^{k-1}(f^p(w))$. By definition $f^p(wa) = u_1u_1^{-1} \cdot u$, which equals $u_1 \cdot \mathtt{Suff}^{k-1}(u_1)^{-1}\mathtt{Suff}^{k-1}(u_1)u_1^{-1} \cdot u$. Hence $f^p(wa) = vqu'$, with $u' = u_1^{-1} \cdot u$ and $v = u_1 \cdot \mathtt{Suff}^{k-1}(u_1)^{-1}$, which equals $f^p(w) \cdot \mathtt{Suff}^{k-1}(f^p(w))^{-1}$. Thus, by construction $(q, a, u', \mathtt{Suff}^{k-1}(qu')) \in \delta$. Since $u_1u' = u_1u_1^{-1} \cdot u = u$, $(q_0, \rtimes wa, u, \mathtt{Suff}^{k-1}(qu')) \in \delta^*$. □

We can now prove Theorem 3.

Proof. Let $\mathcal{T}$ be the transducer defined in Theorem 3. We show that $\forall w \in \Sigma^*$,

$$(w, u) \in R(\mathcal{T}) \iff f(w) = u.$$

By definition of $R(\mathcal{T})$, we know that $(q_0, \rtimes w \ltimes, u, q_f) \in \delta^*$. By definition of δ^*, there exists $u_1, u_2 \in \Delta^*$ and $q \in Q \backslash \{q_0, q_f\}$ such that $(q_0, \rtimes w, u_1, q) \in \delta^*$, $(q, \ltimes, u_2, q_f) \in \delta$, and $u = u_1u_2$. By Lemma 5 we know that $f^p(w) = u_1$. By construction of the DSFST, we have $u_2 = f^p(w_q)^{-1} \cdot f(w_q)$ where $w_q = \min_{\triangleleft}\{w' \mid \exists u, (q_0, \rtimes w', u', q) \in \delta^*\}$. Therefore $(w_q, u'u_2) \in R(\mathcal{T})$. Again, by Lemma 5, $f^p(w_q) = u'$ and so $u'u_2 = f^p(w_q)u_2 = f^p(w_q)f^p(w_q)^{-1} \cdot f(w_q) = f(w_q)$.

We have $\mathtt{Suff}^{k-1}(u_1) = \mathtt{Suff}^{k-1}(f^p(w_q)) = \mathtt{Suff}^{k-1}(f^p(w))$. As f is k-OSL, we know $\mathtt{tails}_f(w_q) = \mathtt{tails}_f(w)$, which implies that $(\lambda, f^p(w_q)^{-1} \cdot f(w_q)) \in \mathtt{tails}_f(w)$. Thus $f(w) = f^p(w)f^p(w_q)^{-1} \cdot f(w_q) = u_1u_2 = u$. □

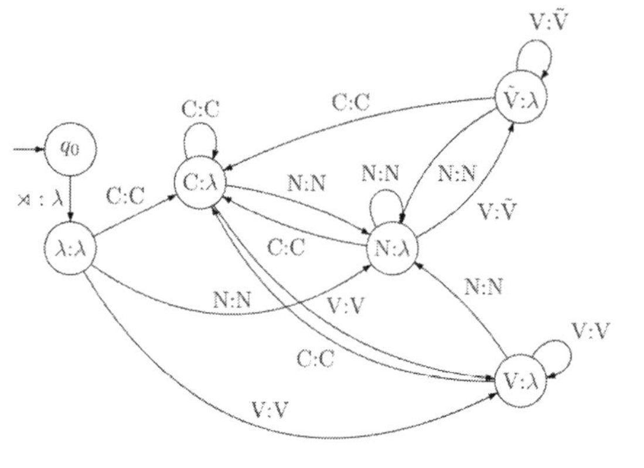

Figure 3: A 2-OSL DSFST that models Johore Malay nasal spreading. Σ={C, N, V} and Δ = {C, N, V, Ṽ}.

Figure 3 presents a 2-OSL transducer that models the nasal spreading example from §2. Note C = obstruent, V = vowels and glides, Ṽ = nasalized V, and N = nasal consonant.

6 Learning OSL functions

6.1 Learning criterion

We adopt the identification in the limit learning paradigm (Gold, 1967), with polynomial bounds on time and data (de la Higuera, 1997). The underlying idea of the paradigm is that if the data available to the algorithm does not contain enough information to distinguish the target from other potential targets, then it is impossible to learn.

We first need to define the following notions. A class $\mathbb{T}$ of functions is represented by a class $\mathbb{R}$ of representations if every $r \in \mathbb{R}$ is of finite size and there is a total and surjective naming function $\mathcal{L} : \mathbb{R} \rightarrow \mathbb{T}$ such that $\mathcal{L}(r) = t$ if and only if for all $w \in \texttt{pre\_image}(t)$, $r(w) = t(w)$, where $r(w)$ is the output of representation r on the input w. We observe that the class of k-OSL functions can be represented by the class of k-OSL DSFSTs.

Definition 8. *Let $\mathbb{T}$ be a class of functions represented by some class $\mathbb{R}$ of representations.*

1. *A* sample *S for a function $t \in \mathbb{T}$ is a finite set of data consistent with t, that is to say $(w, v) \in S$ iff $t(w) = v$. The size of a sample S is the sum of the length of the strings it is composed of: $|S| = \sum_{(w,v)\in S} |w| + |v|$.*
2. *A $(\mathbb{T}, \mathbb{R})$-*learning algorithm *$\mathfrak{A}$ is a program that takes as input a sample for a function $t \in \mathbb{T}$ and outputs a representation from $\mathbb{R}$.*

The paradigm relies on the notion of characteristic sample, adapted here for functions:

Definition 9 (Characteristic sample)**.** *For a $(\mathbb{T}, \mathbb{R})$-learning algorithm $\mathfrak{A}$, a sample CS is a* characteristic sample *of a function $t \in \mathbb{T}$ if for all samples S for t it is the case that $CS \subseteq S$ and $\mathfrak{A}$ returns a representation r such that $\mathcal{L}(r) = t$.*

This definition is the one used in the proof of the OSTIA algorithm. The learning paradigm can now be defined as follows.

Definition 10 (Identification in polynomial time and data)**.** *A class $\mathbb{T}$ of functions is identifiable in polynomial time and data if there exists a $(\mathbb{T}, \mathbb{R})$-learning algorithm $\mathfrak{A}$ and two polynomials $p()$ and $q()$ such that:*

1. *For any sample S of size m for $t \in \mathbb{T}$, $\mathfrak{A}$ returns a hypothesis $r \in \mathbb{R}$ in $\mathcal{O}(p(m))$ time.*
2. *For each representation $r \in \mathbb{R}$ of size n, with $t = \mathcal{L}(r)$, there exists a characteristic sample of t for $\mathfrak{A}$ of size at most $\mathcal{O}(q(n))$.*

6.2 Learning algorithm

We show here that Algorithm 1 learns the OSL functions under the criterion introduced. We call this the Output Strictly Local Function Inference Algorithm (OSLFIA). We assume Σ, Δ, and k are fixed and not part of the input to the learning problem.

Data: Sample $S \subset \{\rtimes\}\Sigma^*\{\ltimes\} \times \Delta^*$ and $k \in \mathbb{N}$
Let q_0, q_f be states with $\{q_0, q_f\} \cap \Delta^{\leq k-1} = \emptyset$
$s \leftarrow \texttt{lcp}(\{y \mid (x, y) \in S\})$; $q \leftarrow \texttt{Suff}^{k-1}(s)$;
$\texttt{smallest}(q) = \rtimes$; $\texttt{out}(q) = s$;
$\delta \leftarrow \{(q_0, \rtimes, s, q)\}$; $R \leftarrow \{q\}$; $C \leftarrow \{q_0, q_f\}$;
while $R \neq \emptyset$ **do**
  $q \leftarrow first(R)$; $s \leftarrow \texttt{smallest}(q)$;
  for *all* *$a \in \Sigma$ in alphabetical order* **do**
    if $\exists (w, u) \in S, x \in \Sigma^*$ *s.t.* $w = sax$ **then**
      $v \leftarrow \texttt{lcp}(\{y \mid \exists x, (sax, y) \in S\})$;
      $r \leftarrow \texttt{Suff}^{k-1}(qv)$;
      $\delta \leftarrow \delta \cup \{(q, a, \texttt{out}(q)^{-1} \cdot v, r)\}$;
      if $r \notin R \cup C$ **then**
        $R \leftarrow R \cup \{r\}$;
        $\texttt{smallest}(r) \leftarrow sa$;
        $\texttt{out}(r) \leftarrow v$;
  if $\exists u, (s, u) \in S$ **then**
    $\delta \leftarrow \delta \cup \{(q, \ltimes, \texttt{out}(q)^{-1} \cdot u, q_f)\}$
  $R \leftarrow R \setminus \{q\}$;
  $C \leftarrow C \cup \{q\}$;
return $\langle C, q_0, q_f, \Sigma, \Delta, \delta \rangle$;

Algorithm 1: OSLFIA

Essentially, the algorithm computes a breadth-first search through the states that are reachable given the learning sample: the set C contains the states already checked while R is a queue made of the states that are reachable but have not been treated yet. Initially, the only transition leaving the initial state is writing the lcp of the output strings of the sample and reaches the state corresponding to the $k - 1$ suffix of this lcp. At each step of the main loop, OSLFIA treats the first state that is in the queue R and computes whenever possible the transitions that leave that state. The outputs associated with each added transition are the longest common prefixes of the outputs associated with the smallest input prefix in the sample that allows the state to be reachable. We show that provided the algorithm is given a sufficient sample the transducer outputted by OSLFIA is onward and in fact a k-OSL transducer. After adding transitions with input letters from Σ to a state, the transition to the final state is added, provided it can be calculated.

6.3 Theoretical results

Here we establish the theoretical results, which culminate in the theorem that OSLFIA identifies the k-

OSL functions in polynomial time and data.

Lemma 6. *For any input sample S, OSLFIA produces its output in time polynomial in the size of S.*

Proof. The main loop is used at most $|\Delta|^{k-1}$ which is constant since both Δ and k are fixed for any learning sample. The smaller loop is executed $|\Sigma|$ times. At each execution: the first conditional can be tested in time linear in n, where $n = \sum_{(w,u)\in S} |w|$; the computation of the `lcp` can be done in nm steps where $m = max\{|u| : (w, u) \in S\}$ with an appropriate data structure (for instance a prefix tree); computing the suffix requires at most m steps. The second conditional can be tested in at most $\texttt{card}(S) \cdot m$ steps; the computation of the final transitions can be done in less than m steps; all the other instructions can be done in constant time. The overall computation time is thus $\mathcal{O}(|\Delta|^{k-1}|\Sigma|(n + nm + \texttt{card}(S) \cdot m + 2m)) = \mathcal{O}(n + m(n + \texttt{card}(S))$ which is polynomial (in fact bounded by a quadratic function) in the size of the learning sample. □

Next we show that for each k-OSL function f, there is a finite kernel of data consistent with f (a 'seed') that is a characteristic sample for OSLFIA.

Definition 11 (A OSLFIA seed). *Given a k-OSL transducer $\langle Q, q_0, q_f, \Sigma, \Delta, \delta\rangle$ computing a k-OSL function f, a sample S is a OSLFIA* seed *for f if*

- *For all $q \in Q$ such that $\exists v \in \Delta^*$ $(q, \ltimes, v, q_f) \in \delta$, $(\rtimes w_q \ltimes, f(w_q)) \in S$, where $w_q = \min_{\triangleleft}\{w \mid \exists u, (q_0, \rtimes w, u, q) \in \delta^*\}$*
- *For all $(q, a, u, q') \in \delta$ with $q' \neq q_f$ and $a \in \{\rtimes\} \cup \Sigma$, for all $b \in \Sigma$ such that there exists $(q', b, u', q'') \in \delta$, there exists $(\rtimes w \ltimes, f(w)) \in S$ and $x \in \Sigma^*$ such that $w = w_q abx$ and $f(w)$ is defined. Also, if there exists v such that $(q', \ltimes, v, q_f) \in \delta$ then $(\rtimes w_q a \ltimes, f(w_q a)) \in S$.*

In what follows, we set $\mathcal{T}^\diamond = \langle Q_\diamond, q_{0_\diamond}, q_{f_\diamond}, \Sigma, \Delta, \delta_\diamond\rangle$ be the target k-OSL transducer, f the function it computes, and $\mathcal{T} = \langle Q, q_0, q_f, \Sigma, \Delta, \delta\rangle$ be the transducer OSLFIA constructs on a sample that contains a seed.

Lemma 7. *If a learning sample S contains a seed then $(q_0, \rtimes w, u, r) \in \delta^* \iff (q_{0_\diamond}, \rtimes w, u, r) \in \delta^*_\diamond$.*

Proof. ($\Rightarrow$). By induction on the length of w. If $|w| = 0$ then $(q_0, \rtimes, u, r) \in \delta$ and so $u = \texttt{lcp}(\{y \mid (x, y) \in S\})$ and $r = \texttt{Suff}^{k-1}(u)$ (initial steps of the algorithm). As S is a seed there is an element $(\rtimes bx \ltimes, f(bx)) \in S$ for all $b \in \Sigma$ and $(\rtimes\ltimes, f(\lambda)) \in S$ if $\lambda \in \texttt{pre\_image}(f)$, which implies that $u = \texttt{lcp}(f(\lambda\Sigma^*))$. As the target is onward, we have $(q_{0_\diamond}, \rtimes, \texttt{lcp}(f(\lambda\Sigma^*)), r') \in \delta_\diamond$ and since it is a k-OSL DSFST $r' = \texttt{Suff}^{k-1}(\texttt{lcp}(f(\lambda\Sigma^*))) = \texttt{Suff}^{k-1}(u) = r$.

Suppose the lemma is true for strings of length less than or equal to n. We refer to this as the first Inductive Hypothesis (IH1). Let wa be of size $n + 1$ such that $(q_0, \rtimes wa, u, r) \in \delta^*$. By definition of δ^*, there exist u_1, u_2, q such that $(q_0, \rtimes w, u_1, q) \in \delta^*$, $(q, a, u_2, r) \in \delta$, and $u = u_1u_2$. By IH1 $(q_{0_\diamond}, \rtimes w, u_1, q) \in \delta^*_\diamond$. We want to show $(q_{0_\diamond}, \rtimes wa, u, r) \in \delta^*_\diamond$ (i.e., $(q, a, u_2, r) \in \delta_\diamond$).

First we show that IH1 also implies that $s = \texttt{smallest}(q)$ such that $s = \rtimes w_q$. Since the algorithm searches breadth-first, s is the smallest input that reaches q in $\mathcal{T}$. If $\rtimes w_q \triangleleft s$ then $\exists q' \neq q$ such that $(q_0, \rtimes w_q, u', q') \in \delta^*$ because $\rtimes w_q$ is a prefix of an input string of the sample S (since S contains a seed). Since $\rtimes w_q \triangleleft s$ and $|s| \leq n$, by IH1 then $(q_{0_\diamond}, \rtimes w_q, u', q') \in \delta^*_\diamond$ which implies $q = q'$ which contradicts the supposition that $\rtimes w_q \triangleleft s$. If $s \triangleleft \rtimes w_q$, then again since $(q_0, \rtimes s, u', q) \in \delta^*$ then by IH1 $(q_{0_\diamond}, \rtimes s, u', q) \in \delta^*_\diamond$. This contradicts the definition of w_q. Therefore $s = \rtimes w_q$.

Next we show that IH1 implies $f^p(w_q) = \texttt{out}(q)$. By construction of the seed, $(\rtimes w_q \ltimes, f(w_q)) \in S$ if $\exists v\, (q, \ltimes, v, q_{f_\diamond}) \in \delta_\diamond$ and $(\rtimes w_q a w' \ltimes, f(w_q a w')) \in S$ for all transitions (q, a, x, q') leaving q in $\mathcal{T}^\diamond$. As the target is onward, $\texttt{lcp}(\{x \mid (q, \sigma, x, q') \in \delta_\diamond, \sigma \in \Sigma \cup \{\ltimes\}\} = \lambda$ (Lemma 1). This implies $\texttt{out}(q) = \texttt{lcp}(\{y \mid \exists a \in \Sigma, x \in \Sigma^*\{\ltimes\}, (\rtimes sax, y) \in S\}) = \texttt{lcp}(\{y \mid \exists a \in \Sigma, x \in \Sigma^*\{\ltimes\}, (\rtimes w_q ax, y) \in S\}) = \texttt{lcp}(\{f(w_q\Sigma^*)\}) = f^p(w_q)$.

Recalling that $(q, a, u_2, r) \in \delta$, we now characterize u_2 to help establish $(q, a, u_2, r) \in \delta_\diamond$. By construction of a seed, there exist elements $(\rtimes w_q abx \ltimes, f(w_q abx))$ in S for all possible $b \in \Sigma$ and an element $(\rtimes w_q a \ltimes, f(w_q a)) \in S$ if $f(w_q a)$ is defined. By the onwardness of the target, this implies that $v = \texttt{lcp}(\{y \mid \exists b, x, (\rtimes sabx \ltimes, y) \in S\} \cup \{f(sa)\}) = \texttt{lcp}(f(sa\Sigma^*)) = f^p(sa)$. Therefore $u_2 = \texttt{out}(q)^{-1} \cdot v = f^p(s)^{-1} \cdot f^p(sa) = f^p(w_q)^{-1} \cdot f^p(w_q a)$.

Finally we identify r to complete this part

of the proof. As the target is OSL, we have $(q_{0_\diamond}, \rtimes w_q a, f^p(w_q a), r') \in \delta_\diamond^*$ (Lemma 3). The fact that $(q_{0_\diamond}, \rtimes w_q, f^p(w_q), q) \in \delta_\diamond^*$ by IH1 and the fact the target is OSL implies $(q, a, f^p(w_q)^{-1} \cdot f^p(w_q a), r') = (q, a, \mathtt{out}(q)^{-1} \cdot s, r') \in \delta_\diamond$. As $\mathcal{T}^\diamond$ is k-OSL, $r' = \mathtt{Suff}^{k-1}(q\mathtt{out}(q)^{-1} \cdot s)$ which is r by construction of the transition in the algorithm. Therefore, as $(q_{0_\diamond}, \rtimes w, u_1, q) \in \delta_\diamond^*$ by IH1, we have $(q_{0_\diamond}, \rtimes wa, u_1\mathtt{out}(q)^{-1} \cdot s, r) = (q_{0_\diamond}, \rtimes wa, u_1 u_2, r) = (q_{0_\diamond}, \rtimes wa, u, r) \in \delta_\diamond^*$

($\Leftarrow$). This is also by induction on the length of w. If $|w| = 0$, as $\mathcal{T}^\diamond$ is onward we have $\mathtt{lcp}(\mathtt{outputs}(q_{0_\diamond})) = \mathtt{lcp}(f(\Sigma^*))$ (Lemma 1) and thus $(q_{0_\diamond}, \rtimes, \mathtt{lcp}(f(\Sigma^*)), r) \in \delta_\diamond$ with $r = \mathtt{Suff}^{k-1}(\mathtt{lcp}(f(\Sigma^*)))$ as $\mathcal{T}^\diamond$ is k-OSL. By construction of the seed, there is at least one element in S using each transition leaving r. As $\mathtt{lcp}(\mathtt{outputs}(r)) = \lambda$ (Lemma 1), this implies $\mathtt{lcp}(\{y \mid (x, y) \in S\}) = \mathtt{lcp}(f(\Sigma^*))$. Therefore $(q_0, \rtimes, \mathtt{lcp}(f(\Sigma^*)), r) \in \delta$.

Suppose the lemma is true for all strings up to length n. We refer to this as the second Inductive Hypothesis (IH2). Pick wa of length $n + 1$ such that $(q_{0_\diamond}, \rtimes wa, u, r) \in \delta_\diamond^*$. By definition of δ^*, q, u_1, u_2 exist such that $(q_{0_\diamond}, \rtimes w, u_1, q) \in \delta_\diamond^*$ and $(q, a, u_2, r) \in \delta_\diamond$, with $u_1 u_2 = u$. By IH2, we have $(q_0, \rtimes w, u_1, q), (q_0, \rtimes w_q, u_1', q) \in \delta^*$ (since $w_q \lhd w$). We want to show $(q, a, u_2, r) \in \delta$.

We first show that $s = \mathtt{smallest}(q) = \rtimes w_q$. Suppose $s \lhd \rtimes w_q$. By construction of the SFST s is a prefix of an element of S which means there exists q' such that $(q_0, s, f^p(s), q') \in \delta_\diamond^*$. But by IH2, this implies that $q' = q$ and the definition of w_q contradicts $s \lhd \rtimes w_q$. Suppose now that $\rtimes w_q \lhd s$. By the construction of the seed, $\rtimes w_q$ is a prefix of an element of the sample, which implies it is considered by the algorithm. As $(q_0, \rtimes w_q, u_1', q) \in \delta^*$ by IH2, $\rtimes w_q$ is a smaller prefix than s that reaches the same state which is impossible as s is the earliest prefix that makes the state q reachable. Therefore $\rtimes w_q = s$ and thus the transition from state q reading a is created when $s = \rtimes w_q$.

Next we show that $f^p(w_q) = \mathtt{out}(q)$. By construction of the seed, there is an element $(\rtimes w_q a w' \ltimes, f(w_q a w')) \in S$ for all transitions $(q, a, x, q') \in \delta_\diamond$ leaving q and $(\rtimes w_q \ltimes, f(w_q)) \in S$ if $\exists v, (q, \ltimes, v, q_f) \in \delta_\diamond$. As the target is onward, $\mathtt{lcp}(\{x \mid (q, \sigma, x, q) \in \delta^*, \sigma \in \Sigma \cup \ltimes\} = \lambda$ (Lemma 1). This implies $\mathtt{out}(q) = \mathtt{lcp}(\{y \mid \exists a, x, (sax, y) \in S\}) = \mathtt{lcp}(\{y \mid \exists a, x, (w_q ax, y) \in S\}) = \mathtt{lcp}(f(w_q\Sigma^*)) = f^p(w_q) = f^p(s)$.

Now let $v = \mathtt{lcp}(\{y \mid \exists b, x, (sabx, y) \in S\})$. Since $s = \rtimes w_q$, $(q_{0_\diamond}, \rtimes w_q a, v, r) \in \delta_\diamond^*$ since, as before, the onwardness of the target implies the $\mathtt{lcp}$ of the output written from r is λ. This is because each possible output from r is in S (because it is in the seed according to the second item of Definition 11). Consequently $v = f^p(w_q a) = f^p(sa)$.

Together these results imply that $u_2 = f^p(w_q)^{-1} \cdot f^p(w_q a) = f^p(s)^{-1} \cdot f^p(sa) = \mathtt{out}(q)^{-1} \cdot v$.

As the target is a k-OSL transducer (and thus deterministic) $\mathtt{Suff}^{k-1}(qu_2) = r$. Therefore the transition $(q, a, \mathtt{out}(q)^{-1} \cdot v, r)$ that is added to δ is the same as the transition (q, a, u_2, r) in $\delta_\diamond$. This implies $(q_0, \rtimes wa, u, r) \in \delta^*$ and proves the lemma. □

Lemma 8. *Any seed for the OSL Learner is a characteristic sample for this algorithm.*

Proof. A corollary of Lemma 7 is that if a seed is contained in a learning sample we have $(q_0, \rtimes w, u, q) \in \delta^* \iff f^p(w) = u$ (Lemma 3) as the target transducer is k-OSL. For all states q where $\exists v, (q, \ltimes, v, q_{f_\diamond}) \in \delta_\diamond$, we have $(\rtimes w_q \ltimes, f(w_q))$ in the seed, which implies the algorithm will add $(q, \ltimes, f^p(w_q)^{-1} \cdot f(w_q), q_f)$ to δ which is exactly the output function of the target. As every state is treated only once, this holds for any learning set containing a seed. Therefore, from any superset of a seed, for any w, the function computed by the outputted transducer of Algorithm 1 is equal to $f^p(w) f^p(w)^{-1} \cdot f(w) = f(w)$. □

Observe that OSLFIA is designed to work with seeds, which contains *minimal* strings. We believe both the seed and algorithm can be adjusted to relax this requirement, though this is left for future work.

Lemma 9. *Given any k-OSL transducer $\mathcal{T}^\diamond$, there exists a seed for the OSL learner that is of size polynomial in the size of $\mathcal{T}^\diamond$.*

Proof. Let $\mathcal{T}^\diamond = \langle Q_\diamond, q_{0_\diamond}, q_{f_\diamond} \Sigma, \Delta, \delta_\diamond, \rangle$. There are at most $\mathtt{card}(Q_\diamond)$ pairs $(\rtimes w_q \ltimes, f(w_q))$ in a seed that corresponds to the first item of Definition 11, each of which is such that $| \rtimes w_q \ltimes | \leq \mathtt{card}(Q^\diamond)$

and $|f(w_q)| \leq \sum_{(q,\sigma,u,q')\in\delta_\diamond} |u|$. We denote by $m_\diamond$ this last quantity and note that $m_\diamond = \mathcal{O}(|\mathcal{T}^\diamond|)$.

For the elements of the second item of Definition 11 we restrict ourselves without loss of generality to pairs $(\rtimes w_q abw' \ltimes, f(w_q abw'))$ where $w' = min_\triangleleft\{x : f(w_q abx) \text{ is defined}\}$. We have $|w'| \leq \mathtt{card}(Q_\diamond)$ and $|f(w_q abw')|$ is in $\mathcal{O}(\mathtt{card}(Q_\diamond)m_\diamond)$. There are at most $|\Sigma|$ pairs $(\rtimes w_q abw' \ltimes, f(w_q abw'))$ for a given transition (q, a, u, q') which implies that the overall bound on the number of such pairs is in $\mathcal{O}(|\Sigma|\mathtt{card}(\delta))$. The overall length of the elements in the seed that fulfill the second item of the definition is in $\mathcal{O}(\mathtt{card}(Q_\diamond)(\mathtt{card}(Q_\diamond) + m_\diamond + |\Sigma|\mathtt{card}(\delta)m_\diamond))$.

The size of the seed studied in this proof is thus in $\mathcal{O}((m_\diamond + |Q_\diamond|)(|Q_\diamond| + |\Sigma|\mathtt{card}(\delta))$ which is polynomial (in fact quadratic) in the size of the target transducer. □

Theorem 4. *OSLFIA identifies the k-OSL functions in polynomial time and data.*

Proof. Immediate from Lemmas 6, 7, 8, and 9. □

We conclude this section by comparing this result to other subsequential function-learning algorithms.

OSTIA (Oncina et al., 1993) is a state-merging algorithm which can identify the class of total subsequential functions in cubic time. (Partial subsequential functions cannot be learned exactly; for a partial function, OSTIA will learn some superset of it.) k-OSL functions include both partial and total functions, so the classes exactly learnable by OSTIA and OSLFIA are, strictly speaking, incomparable.

SOSFIA (Jardine et al., 2014) identifies subclasses of subsequential functions in linear time and data. These subclasses are determined by fixing the structure of a transducer in advance. For every input string, SOSFIA knows exactly which state in the transducer is reached. The sole carrier of information regarding reached states is the input string. But for k-OSL functions, the output strings carry the information about the states reached. As the theorems demonstrate, the destination of a transition is only determined by the output of the transition. Thus no class learned by SOSFIA contains any k-OSL class.

OSTIA-D (OSTIA-R) (Oncina and Varò, 1996; Castellanos et al., 1998) identify a class of subsequential functions with a given domain D (range R) in at least cubic time because it adds steps to OSTIA to prevent merging states that would result in a transducer whose domain (range) is not compatible with D (R). OSTIA-D cannot represent k-OSL functions for the same reasons SOSFIA cannot: domain information is about input strings, not output strings. On the other hand, the range of a k-OSL function is a k-OSL stringset which can be represented with a single acceptor, and thus OSL functions may be learned by OSTIA-R. However, OSLFIA is more efficient both in time and data.[3]

To sum up, OSLFIA is the most efficient algorithm for learning k-OSL functions.

7 Phonology

The example of Johore Malay nasal spreading given in §2 is an example of progressive spreading, since it proceeds from a triggering segment (the nasal) to vowels and glides that *follow* it. There also exist *regressive* spreading processes, in which the trigger follows the target(s). An example from the Mòbà dialect of Yoruba (Ajíbóyè, 2001; Ajíbóyè and Pulleyblank, 2008; Walker, 2014) is shown in (2). An underlying nasalized vowel spreads its nasality to preceding oral vowels and glides.

(2) /ujĩ/ ↦ [ũj̃ĩ], 'praise(n.)'

The difference between progressive and regressive spreading corresponds to reading the input from left-to-right or right-to-left, respectively (Heinz and Lai, 2013). Regressive spreading cannot be modeled with OSL in a left-to-right fashion, because the output of the preceding vowels and glides depends on the presence or absence of a following nasal that could be an unbounded number of segments away. By reading from right-to-left, that nasal trigger will always be read before the target(s), making it akin to progressive spreading. Thus there are two overlapping but non-identical classes, which we call left(-to-right) OSL and right(-to-left) OSL.

There are other types of phonological maps that are neither ISL nor OSL. Consider the optional process of French ə-deletion shown in (3) (Dell, 1973; Dell, 1980; Dell, 1985; Noske, 1993).

[3] To our knowledge no analysis of data complexity for OSTIA, OSTIA-D, and OSTIA-R has been completed (probably because they predate de la Higuera (1997)). Also, an analysis of the data complexity of OSTIA appears daunting.

(3) ə → $\emptyset$ / VC__CV

At issue is how this rule applies. There are two licit pronunciations of /ty dəvənɛ/ 'you became' which are [ty dvənɛ] and [ty dəvnɛ]. The form *[ty dvnɛ] is considered ungrammatical. As Kaplan and Kay (1994) explain, these outputs can be understood as the rule in (3) applying left-to-right ([ty dvənɛ]), right-to-left ([ty dvənɛ]) or simultaneously (*[ty dvnɛ]). What matters is whether the left and right contexts of the rule match the *input* or *output* string: if both match the input it is simultaneous application, and if one side matches the input and the other the output it is left-to-right or right-to-left.

ISL functions always match contexts against the input and therefore they cannot model ə-deletion. In this respect, ISL functions model simultaneous rule application. But there is also a problem with modeling the process as OSL, which is what to output when the ə that will be deleted is read. Consider the input VCəCV. When the DSFST reads the ə, it cannot decide what to output, because whether or not that ə is deleted depends on whether or not the next two symbols in the input are CV. But since the DSFST is deterministic, it must make a decision at this point. It could postpone the decision and output λ. But that would require it to loop at the current state (Lemma 2), which in turn means it cannot distinguish VCəCV from VCəəəCV, a significant problem since only the former meets the context for deletion.

Thus the range of phonological processes that can be modeled with OSL functions is limited to those with one-sided contexts (e.g., either C__ or __D, the former being left OSL and the latter right OSL). In such cases the entire triggering context will be read before the potential target, so there is never a need to delay the decision about what to output. To summarize, phonological rules that apply simultaneously are ISL, and phonological rules with one-sided contexts that apply left-to-right or right-to-left are OSL.

In addition to iterative rules with two-sided contexts, long-distance processes like vowel harmony and consonant agreement and dissimilation are also excluded from the current analysis. While such process have been shown to be subsequential and therefore subregular (see Gainor et al. (2012; Luo (2014; Payne (2013; Heinz and Lai (2013)) they are neither ISL nor OSL because the target and triggering context are not within a fixed window of length k in either the input or output. An example is the long-distance nasal assimilation process in Kikongo (Rose and Walker, 2004), as in (4).

(4) /tu+nik+idi/ ↦ [tunikini] 'we ground'

In Kikongo, the alveolar stop in the suffix /-idi/ surfaces as a nasal when joined to a stem containing a nasal. Since stem nasals appear to occur arbitrarily far from the suffix, there is no k such that the target /d/ and the trigger /n/ are within a window of size k. Thus the process is neither ISL nor OSL.

8 Future Work

Processes like French ə-deletion that have two-sided contexts, with one being on the output side, suggest a class that combines the ISL and OSL properties. We are tentatively calling this class 'Input-Output SL' and are currently working on its properties, FST characterization, and learning algorithm. For long-distance processes, we expect other functional subclasses will strongly characterize these. SL stringsets are just one region of the Subregular Hierarchy (Rogers and Pullum, 2011; Rogers et al., 2013), so we expect functional counterparts of the other regions can be defined. Some of these other regions model long-distance phonotactics (Heinz, 2007; Heinz, 2010; Rogers et al., 2010), so their functional counterparts may prove equally useful for modeling and learning long-distance phonology.

9 Conclusion

We have defined a subregular class of functions called the OSL functions and provided both language-theoretic and automata-theoretic characterizations. The structure of this class is sufficient to allow any k-OSL function to be efficiently learned from positive data. It was shown that the OSL functions—unlike the ISL functions—can model local iterative spreading processes. Future work will aim to combine the results for both ISL and OSL to model iterative processes with two-sided contexts.

Acknowledgments

We thank three reviewers for useful comments, especially the third, who caught a significant error in the first version of this paper.

References

Oládiípò Ajíbóyè and Douglas Pulleyblank. 2008. Mòbà nasal harmony. Ms., University of Lagos and University of British Columbia.

Oládiípò Ajíbóyè. 2001. Nasalization in Mòbà. In Sunyoung Oh, Naomi Sawai, Kayono Shiobara, and Rachel Wojdak, editors, *Proceedings of the Northwest Linguistics Conference*, pages 1–18. University of British Columbia Working Papers in Linguistics 8. Vancouver: University of British Columbia, Department of Linguistics.

William Bennett. 2013. *Dissimilation, Consonant Harmony, and Surface Correspondence*. Ph.D. thesis, Rutgers.

Antonio Castellanos, Enrique Vidal, Miguel A. Varó, and José Oncina. 1998. Language understanding and subsequential transducer learning. *Computer Speech and Language*, 12:193–228.

Jane Chandlee and Jeffrey Heinz. 2012. Bounded copying is subsequential: Implications for metathesis and reduplication. In *Proceedings of the 12th Meeting of the ACL Special Interest Group on Computational Morphology and Phonology*, pages 42–51, Montreal, Canada, June. Association for Computational Linguistics.

Jane Chandlee and Jeffrey Heinz. to appear. Strictly local phonological processes. *Linguistic Inquiry,*. under revision.

Jane Chandlee, Angeliki Athanasopoulou, and Jeffrey Heinz. 2012. Evidence for classifying metathesis patterns as subsequential. In *The Proceedings of the 29th West Coast Conference on Formal Linguistics*, pages 303–309. Cascadilla Press.

Jane Chandlee, Rémi Eyraud, and Jeffrey Heinz. 2014. Learning strictly local subsequential functions. *Transactions of the Association for Computational Linguistics*, 2:491–503, November.

Jane Chandlee. 2014. *Strictly Local Phonological Processes*. Ph.D. thesis, The University of Delaware.

Noam Chomsky and Morris Halle. 1968. *The Sound Pattern of English*. New York: Harper & Row.

Colin de la Higuera. 1997. Characteristic sets for polynomial grammatical inference. *Machine Learning Journal*, 27:125–138.

François Dell. 1973. *Les régles et les sons*. Paris: Hermann.

François Dell. 1980. *Generative phonology and French phonology*. Cambridge: Cambridge University Press.

François Dell. 1985. *Les régles et les sons*. Paris: Hermann, 2 edition.

Jason Eisner. 2003. Simpler and more general minimization for weighted finite-state automata. In *Proceedings of the Joint Meeting of the Human Language Technology Conference and the North American Chapter of the Association for Computational Linguistics (HLT-NAACL 2003)*, pages 64–71.

Robert Frank and Giorgo Satta. 1998. Optimality Theory and the generative complexity of constraint violability. *Computational Linguistics*, 24(2):307–315.

Brian Gainor, Regine Lai, and Jeffrey Heinz. 2012. Computational characterizations of vowel harmony patterns and pathologies. In Jaehoon Choi, E. Alan Hogue, Jeffrey Punske, Deniz Tat, Jessamyn Schertz, and Alex Trueman, editors, *WCCFL 29: Proceedings of the 29th West Coast Conference on Formal Linguistics*, pages 63–71, Somerville, MA. Cascadilla.

E.Mark Gold. 1967. Language identification in the limit. *Information and Control*, 10:447–474.

Thomas Graf. 2010. Logics of phonological reasoning. Master's thesis, University of California, Los Angeles.

Gunnar Hansson. 2010. *Consonant Harmony: Long-Distance Interaction in Phonology*. Number 145 in University of California Publications in Linguistics. University of California Press, Berkeley, CA. Available on-line (free) at eScholarship.org.

Jeffrey Heinz and Regine Lai. 2013. Vowel harmony and subsequentiality. In Andras Kornai and Marco Kuhlmann, editors, *Proceedings of the 13th Meeting on the Mathematics of Language (MoL 13)*, pages 52–63, Sofia, Bulgaria.

Jeffrey Heinz. 2007. *The Inductive Learning of Phonotactic Patterns*. Ph.D. thesis, University of California, Los Angeles.

Jeffrey Heinz. 2009. On the role of locality in learning stress patterns. *Phonology*, 26(2):303–351.

Jeffrey Heinz. 2010. Learning long-distance phonotactics. *Linguistic Inquiry*, 41(4):623–661.

Adam Jardine, Jane Chandlee, Rémi Eyraud, and Jeffrey Heinz. 2014. Very efficient learning of structured classes of subsequential functions from positive data. In Alexander Clark, Makoto Kanazawa, and Ryo Yoshinaka, editors, *Proceedings of the Twelfth International Conference on Grammatical Inference (ICGI 2014)*, volume 34, pages 94–108. JMLR: Workshop and Conference Proceedings, September.

Adam Jardine. 2014. Computationally, tone is different. Under review with Phonology.

C. Douglas Johnson. 1972. *Formal Aspects of Phonological Description*. The Hague: Mouton.

Ronald Kaplan and Martin Kay. 1994. Regular models of phonological rule systems. *Computational Linguistics*, 20(3):331–378.

Michael Kenstowicz and Charles Kisseberth. 1979. *Generative Phonology*. Academic Press, Inc.

Huan Luo. 2014. Long-distance consonant harmony and subsequantiality. Qualifying paper for the University of Delaware's Linguistics PhD Progam.

Robert McNaughton and Seymour Papert. 1971. *Counter-Free Automata*. MIT Press.

Mehryar Mohri. 1997. Finite-state transducers in language and speech processing. *Computational Linguistics*, 23(2):269–311.

Andrew Nevins. 2010. *Locality in Vowel Harmony*. MIT Press.

Roland Noske. 1993. *A theory of syllabification and segmental alternation*. Niemeyer, Tübingen.

David Odden. 2014. *Introducing Phonology*. Cambridge University Press, 2nd edition.

Jose Oncina and Pedro Garcia. 1991. Inductive learning of subsequential functions. Technical Report DSIC II-34, University Politécnia de Valencia.

José Oncina and Miguel A. Varò. 1996. Using domain information during the learning of a subsequential transducer. *Lecture Notes in Artificial Intelligence*, pages 313–325.

José Oncina, Pedro García, and Enrique Vidal. 1993. Learning subsequential transducers for pattern recognition tasks. *IEEE Transactions on Pattern Analysis and Machine Intelligence*, 15:448–458, May.

Farid M. Onn. 1980. *Aspects of Malay Phonology and Morphology: A Generative Approach*. Kuala Lumpur: Universiti Kebangsaan Malaysia.

Amanda Payne. 2013. Dissimilation as a subsequential process. Qualifying paper for the University of Delaware's Linguistics PhD Progam.

James Rogers and Geoffrey Pullum. 2011. Aural pattern recognition experiments and the subregular hierarchy. *Journal of Logic, Language and Information*, 20:329–342.

James Rogers, Jeffrey Heinz, Gil Bailey, Matt Edlefsen, Molly Visscher, David Wellcome, and Sean Wibel. 2010. On languages piecewise testable in the strict sense. In Christian Ebert, Gerhard Jäger, and Jens Michaelis, editors, *The Mathematics of Language*, volume 6149 of *Lecture Notes in Artifical Intelligence*, pages 255–265. Springer.

James Rogers, Jeffrey Heinz, Margaret Fero, Jeremy Hurst, Dakotah Lambert, and Sean Wibel. 2013. Cognitive and sub-regular complexity. In Glyn Morrill and Mark-Jan Nederhof, editors, *Formal Grammar*, volume 8036 of *Lecture Notes in Computer Science*, pages 90–108. Springer.

Sharon Rose and Rachel Walker. 2004. A typology of consonant agreement as correspondence. *Language*, 80:475–531.

Jaques Sakarovitch. 2009. *Elements of Automata Theory*. Cambridge University Press. Translated by Reuben Thomas from the 2003 edition published by Vuibert, Paris.

Keiichiro Suzuki. 1998. *A Typological Investigation of Dissimilation*. Ph.D. thesis, University of Arizona.

Wolfgang Thomas. 1997. Languages, automata, and logic. volume 3, chapter 7. Springer.

Rachel Walker. 2011. *Vowel Patterns in Language*. Cambridge: Cambridge University Press.

Rachel Walker. 2014. Nonlocal trigger-target relations. *Linguistic Inquiry*, 45(3):501–523.

How to Choose Successful Losers in Error-Driven Phonotactic Learning

Giorgio Magri
SFL UMR 7023 (CNRS and University of Paris 8)
59/61 rue Pouchet, 75017 Paris
France
magrigrg@gmail.com

René Kager
UiL-OTS (Utrecht University)
Trans 10, 3512 JK Utrecht
The Netherlands
RWJKager@uu.nl

Abstract

An error-driven phonotactic learner is trained on a stream of licit phonological forms. Each piece of training data counts as a *winner* in terms of Optimality Theory. In order to test its current grammar, the learner needs to compare the current winner with a properly chosen *loser*. This paper advocates a new subroutine for the choice of the loser, based on the idea of minimizing the "distance" from the given winner.

1 Error-driven phonotactic learning and the problem of the choice of the loser

Phonotactics is knowledge of the distinction between licit and illicit phonological forms (Chomsky and Halle, 1965). We adopt the model of phonotactics developed within *Optimality Theory* (Prince and Smolensky, 2004), briefly reviewed here. A *candidate set* is a collection of pairs (x, y); the first element x is called the *underlying* form; the second element y is called the *candidate surface* form. A *constraint* assigns to each candidate pair (x, y) a non-negative number of *violations*, which measures how the mapping of the underlying form x to the surface form y deviates from the ideal relative to the specific perspective valued by that constraint. Constraints come in two types. *Faithfulness constraints* punish a candidate pair (x, y) based on the discrepancy between the underlying form x and the surface form y. For instance, the faithfulness constraint IDENT[VOICE] is violated by a candidate pair of segments differing in voicing. *Markedness constraints* punish a candidate pair (x, y) based on the ill-formedness of the surface form y. For instance, the markedness constraint NOVOICEOBS is violated by a candidate pair of segments whose surface segment is a voiced obstruent.

A *constraint ranking* $\gg$ is a linear order over the constraint set. The *grammar* $G_{\gg}$ corresponding to the ranking $\gg$ takes an underlying form x and returns a candidate (x, y) which $\gg$*-beats* any other candidate $(x, \not{z})$ whose underlying form is x and whose surface form $\not{z}$ is different from y, in the sense of condition (1). The candidate (x, y) is then called the *winner* while $(x, \not{z})$ is called a *loser*. Losers are stricken out as a mnemonic.

(1) There exists a constraint which is *winner-preferring* (i.e., assigns *less* violations to the winner candidate (x, y) than to the loser candidate $(x, \not{z})$) which is $\gg$-ranked above every constraint which is *loser-preferring* (i.e., assigns *more* violations to the winner candidate (x, y) than to the loser candidate $(x, \not{z})$).

A surface form y is *phonotactically licit* according to the grammar $G_{\gg}$ provided there exists some underlying form x which is mapped to y by that grammar.

An *error-driven* phonotactic learner is trained on a sequence of phonological forms all phonotactically licit according to a target grammar and it tries to infer that grammar as follows. It maintains a current hypothesis of the target grammar, which is initialized to a most restrictive grammar, namely one which deems illicit as many forms as possible. The current grammar is then slightly updated in the direction of a looser phonotactics, whenever it incorrectly predicts the current piece of data to be il-

Proceedings of the 14th Meeting on the Mathematics of Language (MoL 14), pages 126–138,
Chicago, USA, July 25–26, 2015. ©2015 Association for Computational Linguistics

licit. This learning scheme is formalized in OT as the *error-driven ranking algorithm* (EDRA) outlined in the following pseudo-code and detailed below (Tesar and Smolensky, 1998; Boersma, 1998).

```
1: Initialize the current ranking vector θ
2: repeat
3:     get a licit winner surface form y
4:     choose a corresponding underlying form x
5:     choose a loser form z̸ to compare to y
6:     If the winner (x, y) does not beat the loser
           (x, z̸) according to the ranking vector θ:
7:         update the current ranking vector θ
8: until no more errors are made at line 6
```

The EDRA knows the underlying constraint set $C_1, \ldots, C_n$. The current constraint ranking is represented by assigning to each constraint C_k a numerical *ranking value* θ_k, with the understanding that high ranking values correspond to high ranked constraints. The ranking values are collected into a *ranking vector* $\boldsymbol{\theta} = (\theta_1 \ldots, \theta_n)$. At line 1, the ranking values of the faithfulness and the markedness constraints are initialized to 0 and to a large positive constant $\theta > 0$, respectively. Thus, the markedness constraints start out above the faithfulness ones, yielding a grammar which is phonotactically maximally restrictive (Smolensky, 1996).

At line 3, the EDRA is fed a piece of training data, consisting of a surface form y licit according to the target constraint ranking. No assumptions are made on the sequence of training data (e.g., no assumptions are made on the frequency with which various licit forms are fed to the learner). At line 4, the EDRA needs to reconstruct an underlying form x corresponding to the current winner surface form y. A common choice is to set x identical to y, under the assumption that the underlying OT typology is *idempotent*, namely it maps every phonotactically licit form faithfully into itself (Magri, 2015b). The proper definition of the subroutine for the choice of the loser form $\not{z}$ at line 5 is the topic of this paper and will thus be discussed in detail below.

At line 6, the EDRA checks whether the current ranking vector $\boldsymbol{\theta}$ satisfies condition (2), where W and L are the sets of winner- and loser-preferring constraints relative to the intended winner and loser candidates (x, y) and $(x, \not{z})$.

(2) $$\max_{h \in W} \theta_h > \max_{k \in L} \theta_k$$

If condition (2) holds, any ranking $\gg$ which respects the current ranking vector $\boldsymbol{\theta}$ (in the sense that $C_h \gg C_k$ whenever $\theta_h > \theta_k$) satisfies the OT condition (1), namely succeeds at making the intended winner (x, y) beat the intended loser $(x, \not{z})$ (Boersma, 2009). In this case, the learner has nothing to learn from the comparison between the current winner and loser forms.

Failure of condition (2) instead suggests that the current ranking values of the loser-preferring (winner-preferring) constraints are too large (too small, respectively) and thus need to be updated at line 7. We assume the re-ranking rule (3) (Tesar and Smolensky, 1998; Boersma, 1998; Magri, 2012).

(3) a. Increase the ranking values of the w winner-preferring constraints by $\frac{1}{w+1}$;

b. decrease the ranking values of the *undominated* loser-preferring constraints by 1.

Each winner-preferring constraint is promoted by $\frac{1}{w+1}$, where w is the total number of winner-preferring constraints. The loser-preferring constraints are demoted by 1. Only those loser-preferring constraints that really need to be demoted are indeed demoted, namely those which are not currently ranked underneath a winner-preferring constraint and are therefore called *undominated*.

The only implementation detail which has been left open in this outline of the EDRA model is the proper definition of subroutine for the choice of the loser at line 5. This is the topic of this paper.

2 Two test cases to evaluate subroutines for the choice of the current loser

The EDRA model is guaranteed to *converge* (under the assumption that the target grammar is idempotent): after a finite (small) number of iterations, it is not possible to sample from the set of target licit forms any surface form y which would force the learner to make an update in the if-loop at lines 6-8. Convergence holds irrespectively of the subroutine for the choice of the current loser used at line 5. Suppose now that this subroutine satisfies the basic condition (4). This condition says that the EDRA never wastes data (Tesar and Smolensky, 1998): if there is an opportunity to learn something from the current winner (i.e., if there exists at least a loser which is able to trigger an update), the EDRA will not "waste" that opportunity (i.e., the chosen loser indeed triggers an update).

(4) The subroutine for the choice of the loser returns a loser which triggers an update at line 7, whenever such a loser exists.

This condition (4) ensures that, if a surface form y is licit according to the target grammar the EDRA has been trained on, then it is also licit according to any ranking $\gg$ which respects the final ranking vector $\boldsymbol{\theta}^{\text{fin}}$ entertained by the EDRA at convergence (in the sense that $C_h \gg C_k$ whenever $\theta_h^{\text{fin}} > \theta_k^{\text{fin}}$). In other words, the EDRA succeeds at half of the learning problem: it has learned to recognize licit forms as such. The ranking learned by the EDRA could nonetheless deem licit too many forms. In other words, it could describe a phonotactics which, although *consistent* with the target one, is not sufficiently *restrictive*. Are there guarantees that the EDRA also learns to recognize illicit forms as such?

Consider a phonotactic pattern which has the following property: there exists a subset of the markedness constraints which punish exactly all and only the illicit forms. This phonotactic pattern can thus be analyzed in terms of a constraint ranking such as (5): the designated subset of markedness constraints hold sway at the top while the remaining markedness constraints are silent at the bottom. The relative ranking of the faithfulness constraints sandwiched in between is irrelevant. We therefore refer to these phonotactic patterns as *$\mathcal{F}$-irrelevant*.

(5) *a subset of $\mathcal{M}$ constraints*

all $\mathcal{F}$ constraints

the remaining $\mathcal{M}$ constraints

For instance, suppose that the constraint set contains a markedness constraint against voiced velar obstruents and a markedness constraint against dorsal fricatives. The velar inventory [~~g~~ k ~~ɣ~~ ~~x~~], which only admits the voiceless velar stop (illicit segments are stricken out), follows by just letting those two markedness constraints hold sway at the top.

The EDRA model described in section 1 has been shown to be restrictive when the target phonotactics is $\mathcal{F}$-irrelevant (Magri, 2013a; Magri, 2014a; Magri, 2015c; Magri, 2015a). In other words, it succeeds at learning the target phonotactics. This success holds irrespectively of the details of the phonological analysis (e.g., the content of the markedness and faithfulness constraints or any other properties of the target ranking, besides it being $\mathcal{F}$-irrelevant). It also holds irrespectively of how the current loser is chosen at line 5 of the pseudo-code—as long as condition (4) is respected. In order to tackle the issue of the proper definition of the subroutine for the choice of the loser at line 5, we thus need to look at the behavior of the EDRA model on phonotactic patterns which are *not* $\mathcal{F}$-irrelevant. Let's briefly recall two two examples of such phonotactic patterns which are not $\mathcal{F}$-irrelevant (Magri and Kager, 2015).

Voicing is especially effortful at the velar place: due to the small oral volume behind the velar constriction, the supra-glottal pressure quickly equalizes the sub-glottal pressure, hindering vocal cords vibration (Ohala, 1983). Many attested velar inventories comply with phonetic markedness, namely have voiceless stops or fricatives without the voiced ones. Yet, UPSID (Maddieson, 1984) documents two inventories [~~g~~ k ɣ ~~x~~] and [g k ɣ ~~x~~] which are phonetically *counterintuitive*, as they admit the voiced fricative at the exclusion of the voiceless one. If we could posit a markedness constraint which punishes [x] at the exclusion of [ɣ], these inventories could be generated by letting that markedness constraint hold sway at the top of the ranking (5). Yet, such a markedness constraint would be incompatible with the *grounding hypothesis* (Hayes and Steriade, 2004): [x] is not any worse than [ɣ] from any phonetic perspective. Fortunately, the desired inventory can be generated in compliance with the grounding hypothesis whenever /ɣ/ is harder to neutralize than /x/, so that the former surfaces at the exclusion of the latter. This neutralization pattern requires some faithfulness constraints (which preserve /ɣ/ from neutralizing) to be ranked above some other faithfulness constraints (those violated by the neutralization of /x/). In conclusion, the grounding hypothesis forces us to posit a crucial relative ranking among the faithfulness constraints. The learnability guarantees recalled above for $\mathcal{F}$-irrelevant target rankings (5) thus do not apply in these cases.

Let's look closer at the inventory [~~g~~ k ɣ ~~x~~], which lacks the voiced stop and the voiceless fricative. We analyze this inventory as follows: /x/ can be neutralized to [k] preserving voicing, while /ɣ/ cannot be neutralized preserving voicing, because [g] is independently ruled out by a dedicated constraint. This intuition can be cashed out as follows. We assume that only velar obstruents are candidates of the velar obstruents, as stated in (6a). The ranking (6b) then yields the target inventory [~~g~~ k ɣ ~~x~~].

(6) a. Gen(/g k ɣ x/) = [g k ɣ x]

b.

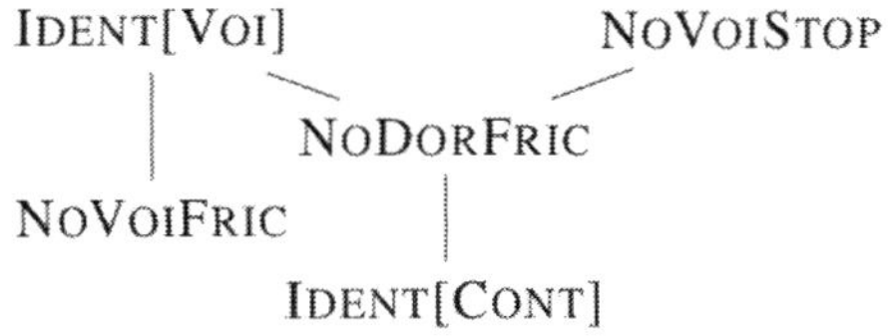

The inventory [g k ɣ ~~x~~] only lacks the voiceless fricative and thus differs from the inventory considered above only because the voiced velar stop [g] is now licit. We analyze this inventory as follows: /x/ can be neutralized to [h] preserving voicing, while /ɣ/ cannot be neutralized while preserving voicing, because [ɦ] is independently ruled out by a dedicated constraint. This intuition can be cashed out as follows. We assume place impermeability apart from the velar/glottal border: only the velar and glottal obstruents are candidates of the velar and glottal obstruents, as stated in (7a). The ranking (7b) then generates the target inventory [g k ɣ ~~x~~].

(7) a. Gen(/g k ɣ x ʔ ɦ h/) = [g k ɣ x ʔ ɦ h]

b.

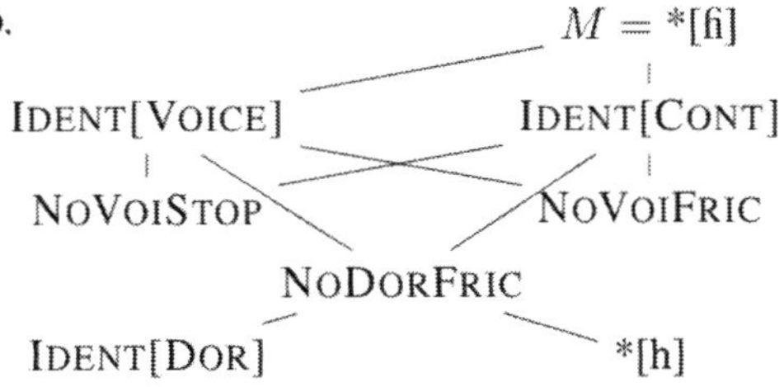

From the perspective of the phonological analysis, the assumption (6a) of velar place impermeability and the assumption (7a) of place impermeability apart from the velar/glottal border are not restrictive: they can be reinterpreted as the assumption that the faithfulness constraints for place features are high ranked. Yet, this interpretation introduces additional relative ranking conditions among the faithfulness constraints which would need to be carefully considered in the learnability analyses. To start from the simplest case, the learnability analyses developed in the rest of the paper explicitly adopt the restrictive assumptions (6a) and (7a) on candidacy.

The only detail in the description of the EDRA model which has been left open in section 1 concerns the proper definition of subroutine for the choice of the loser at line 5. The rest of this paper tackles this issue using the test cases of the two velar inventories just discussed.

3 Motivating a new subroutine for the choice of the current loser

Suppose that the EDRA model is trained on the velar inventory [~~g~~ k ɣ ~~x~~] and thus needs to learn the corresponding constraint ranking (6b). At line 3, the model is effectively always fed the surface form [ɣ], since [ɣ] and [k] are the only two licit forms and the latter can be ignored, because it is unmarked relative to the constraints assumed (it violates none) and thus never prompts an update. At line 4, the model assumes the faithful underlying form /ɣ/. At line 5, the model has to choose the current loser candidate among [g], [k], or [x]. According to the *classical subroutine* for the choice of the loser, the learner chooses as the current loser the candidate predicted to win by the current ranking values—more precisely, by an arbitrary ranking consistent with the current ranking values (Tesar and Smolensky, 1998; Magri, 2013b). This classical subroutine for the choice of the loser satisfies condition (4), namely it never wastes data: if there exists at least a loser which is able to trigger an update, the chosen loser can be shown to indeed trigger an update.

Unfortunately, the classical subroutine for the choice of the loser leads to trouble when the EDRA model is trained on the inventory [~~g~~ k ɣ ~~x~~]. Here is why. The comparison between the winner mapping (/ɣ/, [ɣ]) and the three loser mappings (/ɣ/, ~~[g]~~), (/ɣ/, ~~[k]~~), and (/ɣ/, ~~[x]~~) sorts the constraints into winner-/loser-preferring as represented in (8) with Elementary Ranking Conditions (Prince, 2002).

(8)

	IDENT[VOI]	ID[CON]	NODORFRIC	NOVOISTOP	NOVOIFRI
(/ɣ/,[ɣ])~(/ɣ/,~~[g]~~)		W	L	W	L
(/ɣ/,[ɣ])~(/ɣ/,~~[k]~~)	W	W	L		L
(/ɣ/,[ɣ])~(/ɣ/,~~[x]~~)	W				L

The loser [ɡ] will (almost) never be chosen, because the constraint NoVoicedStop is winner-preferring in the corresponding first ERC in (8), starts ranked at the top, and is never demoted (because never loser-preferring). The choice of the current loser thus effectively boils down to [k] and [x]. The markedness constraints start out above the faithfulness constraints. That ranking configuration is preserved for a large initial portion of the run. Throughout that portion, the choice of the current loser is thus completely determined by the markedness constraints. Since [k] is unmarked, the classical subroutine for the choice of the loser always chooses [k]. Unfortunately, the corresponding second ERC in (8) promotes the two faithfulness constraints Ident[Voice] and Ident[Cont] on a par. If the EDRA stubbornly always chooses [k] as the current loser, it will always promote the two faithfulness constraints on a par until they reach the top of the ranking, thus failing at learning the target ranking (6b) which instead requires Ident[Voice] to be ranked above Ident[Cont].

We thus replace the classical subroutine with the *new subroutine* described below in pseudo-code. Here, we consider an arbitrary underlying form x (while the EDRA model always chooses x equal to the winner form y). For a related proposal, see (Riggle, 2004). Three remarks are in order. First,

---

Require: a current winner (x, y) candidate:

1: construct the ERC matrix corresponding to the comparisons $(x, y) \sim (x, \not{z})$ for all possible loser candidates $\not{z}$ for the underlying form x

2: split any ERC with multiple L's into multiple ERCs with a single L;

3: determine the smallest number $\hat{n}$ such that there exists an ERC with $\hat{n}$ winner-preferrers which is inconsistent with the current ranking values;

4: pick at random among the inconsistent ERCs with $\hat{n}$ winner-preferrers.

---

the new subroutine satisfies condition (4): if there exists a loser which is able to prompt an update, the new subroutine will return one such loser, as it only searches among losers whose corresponding ERC is inconsistent with the current ranking values. Second, the new subroutine chooses among such losers the one(s) which minimizes the "difference" between the winner and the loser, as measured in terms of the number of winner-preferring constraints which distinguish between them. Third, the new subroutine is computationally less expensive than the classical one, because it circumvents the computation of the predicted optimal candidate.

When the new subroutine for the choice of the loser is deployed on the surface form y = [ɣ] corresponding to the ERC block (8), it prevents the EDRA from choosing the loser [~~ɡ~~], because the corresponding first ERC is already consistent with the current ranking values (through the high ranked winner-preferring markedness constraint NoVoiStop). And it also prevents it from choosing the loser [~~k~~], because the corresponding second ERC has "too many" W's. The EDRA is thus biased towards choosing the loser [~~x~~]. The corresponding third ERC promotes Ident[Voice] but not Ident[Cont], leading to the target ranking (6b). We thus obtain the following

Theorem 1 *When trained on an arbitrary sequence of data sampled from the velar inventory [~~ɡ~~ k ɣ ~~x~~], the EDRA model with the new subroutine for the choice of the loser succeeds at learning the target ranking (6b).*

4 How to analyze the new subroutine

The preceding section has motivated a new subroutine for the choice of the current loser. This section highlights some formal properties of this new subroutine which turn out useful in the analysis of EDRA's restrictiveness. For concreteness, we focus on the velar inventory [ɡ k ɣ ~~x~~], with the analysis (7) recalled in section 2. This analysis requires the voiceless glottal fricative [h] to be licit and the voiced glottal fricative [ɦ] to be instead illicit. To start from the simplest case, we minimize the number of licit forms that the model is trained on, by assuming that the glottal stop [ʔ] is illicit as well (say, because of a dedicated high ranked markedness constraint *[ʔ]). Thus, suppose that the EDRA model is trained on the velar/glottal inventory [ɡ k ɣ ~~x~~ ~~ʔ~~ ~~ɦ~~ h]. At line 3, the model is effectively always fed one of the surface forms [ɡ], [ɣ], and [h], since the only other licit form [k] is unmarked and can thus be ignored. At line 4, the model assumes the corresponding faithful underlying forms. At line 5, the model chooses a current loser form. The loser candidates [ʔ] and [ɦ] can be ignored, because the corresponding constraints *[ʔ] and *[ɦ] are never violated and thus stay ranked at the top—just like NoVoiStop in (8) above. The ERCs corresponding to the other losers are listed in (9).

(9)

	ID[DOR]	IDENT[VOI]	ID[CON]	NODORFRIC	NOVOISTOP	NOVOIFRI	*[h]
(/g/,[g])∼(/g/,~~[k]~~)		W			L		
(/g/,[g])∼(/g/,~~[ɣ]~~)			W	W	L	W	
(/g/,[g])∼(/g/,~~[x]~~)		W	W	W	L		
(/g/,[g])∼(/g/,~~[h]~~)	W	W	W		L		W
(/ɣ/,[ɣ])∼(/ɣ/,~~[g]~~)			W	L	W	L	
(/ɣ/,[ɣ])∼(/ɣ/,~~[k]~~)		W	W	L		L	
(/ɣ/,[ɣ])∼(/ɣ/,~~[x]~~)		W				L	
(/ɣ/,[ɣ])∼(/ɣ/,~~[h]~~)	W	W		L		L	W
(/h/,[h])∼(/h/,~~[g]~~)	W	W	W		W		L
(/h/,[h])∼(/h/,~~[k]~~)	W		W				L
(/h/,[h])∼(/h/,~~[ɣ]~~)	W	W		W		W	L
(/h/,[h])∼/h/,~~[x]~~)	W			W			L

Because of the new subroutine for the choice of the loser, the original ERC matrix (9) can effectively be simplified block-by-block as in (10). To illustrate, consider for instance the top block, corresponding to the surface form [g]. The bottom two ERCs of the block corresponding to the losers ~~[x]~~ and ~~[h]~~ can be ignored: since these losers are too different from the intended winner [g], their corresponding ERCs are entailed by the other ERCs in the block and will therefore never be selected by the new subroutine for the choice of the current loser.

(10)

	ID[DOR]	IDENT[VOI]	ID[CON]	NODORFRIC	NOVOISTP	NOVOIFRI	*[h]
ERC 1		W			L		
ERC 2			W	W	L	W	
ERC 3			W	L	W		
ERC 4			W		W	L	
ERC 5		W	W	L			
ERC 6		W				L	
ERC 7	W	W		L			W
ERC 8	W		W				L
ERC 9	W			W			L

This illustrates an important formal property of the new subroutine: ERCs entailed by other ERCs in the same block (namely ERCs which correspond to too dissimilar losers) can be ignored. That's instead not always the case with the classical subroutine for the choice of the loser, as shown above with the choice of the loser [k] in the case of the ERC matrix (8).

Let S be the first time when the current ranking vector entertained by the EDRA becomes consistent with either ERC 1 *or* ERC 6 in (10)—convergence ensures that such a time S exists. Because of the new subroutine for the choice of the current loser, ERC 2 cannot trigger any update before time S, because ERC 2 belongs to the same block as ERC 1, because ERC 2 has more W's than ERC 1, and because the current ranking vector is never consistent with ERC 1 before time S. Analogously, ERCs 3, 4, 5, and 7 cannot trigger any update before time S. In other words, the run up to time S is determined by ERCs 1, 6, 8 and 9 alone. Consider next ERC 7. Since it has more W's than the other ERCs 3-6 in the same block, it cannot trigger any update until the current ranking values have become consistent with the other ERCs 3-6. In other words, if ERC 7 triggers updates at all in the run considered, it will start triggering updates only late into the run. Thus, let the time T be defined as follows. If ERC 7 triggers at least an update in the run considered, T is the smallest time such that ERC 7 triggers an update between times T and $T+1$; if ERC 7 triggers no updates in the run considered, T is the final time of the run. Of course, $S \leq T$ (before time S, the current ranking values are still inconsistent with ERC 6 and the EDRA is thus forbidden to consider ERC7, which has more W's). In conclusion, a generic run of the EDRA model on the current test case can be split into three stages, as in (11).

(11)

start		S		T		end
	only ERCs 1, 6, 8, 9		all ERCs but 7		all ERCs	

This reasoning illustrates another formal property of the new subroutine for the choice of the loser: that it makes the various ERCs enter the scene in stages, ordered by their complexity, namely by the number of winner-preferring constraints they re-rank. This means in turn that the analysis of a generic run can be split into different stages, with an increasing number of ERCs active at each stage.

This turns out to be very useful for the analysis of EDRA's restrictiveness. In fact, establishing restrictiveness requires a characterization of the final ranking vector entertained by the EDRA at convergence. To obtain that characterization, we start from the initial stage and work towards the end. For each stage, we characterize the ranking vectors the learner can end up with at the end of that stage. Obviously, we have to do that for each ranking vector the learner could end up with at the end of the preceding stage. This logics is illustrated in (12). Suppose that the analysis of the first stage (in between the beginning of the run and time S) concludes that the EDRA can end up with one of two ranking vectors $\boldsymbol{\theta}_1^S$ and $\boldsymbol{\theta}_2^S$ at the time S when that stage ends. The analysis at the second stage (in between times S and T) will then have to be repeated twice, for each of the two ranking vectors viable at time S. And so on.

(12)

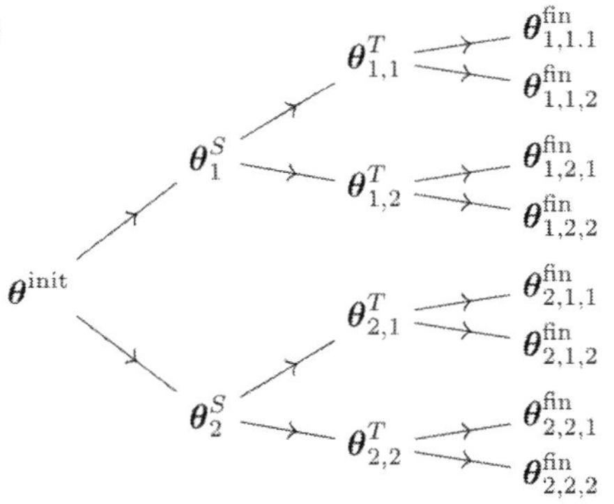

These considerations suggest that we aim for particularly tight analyses of the ranking vectors entertained at the end of the initial stages, in order to avoid a combinatorial explosion of the analyses required at later stages. Of course, tight analyses are readily possible when only a few training ERCs can trigger updates and thus mold the current ranking vector. As we increase the number of training ERCs which trigger updates, the analysis becomes more involved, and the characterization of the stage-final ranking vectors becomes looser. As illustrated in (11), the new subroutine for the choice of the loser thus comes very handy for the analysis of restrictiveness, as it ensures that the EDRA is trained on few ERCs at the beginning of the run, with additional ERCs entering the scene only at later stages.

The final appendix makes these considerations concrete through a detailed analysis of the behavior of the EDRA model with the new subroutine for the choice of the loser trained on the ERC matrix (10) corresponding to the inventory [ɡ k ɣ x ʔ ɦ h]. The resulting analysis establishes the following result.

Theorem 2 *When trained on an arbitrary sequence of data sampled from the glottal/velar inventory* [ɡ k ɣ x ʔ ɦ h]*, the EDRA model with the new subroutine for the choice of the loser succeeds at learning the ranking (7b).*

5 Conclusion

This paper has motivated a new subroutine for the choice of the current loser in phonotactic error-driven learning. Informally, the new subroutine chooses a loser which is as similar as possible to the intended winner, while being able to trigger an update. Similarity in measured in terms of the number of winner-preferring constraints in the corresponding ERC. Crucially, this new subroutine allows the various training ERCs to become active in stages, ordered by their complexity, measured in terms of the number of winner-preferring constraints. This allows for careful restrictiveness guarantees, such as the one provided by theorem 2. The proof of the theorem illustrates a number of techniques for the restrictiveness analysis of the EDRA model with the new subroutine for the choice of the loser.

A Proof of theorem 2

A.1 Analysis at an arbitrary time in the run

The markedness constraint *[h] starts high and the faithfulness constraint IDENT[DOR] starts low. Lemma 1 says that *[h] can never drop by more than 5/4 underneath IDENT[DOR]. This follows from the fact that only ERCs 7, 8, and 9 in (10) re-rank these two constraints. And that ERC 7 promotes them in tandem, and thus does not contribute to their separation. While ERCs 8 and 9 cannot demote *[h] a long way underneath IDENT[DOR], as IDENT[DOR] is winner-preferring in both ERCs 8 and 9.

Lemma 1 *The ranking values of the markedness constraint* *[h] *and the faithfulness constraint* IDENT[DOR] *satisfy the following inequality:*

(13) $$\theta^t_{*[\text{h}]} \geq \theta^t_{\text{ID[DOR]}} - \frac{5}{4}$$

at any time t in any run.

Proof. The proof is by induction on time t. The inequality (13) trivially holds at the initial time $t = 0$, because of the choice of the initial ranking values

$\theta^{t=0}_{*[h]} = \theta > 0$ and $\theta^{t=0}_{\text{ID[DOR]}} = 0$. Assume that the inequality holds at time t and let me show that it then holds at time $t+1$ as well. If the update between times t and $t+1$ has been triggered by the ERCs 1 through 6, then the inequality holds at time $t+1$ because it held at time t and the two constraints *[h] and IDENT[DOR] have not been re-ranked between times t and $t+1$. If the update between times t and $t+1$ has been triggered by the ERC 7, then the inequality holds at time $t+1$ because it held at time t and both constraints *[h] and IDENT[DOR] have been promoted by the same amount in between times t and $t+1$. Finally, if the update between times t and $t+1$ has been triggered by the ERCs 8 or 9, then the inequality holds at time $t+1$ because of the following chain of inequalities.

(14) $$\theta^{t+1}_{*[h]} \overset{(a)}{=} \theta^{t}_{*[h]} - 1 \overset{(b)}{\geq} \theta^{t}_{\text{ID[DOR]}} - 1 \overset{(c)}{=} \theta^{t+1}_{\text{ID[DOR]}} - \frac{1}{4} - 1$$

At step (14a), I have used the fact that the update by ERCs 8 or 9 in between times t and $t+1$ has demoted the constraint *[h] by 1, according to the re-ranking rule (3b). At step (14b), I have used the fact that, in order for ERCs 8 or 9 to have been able to trigger an update in between times t and $t+1$, the current ranking value $\theta^{t}_{*[h]}$ of the loser-preferring constraint *[h] must have been larger than or equal to the ranking value $\theta^{t}_{\text{ID[DOR]}}$ of the winner-preferring constraint IDENT[DOR]. Finally at step (14c), I have used the fact that the update by ERCs 8 or 9 in between times t and $t+1$ has promoted the constraint IDENT[DOR] by $1/4$, as that ERC has $w=3$ winner-preferring constraints and the re-ranking rule (3a) set the promotion amount equal to $\frac{1}{w+1}$. ■

If ERCs 8 and 9 were to trigger lots of updates, IDENT[DOR] would be promoted a lot and *[h] would be demoted a lot. In the end, *[h] would thus find itself underneath IDENT[DOR] separated by a large distance. But lemma 1 says that is impossible. Hence, ERCs 8 and 9 can never trigger too many updates, as stated by lemma 2.

Lemma 2 *The numbers α^t_8 and α^t_9 of updates triggered by ERCs 8 and 9 up to an arbitrary time t in an arbitrary run can be bound as follows:*

(15) $$\alpha^t_8 + \alpha^t_9 \leq \frac{4}{5}\theta + 1$$

where θ is the initial ranking value of the markedness constraints.

Proof. The ranking values $\theta^{t}_{*[h]}$ and $\theta^{t}_{\text{ID[DOR]}}$ of the constraints *[h] and IDENT[DOR] at an arbitrary time t can be expressed as follows in terms of the numbers of updates $\alpha^t_7, \alpha^t_8, \alpha^t_9$ triggered by the ERCs 7, 8, and 9 up to time t.

(16) a. $$\theta^{t}_{*[h]} = \theta + \frac{1}{5}\alpha^t_7 - \alpha^t_8 - \alpha^t_9$$

b. $$\theta^{t}_{\text{ID[DOR]}} = \frac{1}{5}\alpha^t_7 + \frac{1}{4}\alpha^t_8 + \frac{1}{4}\alpha^t_9$$

The inequality (15) follows by plugging the expressions (16a) and (16b) into (13). ■

A.2 Analysis up to time T

Recall from subsection 4 that time T is the smallest time such that ERC 7 triggers an update between times T and $T+1$ (or the time when the run ends, in case ERC 7 triggers no updates). Constraint IDENT[DOR] is only promoted by ERCs 8 and 9 up to time T (ERC 7 triggers no updates before time T). Since these two ERCs cannot trigger too many updates by lemma 2, IDENT[DOR] cannot raise too high up to time T, as stated by the following lemma.

Lemma 3 *The ranking value of the faithfulness constraint* IDENT[DOR] *satisfies*

(17) $$\theta^{t}_{\text{ID[DOR]}} \leq \frac{1}{5}\theta + \frac{1}{4}$$

at an arbitrary time $t \leq T$.

Proof. The faithfulness constraint IDENT[DOR] is only promoted by ERCs 8 and 9 up to time T (ERC 7 triggers no updates before time T). The ranking value of IDENT[DOR] at an arbitrary time $t \leq T$ can then be expressed as follows in terms of the numbers α^t_8, α^t_9 of updates triggered by ERCs 8 and 9 up to time t.

(18) $$\theta^{t}_{\text{ID[DOR]}} = \frac{1}{4}(\alpha^t_8 + \alpha^t_9)$$

Plugging (15) into (18) yields (17). ■

The following lemma says that the markedness constraint NODORFRIC cannot have dropped too much before time T. This follows from the fact that only ERCs 2 and 5 demote NODORFRIC up to time T (ERC 7 has not triggered any update yet). In order for NODORFRIC to have been demoted a long way, these two ERCs must have triggered many updates. Yet, the faithfulness constraint IDENT[CON] is winner-preferring in both ERCs and is thus promoted by each update they trigger. These ERCs thus cannot trigger too many updates, because they cannot demote NODORFRIC a long way underneath IDENT[CON].

Lemma 4 *The ranking value of the markedness constraint* NODORFRIC *satisfies*

$$(19) \quad \theta^t_{\text{NoDorFric}} > \frac{1}{5}\theta + \frac{1}{4}$$

at an arbitrary time $t \leq T$.

Proof. Suppose by contradiction that the claim is false. This means that there exists some time $t < T$ such that the markedness constraint NODORFRIC is demoted in between times $t-1$ and t and its ranking value at time t is smaller than or equal to the forbidden threshold $\theta/5 + 1/4$. Since constraints are demoted by 1, its ranking value at the time $t-1$ preceding the update must have been already smaller than or equal to $\theta/5 + 1/4 + 1$, as stated in (20a). Only ERCs 3 and 5 can have triggered this demotion (ERC 7 triggers no updates before time T). Crucially, the constraint IDENT[CONT] is winner-preferring relative to both ERCs 3 and 5. In order for either ERC 3 or 5 to have been able to demote NODORFRIC in between times $t-1$ and t, the ranking value of the winner-preferring constraint IDENT[CONT] at time $t-1$ must thus have been smaller than or equal to the ranking value of the loser-preferring constraint NODORFRIC at time $t-1$, as stated in (20b). The rest of the proof derives a contradiction from these two inequalities (20).

$$(20) \quad \text{a.} \quad \theta^{t-1}_{\text{NoDorFric}} \leq \frac{1}{5}\theta + \frac{1}{4} + 1$$

$$\text{b.} \quad \theta^{t-1}_{\text{NoDorFric}} \geq \theta^{t-1}_{\text{Id[Con]}}$$

The ranking value of the markedness constraint NODORFRIC at time $t-1$ can be lower bounded as $\theta^{t-1}_{\text{NoDorFric}} \geq \theta - \alpha^{t-1}_3 - \alpha^{t-1}_5$, by only considering the contribution of the ERCs 3 and 5 which demote it, while ignoring the contribution of the ERCs 2 and 9 which promote it. Plugging this bound into (20a) yields the following bound on the number of updates triggered by ERCs 3 and 5 up to time $t-1$.

$$(21) \quad \alpha^{t-1}_3 + \alpha^{t-1}_5 \geq \frac{4}{5}\theta - \frac{5}{4}$$

The ranking value of the faithfulness constraint IDENT[CON] at time $t-1$ can be lower bounded as $\theta^{t-1}_{\text{Id[Con]}} \geq \frac{1}{3}\alpha^{t-1}_3 + \frac{1}{3}\alpha^{t-1}_5$, by only considering the contribution of ERCs 3 and 5. Using the bound (21) on the number of updates triggered by ERCs 3 an 5, we obtain the following lower bound on the ranking value of IDENT[CON].

$$(22) \quad \theta^{t-1}_{\text{Id[Con]}} \geq \frac{4}{15}\theta - \frac{5}{12}$$

The inequalities (20a), (20b), and (22) are contradictory (provided θ is large), because they require the ranking value $\theta^{t-1}_{\text{NoDorFric}}$ to be smaller than $\frac{1}{5}\theta + \frac{5}{4} \simeq \frac{3}{15}\theta$ but larger than $\frac{4}{15}\theta - \frac{5}{12} \simeq \frac{4}{15}\theta$. ■

A.3 Analysis at time S

Recall that time S is the first time when the current ranking vector becomes consistent with either ERC 1 or ERC 6. Because of the new subroutine for the choice of the loser, the run up to time S is determined by ERCs 1, 6, 8, and 9, as noted above. Since the faithfulness constraint IDENT[VOICE] is the only winner-preferring constraint in both ERCs 1 and 6, it must raise a long way in order for the current ranking vector to become consistent with either ERC 1 or ERC 6 at time S, as stated by lemma 5.

Lemma 5 *The ranking value of the faithfulness constraint* IDENT[VOICE] *satisfies the following inequality at time* S*:*

$$(23) \quad \theta^S_{\text{Id[Voi]}} \geq \frac{1}{3}\theta$$

Proof. For concreteness, suppose it is ERC 1 which becomes consistent with the current ranking vector at time S (the reasoning is identical if it is ERC 6 instead). This means that the ranking value of the winner-preferring constraint IDENT[VOICE] is larger than the ranking value of the loser-preferring constraint NOVOICEDSTOP at time S, as stated by the following inequality.

$$(24) \quad \theta^S_{\text{Id[Voi]}} \geq \theta^S_{\text{NoVoiStop}}$$

By definition of time S, only ERCs 1 and 6 have promoted IDENT[VOICE] up to time S and only ERC 1 has demoted NOVOISTOP. Their ranking values can thus be expressed as follows.

$$(25) \quad \text{a.} \quad \theta^S_{\text{Id[Voi]}} = \frac{1}{2}\alpha^S_1 + \frac{1}{2}\alpha^S_6$$

$$\text{b.} \quad \theta^S_{\text{NoVoiStop}} = \theta - \alpha^S_1$$

Plugging (25a) and (25b) into (24) yields the following bound on the number of updates triggered by ERC 1 up to time S.

(26) $\alpha_1^S \geq \frac{2}{3}\theta - \frac{1}{3}\alpha_6^S.$

The following chain of inequalities then yields the desired bound on the ranking value of IDENT[VOICE] at time S.

(27) $$\theta^S_{\text{Id[voi]}} \overset{(a)}{=} \frac{1}{2}\alpha_1^S + \frac{1}{2}\alpha_6^S \overset{(b)}{\geq} \frac{1}{3}\theta - \frac{1}{6}\alpha_6^S + \frac{1}{2}\alpha_6^S \overset{(c)}{\geq} \frac{1}{3}\theta$$

At step (27a), I have used the expression (25a) of the ranking value of IDENT[VOICE]. At step 27b), I have used the bound (26) on α_1^S. At step (27c), I have lower bounded by getting rid of the contribution of α_6^S, which is crucially multiplied by a positive coefficient. ■

A.4 Analysis after time S

The faithfulness constraint IDENT[DOR] is only promoted by ERCs 7, 8, and 9. The latter two ERCs 8 and 9 promote IDENT[DOR] and not IDENT[VOICE]. Yet, they can only trigger few updates by lemma 2, and thus cannot give a substantial advantage to the former constraint over the latter. Furthermore, ERC 7 promotes both IDENT[DOR] and IDENT[VOI], and thus does not give the former any advantage over the latter. The following lemma thus concludes that IDENT[DOR] will never be able to surpass IDENT[VOICE], which already sits high at time S by lemma 5.

Lemma 6 *The ranking values of the faithfulness constraints* IDENT[VOICE] *and* IDENT[DOR] *satisfy the following inequality at any time time $t \geq S$:*

(28) $\theta^t_{\text{ID[VOI]}} \geq \theta^t_{\text{ID[DOR]}} + 2$

Proof. Suppose by contradiction that (28) fails at some time $t \geq S$, as stated in (29).

(29) $\theta^t_{\text{ID[DOR]}} > \theta^t_{\text{ID[VOI]}} - 2$

From now on, let $\alpha_i^{S,t}$ denote the number of updates triggered by the ith ERC in between times S and t. Thus, $\alpha_i^t = \alpha_i^S + \alpha_i^{S,t}$. The ranking value of the faithfulness constraint IDENT[DOR] at time t can be expressed as in (30). At step (30a), I have used the fact that this constraint is promoted only by ERCs 7, 8, and 9. At step (30b), I have used the fact that ERC 7 triggers no updates before time T and thus also no updates before time S (because $S \leq T$), so that $\alpha_7^S = 0$ and thus $\alpha_7^t = \alpha_7^{S,t}$.

(30) $$\theta^t_{\text{ID[DOR]}} \overset{(a)}{=} \frac{1}{4}\alpha_8^t + \frac{1}{4}\alpha_9^t + \frac{1}{5}\alpha_7^t \overset{(b)}{=} \frac{1}{4}\alpha_8^t + \frac{1}{4}\alpha_9^t + \frac{1}{5}\alpha_7^{S,t}$$

The ranking of the faithfulness constraint IDENT[VOICE] at time t can be expressed as in (31). At step (31a), I have expressed the ranking value at time $t \geq S$ as the ranking value at time S plus the increment in the ranking value due to the promotions between times S and t. At step (31b), I have lowered bounded the ranking value of IDENT[VOICE] at time S using (23).

(31) $$\begin{aligned} \theta^t_{\text{ID[VOI]}} &= \\ &\overset{(a)}{=} \theta^S_{\text{ID[VOI]}} + \frac{1}{2}\alpha_1^{S,t} + \frac{1}{3}\alpha_5^{S,t} + \frac{1}{2}\alpha_6^{S,t} + \frac{1}{5}\alpha_7^{S,t} \\ &\geq \theta^S_{\text{ID[VOI]}} + \frac{1}{5}\alpha_7^{S,t} \\ &\overset{(b)}{\geq} \frac{1}{3}\theta + \frac{1}{5}\alpha_7^{S,t} \end{aligned}$$

Plugging (30) and (31) into (29) yields $\alpha_8^t + \alpha_9^t > \frac{4}{3}\theta - 2$, which contradicts (15). ■

A.5 An auxiliary result

The next step in the analysis (namely, the proof of lemma 7 below) rests on theorem 3 (Magri, 2014b).

Theorem 3 *Consider an arbitrary run of the EDRA with the re-ranking rule (3). Assume that each training ERC has a unique* L. *Focus on a specific training ERC, say the $\bar{\imath}$th one. Let C_ℓ be its unique loser-preferring constraint and let C_h be one of its (possibly many) winner-preferring constraints, as in (32).*

(32) $$\begin{array}{lccccccc} & & \dots & C_h & \dots & C_\ell & \dots & \\ \bar{\imath}\textit{th ERC} = [& & \dots & \text{W} & \dots & \text{L} & \dots &] \end{array}$$

Define the coefficient δ_i as follows:

(33) $$\delta_i = \begin{cases} & \quad\quad\quad\quad\quad \dots \quad C_h \quad \dots \quad C_\ell \quad \dots \\ \frac{1}{w_i+1} & \textit{if ith ERC} = [\dots e/\text{W}/\not{\text{L}} \dots \quad \text{W} \quad \dots] \\ \frac{w_i+2}{w_i+1} & \textit{if ith ERC} = [\dots \quad \text{L} \quad \dots \quad \text{W} \quad \dots] \\ 1 & \textit{if ith ERC} = [\dots \quad \text{L} \quad \dots \text{L}/e/\not{\text{W}} \dots] \\ 0 & \textit{otherwise} \end{cases}$$

The number of updates $\alpha_{\bar{\imath}}$ triggered by the $\bar{\imath}$th input ERC is either null or else bounded as follows

$$(34)\ \alpha_{\bar{\imath}} \leq \frac{w_{\bar{\imath}}+1}{w_{\bar{\imath}}+2}\Big(\underbrace{\theta_\ell^{\text{init}} - \theta_h^{\text{init}}}_{(a)} + \underbrace{1}_{(b)} + \underbrace{\sum_i \delta_i \cdot \alpha_i}_{(c)}\Big)$$

where $\theta_\ell^{\text{init}}$ and θ_h^{init} are the initial raking values of the two constraints C_ℓ and C_h; α_i is the number of updates triggered by the ith training ERC; w_i is the number of its winner-preferring constraints; the sum in (34c) runs over all training ERCs.

Here is the intuitive idea. Suppose that the initial ranking value $\theta_\ell^{\text{init}}$ of the loser-preferrer C_ℓ is larger than the initial ranking value θ_h^{init} of the winner-preferrer C_h. A certain number of updates by the $\bar{\imath}$th ERC are thus justified just in order to compensate for this bad choice of the initial ranking values, as quantified by the term (34a). At that point, the two constraints could in principle have exactly the same ranking values. An additional update is thus justified in order to bring the winner-preferring constraint C_h above the loser-preferrer C_ℓ, yielding the term (34b). Further updates by this $\bar{\imath}$th ERC are only justified if this ranking configuration $C_h \gg C_\ell$ is disrupted by updates triggered by some other training ERCs, as quantified by the term (34c). This term sums the number of updates α_i triggered by the generic ith training ERC multiplied by the "amount of disruption" δ_i caused by that ERC to the ranking configuration $C_h \gg C_\ell$. For instance, suppose that the ith training ERC looks like the top ERC listed in (33). The amount δ_i of disruption caused by that ERC is $\delta_i = \frac{1}{w_i+1}$, because that ERC disrupts the ranking configuration $C_h \gg C_\ell$ by promoting C_ℓ by $\frac{1}{w_i+1}$.

A.6 Analysis after time T

Lemma 7 says that IDENT[CONT] is always ranked above IDENT[DOR] after time T, with a sufficient distance in between the two faithfulness constraints (at least 2). The proof of this lemma is more involved than the proof of the preceding lemmas. The difficulty is due to the fact that ERC 2 promotes IDENT[CONT] at the exclusion of IDENT[DOR] while ERC 7 promotes IDENT[DOR] at the exclusion of IDENT[CONT]. In order to compare the ranking values of these two faithfulness constraints, we thus need some connection between the numbers of updates α_2^t and α_7^t triggered by the two ERCs 2 and 7. What allows this connection to be established is the fact that the constraint NODORFRIC is winner-preferring in ERC 2 but loser-preferring in ERC 7. Since ERC 2 thus promotes the constraint NODORFRIC which ERC 7 tries to demote, updates by ERC 2 "buy" extra updates by ERC 7. If ERC 2 happens to trigger few updates (and thus to contribute little to the height of IDENT[CONT]), then it will buy only few updates by ERC 7 (which will therefore contribute little to the height of IDENT[DOR]). Theorem 3 is used to formalize this intuition, yielding the link between α_2^t and α_7^t in (38).

Lemma 7 *Suppose that in the run considered, ERC 7 does trigger at least an update. The ranking values of the faithfulness constraints* IDENT[CON] *and* IDENT[DOR] *satisfy the following inequality:*

$$(35)\qquad \theta^t_{\text{ID[CON]}} \geq \theta^t_{\text{ID[DOR]}} + 2$$

at any time time $t \geq T$.

Proof. Since ERC 7 triggers an update between times T and $T+1$, the loser-preferring constraint NODORFRIC cannot be underneath the winner-preferring constraint IDENT[VOICE] at time T, as stated in (36a). Furthermore, the current ranking vector at time T must be consistent with ERC 5 (otherwise, the algorithm would have chosen ERC 5 instead of ERC 7, as the former has less W's). This means that the loser-preferring constraint NODORFRIC is already underneath IDENT[CON] at time T, as stated in (36b).

(36) a. $\theta^T_{\text{NoDorFric}} \geq \theta^T_{\text{ID[Voi]}}$

b. $\theta^T_{\text{ID[Con]}} > \theta^T_{\text{NoDorFric}}$

The chain of inequalities in (37) thus holds. In step (37a), I have used (36). In step (37b), I have used the fact that $T \geq S$ and that the ranking values of the faithfulness constraints can only grow with time (because they are never demoted). In step (37c), I have used the inequality (23).

$$(37)\qquad \theta^T_{\text{ID[Con]}} \overset{(a)}{>} \theta^T_{\text{ID[Voi]}} \overset{(b)}{\geq} \theta^S_{\text{ID[Voi]}} \overset{(c)}{\geq} \frac{1}{3}\theta$$

Since ERC 7 triggers no updates before time T, the number α_7^t of updates it has triggered up to time t is equal to the number $\alpha_7^{T,t}$ of updates it has triggered between times T and t, as stated in (38a). Applying theorem 3 to ERC 7 pivoting on its winner-preferring constraint IDENT[DOR] and considering time T as the initial time yields the inequality (38b). In step (38c), I have used (36b) together with the fact that $\theta^T_{\text{ID[DOR]}} \geq 0$.

(38)
$$\alpha_7^t \overset{(a)}{=} \alpha_7^{T,t} \leq$$
$$\overset{(b)}{\leq} \frac{5}{6}\left(\theta^T_{\text{NODORFRIC}} - \theta^T_{\text{ID[DOR]}}\right) + 1 + \frac{5}{24}\left(\alpha_2^{T,t} + \alpha_9^{T,t}\right)$$
$$\overset{(c)}{\leq} \frac{5}{6}\theta^T_{\text{ID[CON]}} + 1 + \frac{5}{24}\left(\alpha_2^{T,t} + \alpha_9^{T,t}\right)$$

The inequality (39) completes the proof.

(39)
$$\theta^t_{\text{ID[DOR]}} =$$
$$\overset{(a)}{=} \frac{1}{4}(\alpha_8^t + \alpha_9^t) + \frac{1}{5}\alpha_7^t$$
$$\overset{(b)}{\leq} \underbrace{\frac{1}{4}(\alpha_8^t + \alpha_9^t)} + \frac{1}{6}\theta^T_{\text{ID[CON]}} + \frac{1}{5} + \frac{1}{24}\alpha_2^{T,t} + \underbrace{\frac{1}{24}\alpha_9^{T,t}}$$
$$\overset{(c)}{\leq} \left(\frac{1}{4} + \frac{1}{24}\right)(\alpha_8^t + \alpha_9^t) + \frac{1}{6}\theta^T_{\text{ID[CON]}} + \frac{1}{5} + \frac{1}{24}\alpha_2^{T,t}$$
$$\overset{(d)}{\leq} \frac{7}{24}\left(\frac{4}{5}\theta + 1\right) + \frac{1}{6}\theta^T_{\text{ID[CON]}} + \frac{1}{5} + \frac{1}{24}\alpha_2^{T,t}$$
$$\overset{(e)}{\leq} \frac{7}{24}\left(\frac{4}{5}3\theta^T_{\text{ID[CON]}} + 1\right) + \frac{1}{6}\theta^T_{\text{ID[CON]}} + \frac{1}{5} + \frac{1}{24}\alpha_2^{T,t}$$
$$= \frac{7}{10}\theta^T_{\text{ID[CON]}} + \frac{1}{6}\theta^T_{\text{ID[CON]}} + \left(\frac{1}{5} + \frac{7}{24}\right) + \frac{1}{24}\alpha_2^{T,t}$$
$$= \theta^T_{\text{ID[CON]}} - \frac{8}{60}\theta^T_{\text{ID[CON]}} + \left(\frac{1}{5} + \frac{7}{24}\right) + \frac{1}{24}\alpha_2^{T,t}$$
$$\overset{(f)}{\leq} \theta^T_{\text{ID[CON]}} \underbrace{- \frac{8}{60}3\theta + \left(\frac{1}{5} + \frac{7}{24}\right)} + \frac{1}{24}\alpha_2^{T,t}$$
$$\overset{(g)}{\leq} \theta^T_{\text{ID[CON]}} - 2 + \frac{1}{24}\alpha_2^{T,t} \overset{(h)}{\leq} \theta^t_{\text{ID[CON]}} - 2$$

The ranking value of the faithfulness constraint IDENT[DOR] can be expressed as in (39a) in terms of the contribution of the three ERCs 7, 8, and 9 which promote it. At step (39b), I have upper bounded the number α_7^t of updates triggered by ERC 7 through (38). At step (c), I have upper bounded the sum of the two terms marked with a wiggly line with $\left(\frac{1}{4} + \frac{1}{24}\right)(\alpha_8^t + \alpha_9^t)$. At step (39d), I have upper bounded $\alpha_8^t + \alpha_9^t$ using (15). At steps (39e) and (39f), I have used the inequality $\theta < 3\theta^T_{\text{ID[COR]}}$ obtained in (37). At step (39g), I have used the fact that the quantity marked by the wiggly line is smaller than -2, provided the initial ranking value θ is large enough. Finally at step (39h), I have used the fact that $\theta^t_{\text{ID[CON]}} \geq \theta^T_{\text{ID[CON]}} + \frac{1}{4}\alpha_2^{T,t}$, because ERC 2 promotes IDENT[CONT] by $1/4$. ■

A.7 Putting the pieces together

It is easy to see that, if some rankings values are consistent with the ERC matrix (9b) and furthermore the ranking values of the two constraints sc NoDorFric and IDENT[DOR] satisfy the strict inequality (40), then each of the rankings consistent with those ranking values neutralizes [x] and thus generate the target inventory [g k ɣ x̶].

(40) $\theta_{\text{NODORFRIC}} > \theta_{\text{ID[DOR]}}$

In order to guarantee that the EDRA model succeeds at learning this inventory and thus complete the proof of the theorem, it thus suffices to prove that its final ranking values satisfy this inequality (40), as convergence ensures consistency with the training ERC matrix (9b). To this end, we repeat below some of the inequalities which have been obtained with the preceding lemmas.

(17) $\theta^t_{\text{ID[DOR]}} \leq \frac{1}{5}\theta + \frac{1}{4}$ for any $t \leq T$ (lemma 3)

(19) $\theta^t_{\text{NODORFRIC}} > \frac{1}{5}\theta + \frac{1}{4}$ for any $t \leq T$ (lemma 4)

(28) $\theta^t_{\text{ID[VOI]}} \geq \theta^t_{\text{ID[DOR]}} + 2$ for any $t \geq S$ (lemma 6)

(35) $\theta^t_{\text{ID[CON]}} \geq \theta^t_{\text{ID[DOR]}} + 2$ for any $t \geq T$ (lemma 7)

The two inequalities (17) and (19) guarantee that the markedness constraint NODORFRIC is indeed ranked above IDENT[DOR] up to time T. In other words, that the inequality (40) holds up to time T. If ERC 7 never triggers any update in the run considered, time T is the final time of the run, and the proof is completed. Thus, suppose that T is not the end of the run. The inequalities (28) and (35) ensure that the two faithfulness constraints IDENT[VOICE] and IDENT[CONT] are always above IDENT[DOR] from time T on, with enough space in between (namely at least 2). Since NODORFRIC is ranked above IDENT[DOR] at time T, it can never demoted below it, because any ERC where it is loser-preferring counts at least one of the two constraints IDENT[VOICE] or IDENT[CONT] as winner-preferring.

References

Paul Boersma. 1998. *Functional Phonology*. Ph.D. thesis, University of Amsterdam, The Netherlands. The Hague: Holland Academic Graphics.

Paul Boersma. 2009. Some correct error-driven versions of the constraint demotion algorithm. *Linguistic Inquiry*, 40:667–686.

Noam Chomsky and Morris Halle. 1965. Some controversial questions in phonological theory. *Journal of Linguistics*, 1:97–138.

Bruce Hayes and Donca Steriade. 2004. Introduction: the phonetic bases of phonological markedness. In Bruce Hayes, Robert Kirchner, and Donca Steriade, editors, *Phonetically based phonology*, pages 1–33. Cambridge University Press.

Ian Maddieson. 1984. *Patterns of Sounds*. Cambridge University Press.

Giorgio Magri and René Kager. 2015. How to account for phonetically counterintuitive segment inventories using only phonetically grounded markedness constraints. In Thuy Bui and Deniz Ozyildiz, editors, *Proceedings of NELS 45*, Amherst, MA. GLSA Publications.

Giorgio Magri. 2012. Convergence of error-driven ranking algorithms. *Phonology*, 29(2):213–269.

Giorgio Magri. 2013a. The error-driven ranking model of the early stage of the acquisition of phonotactics: an initial result on restrictiveness. In Hsin-Lun Huang, Ethan Poole, and Amanda Rysling, editors, *Proceedings of NELS 43: the 43rd annual meeting of the North East Linguistic Society*, pages 277–290.

Giorgio Magri. 2013b. A note on the GLA's choice of the current loser from the perspective of factorizability. *Journal of Logic, Language, and Information*, 22:231–247.

Giorgio Magri. 2014a. The EDRA model of the acquisition of phonotactics: the problem of $\mathcal{F}$-controlling. In Özlem Çetinoglu, Jeffrey Heinz, Andreas Maletti, and Jason Riggle, editors, *Proceedings of MORPHFSM 2014*. Association for Computational Linguistics.

Giorgio Magri. 2014b. Tools for the robust analysis of error-driven ranking algorithms and their implications for modeling the child's acquisition of phonotactics. *Journal of Logic and Computation*, 24.1:135–186.

Giorgio Magri. 2015a. How to control the height of the faithfulness constraints. Ms.

Giorgio Magri. 2015b. Idempotency in constraint-based phonology and the formal underpinning of correspondence theory. Submitted manuscript.

Giorgio Magri. 2015c. Restrictiveness and the relative ranking of the markedness constraints. Ms.

John J. Ohala. 1983. The origin of sound patterns in vocal tract constraints. In P. F. MacNeilage, editor, *The production of speech*, pages 189–216. Springer Verlag, New York.

Alan Prince and Paul Smolensky. 2004. *Optimality Theory: Constraint Interaction in generative grammar*. Blackwell, Oxford. As Technical Report CU-CS-696-93, Department of Computer Science, University of Colorado at Boulder, and Technical Report TR-2, Rutgers Center for Cognitive Science, Rutgers University, New Brunswick, NJ, April 1993. Also available as ROA 537 version.

Alan Prince. 2002. Entailed ranking arguments. Ms., Rutgers University, New Brunswick, NJ. Rutgers Optimality Archive, ROA 500. Available at http://www.roa.rutgers.edu.

Jason Riggle. 2004. Contenders and learning. In *Proceedings of the 23rd annual meeting of the West Coast Conference on Formal Linguistics*, pages 101–114.

Paul Smolensky. 1996. The initial state and Richness of the Base in Optimality Theory. John Hopkins Technical Report.

Bruce Tesar and Paul Smolensky. 1998. Learnability in Optimality Theory. *Linguistic Inquiry*, 29:229–268.

A Concatenation Operation to Derive Autosegmental Graphs

Adam Jardine and **Jeffrey Heinz**
University of Delaware
{ajardine,heinz}@udel.edu

Abstract

Autosegmental phonology represents words with graph structures. This paper introduces a way of reasoning about autosegmental graphs as strings of concatenated graph primitives. The main result shows that the sets of autosegmental graphs so generated obey two important, putatively universal, constraints in phonological theory provided that the graph primitives also obey these constraints. These constraints are the Obligatory Contour Principle and the No Crossing Constraint. Thus, these constraints can be understood as being derived from a finite basis under concatenation. This contrasts with (and complements) earlier analyses of autosegmental representations, where these constraints were presented as axioms of the grammatical system. Empirically motivated examples are provided.

1 Introduction

Autosegmental phonology represents words with graph structures. This paper provides a new way of defining the set of valid autosegmental representations through concatenating a finite set of graph primitives with particular properties. This 'bottom-up' approach to formalizing autosegmental representations (henceforth APRs) contrasts with the 'top-down', axiomatic approach of previous formalizations of APRs (Goldsmith, 1976; Bird and Klein, 1990; Coleman and Local, 1991; Kornai, 1995). However, we show that APR graphs constructed in the way we define hold to these axioms. One advantage to this perspective is that it brings out the *string-like* quality of APRs, in that they can be generated by the concatenation of a finite set of primitives. Furthermore, it shows that two putatively universal constraints, the Obligatory Contour Principle and the No Crossing Constraint (see below), are guaranteed to hold of autosegmental representations provided the graph primitives also obey these constraints. In other words, concatenation preserves these properties. Finally, the empirical generalization that languages may exhibit unbounded spreading but not unbounded contours is naturally expressed by this finite set of primitives, as spreading is derivable through concatenation but the only available contours are those found in the set of graph primitives. In short, important properties of autosegmental representations of words can be understood as being derived from a finite basis under concatenation.

Goldsmith (1976) originally defined APRs as *graphs*. Likewise, this paper models APRs using graphs representing both the associations and precedence relations of APRs. We apply established graph-theoretic methods to APRs, in particular *graph concatenation*, as defined by Engelfriet and Vereijken (1997). Engelfriet and Vereijken (1997) generate all graphs from concatenation and sum operations and a finite set of primitives. What is proposed here is a much weaker version of this idea, using concatenation only to build a specific class of graphs from a set of primitives. In doing so, it is shown how the properties of structures in the generated class derive from the operation and the primitives.

As detailed in the next section, there are several properties that most researchers agree are essential to APRs. One is that their composite autosegments

Proceedings of the 14th Meeting on the Mathematics of Language (MoL 14), pages 139–151,
Chicago, USA, July 25–26, 2015.

are divided up into disjoint strings called *tiers*, with associations linking autosegments on different tiers. Second, the No-Crossing Constraint (NCC) (Goldsmith, 1976; Hammond, 1988; Coleman and Local, 1991) states that these associations cannot 'cross'; i.e., they must respect the precedence relations on each tier. Finally, the Obligatory Contour Prinicple (OCP) (Leben, 1973) states that on the melody tier adjacent autosegments cannot be identical.

Formal treatments of these properties, starting with Goldsmith (1976), state these properties as axioms. For example, Bird and Klein (1990) provide a model-theoretic definition of APRs given a particular interpretation of association as overlap, and state axioms restricting the overlap relation. More recently, Jardine (2014) axiomatizes the NCC and one-to-one association in monadic-second order logic. Kornai (1995)'s treatment defines concatenation operations similar to the one given here, but his definition of APRs as *bistrings* does not derive from these operations. As a result, key properties like the NCC must be specified as axioms.

Instead, the current paper shows that the NCC and OCP can be derived by a concatenation operation alone, given a well-defined set of primitives. This paper is structured as follows. §2 details the set of properties phonologists deem important for APRs. §3 gives the relevant mathematical preliminaries, and §4 defines APRs as graphs and how the properties in §2 can be formalized as axioms. §5 defines a concatenation operation over graphs, and §6 proves how APR graphs derived using this concatenation operation obey the relevant axioms from §4. §7 then shows how to describe some common natural language phenomena using concatenation, as well as some phenomena that raise issues for concatenation. §8 reviews the advantages of viewing APRs through concatenation and discusses future work, and §9 concludes.

2 Basics of Autosegmental Phonology

Autosegmental phonology (AP) (Goldsmith, 1976; Goldsmith, 1979; Clements, 1976; McCarthy, 1979; McCarthy, 1985) has been a widely adopted theory of phonological representations in which phonological units, called *autosegments*, appear on one of some finite set of strings, or *tiers*, and related to autosegments on other tiers by *association*. Such *autosegmental representations* (APRs) are usually depicted with the tiers as vertically separated strings of symbols and the association relation shown as lines drawn between autosegments, as in (1) below.

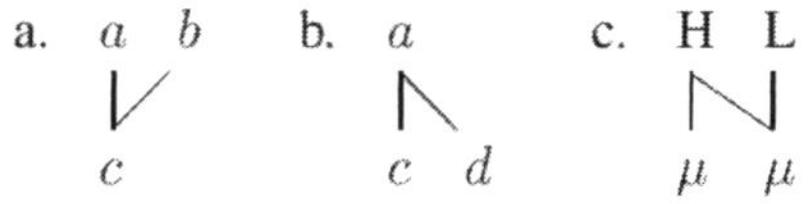

(1)

The core insight APRs express is that a single autosegment on one tier may be associated to multiple autosegments on another tier, as in (1). For purposes of exposition, this paper focuses on two-tiered APRs: a *melody tier*, which carries featural information, and a *timing tier*, which represents how features on the melody tier are pronounced in the linear speech stream. For example, in tonal phonology, APRs often comprise a melody tier over the symbols $\{\text{H}, \text{L}\}$ for high and low tones and a timing tier over $\{\mu\}$ for morae (the timing unit most commonly associated with tone). The APR in (1c) thus represents a high-toned mora followed by a falling tone mora.

Thus, the insights of autosegmental phonology can be studied minimally with two-tier APRs, and so this paper focuses on two-tier APRs. However, in practice, APRs often use more than two tiers. As we explain at the appropriate points throughout the paper, the concepts discussed here can be straightforwardly applied to AP graphs with multiple tiers.

Two principles have been seen as crucial to constraining the theory of APRs. One is the No Crossing Constraint (NCC) (Goldsmith, 1976; Hammond, 1988; Coleman and Local, 1991), which states that if autosegment a is associated to autosegment y, no autosegment b which follows a on its tier may be associated to an autosegment x which precedes y. An example APR violating the NCC is given in (2a). The other principle is the Obligatory Contour Principle (OCP), which states that on each tier, adjacent autosegments must be different (Leben, 1973; McCarthy, 1986). The APR in (2b) violates the OCP.

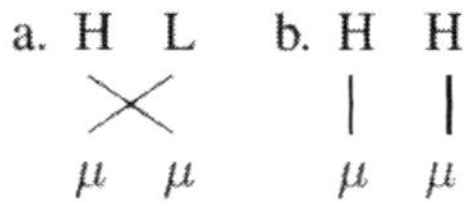

(2)

Formal definitions of the NCC and OCP will be given in the following section, after we have defined

APRs explicitly in terms of graphs. The NCC is usually considered to be inviolable, where the OCP is considered violable by some authors (Odden, 1986). This paper treats the OCP as an inviolable principle, although this point is returned to in §8.

It is often, but not always, assumed that the sets of autosegments which are allowed to appear on each tier are disjoint. This assumption is usually adhered to in tonal and featural APRs, but not always in morphological APRs in which separate tiers represent separate morphemes (a la McCarthy (1979)). Here, we assume that the sets of elements allowed to appear on each tier are disjoint, and leave theories of APRs which allow a particular autosegment to appear on multiple tiers for future work.

3 Preliminaries

Let $\mathbb{N}$ represent the natural numbers. Given a set X of elements, a *partition* P is a set $\{X_0, X_1, ... X_n\}$ of nonempty subsets or *blocks* of X such that X is the union of these blocks and for each $X_i, X_j \in P$, $X_i \cap X_j = \emptyset$. P induces an equivalence relation $\sim_P$ over X such that for all $x, y \in X$, $x \sim_P y$ iff for some $X_i \in P$, $x \in X_i$ and $y \in X_i$. We also say $\sim_P$ *partitions* X into P. A partition P is said to *refine* another partition P' iff every block of P' is a union of blocks of P. We also say $\sim_P$ is then *finer* than $\sim_{P'}$. If R is a relation on X then let $\sim_R$ denote the *finest equivalence relation on* X *containing* R.

If Σ is a finite alphabet of symbols, then Σ^* denotes the set of all strings over that alphabet, including the empty string λ. We consider here alphabets structured by partitions. We refer to a partition $T = \{T_0, T_1, ..., T_n\}$ of Σ as a *tier partition* over Σ, and refer to some block T_i in T as a *tier alphabet*.

A *labeled mixed graph* is a tuple $\langle V, E, A, \ell \rangle$ where V is a set of *nodes*, E is the set of *undirected* edges, A is the set of *directed* edges (or *arcs*), and $\ell : V \to \Sigma$ is a total labeling function assigning each node in V a label in an alphabet Σ. For elements of the set V we will use early elements in $\mathbb{N}$. An undirected edge is a set $\{x, y\}$ of cardinality 2 of nodes $x, y \in V$, and a directed edge is a 2-tuple (x, y) of nodes in V. When not obvious from context, the elements of a graph G will be marked with subscripts; e.g., V_G. Let G_λ, the *empty graph*, refer to the graph $\langle \emptyset, \emptyset, \emptyset, \emptyset \rangle$.

Unless otherwise noted, all graphs in this paper are labeled mixed graphs, and thus will simply be referred to as *graphs*. All graphs are also assumed to be *simple* graphs without multiple edges; $\{x, y\} \in E$ implies $(x, y) \notin A$, and $(x, y) \in A$ implies $\{x, y\} \notin E$. Let $GR(\Sigma)$ denote the union of $\{G_\lambda\}$ with all graphs whose labels are in Σ.

A graph H is a *subgraph* of a graph G if $V_H \subseteq V_G$, $E_H \subseteq E_G$, $A_H \subseteq A_G$, and $\ell_H \subseteq \ell_G$. A subgraph H of G is an *induced subgraph* if for some subset X of V_G, $V_H = X$ and for all $x, y \in X$, $\{x, y\} \in E_G$ iff $\{x, y\} \in E_H$ and $(x, y) \in A_G$ iff $(x, y) \in A_H$. In other words, H has exactly the edges in G that appear between the nodes in X. We say X *induces* H and also write $G[X]$ for H. By a partition of G we refer to some set $\{G[V_0], G[V_1], ..., G[V_n]\}$ where $\{V_0, V_1, ..., V_n\}$ is a partition of V.

4 APRs as graphs

Here we define *autosegmental graphs* (APGs), or explicit graph representations of APRs. In this section, the set of valid APGs is defined axiomatically based on the phonological principles discussed in §2. In §6.2 we show that these principles can all be derived from graph concatenation. For an APG G, A represents the ordering relation on each tier, and E represents the association relations between them.[1]

We first define the tiers as subgraphs of G that are *string graphs* for which A represents the successor relation (Engelfriet and Hoogeboom, 2001).

Figure 1: A string graph

Let $\preccurlyeq$ be the reflexive, transitive closure of A. That is, for any $x, y \in V$, if $x \preccurlyeq y$ then either $x = y$ or there is a directed path from x to y.

Definition 1 *A graph is a* string graph *if* $E = \emptyset$ *and its relation* $\preccurlyeq$ *is a total order on* V.

[1]It should be noted that linguists often leave the precedence relation on each tier as implicit or otherwise distinct from the model of associations (see Coleman and Local (1991) for an overview). However, with precedence directly in the graph, an APG represents all of the information in an APR, and thus this information can be studied by established graph-theoretic techniques, such as the graph concatenation considered here.

Let $\sim_A$ be the smallest equivalence relation that results from the symmetric closure of $\preccurlyeq$. The first axiom says $\sim_A$ partitions V into two tiers.

Axiom 1 *V is partitioned by $\sim_A$ into at most two sets V_0, V_1 such that $G[V_0]$ and $G[V_1]$ are string graphs. V_0 and V_1 are the* tiers *of G.*

The second axiom, related to Axiom 1, is that the partition of G into tiers respects some partition of Σ.

Axiom 2 *There is some tier partition $T = \{T_0, T_1\}$ over Σ such that ℓ forms a morphism from $\sim_T$ to $\sim_A$ such that $\ell(x) \sim_T \ell(y)$* iff *$x \sim_A y$.*

Axiom 2 corresponds to the principle discussed in §2 that each kind of autosegment may only appear on a particular tier. Note that a tier in G thus corresponds to a tier alphabet in T. For notational brevity, we mark this with matching subscripts; e.g., V_0 is the subset of V s.t. for all $v \in V_0$, $\ell(v) \in T_0$.

Axiom 3 governs the general form of associations.

Axiom 3 *For all $\{x, y\} \in E$, $x \not\sim_A y$.*

This simply states that the undirected edges, which again represent associations, must have one end in each tier. Thus, as noted by Coleman and Local (1991), the set of associations between two tiers in an APG forms a bipartite undirected graph $\langle V, E, \ell\rangle$ where the two parts are the tiers V_0 and V_1.

Having defined the structure of APGs in Axioms 1 through 3, we now define the NCC and OCP.

Axiom 4 (NCC) *For all $u, v, x, y \in V$, if $\{u, x\}, \{v, y\} \in E$ and $u \preccurlyeq v$, then $x \preccurlyeq y$.*

Similar axioms have also been defined by Bird and Klein (1990), Kornai (1995), and Jardine (2014).

Finally, Axiom 5 defines the OCP. Recall that the OCP only holds at the *melodic* level, so we choose only one of the tiers V_m for the OCP to hold.

Axiom 5 (OCP) *For one tier V_m, for all $x, y \in V_m$, $(x, y) \in A$ implies $\ell(x) \neq \ell(y)$.*

This concludes the axioms for APGs. For an alphabet Σ and tier partition $T = \{T_m, T_t\}$ over Σ, let $APG(\Sigma, T)$ denote the *class* of APGs obeying the tier partition T of Σ, where for each $G \in APG(\Sigma, T)$, ℓ maps elements in the tier V_m adhering to Axiom 5 to T_m.[2] §6 shows how to derive these axioms from the concatenation, as defined in the following section, of an alphabet of graph primitives with certain properties.

[2]Note that E is not required to be nonempty; that is, APGs with no association lines are allowed. This can model, for example, underlying APRs with unassociated melodies.

These axioms can be extended to graphs with more than two tiers. Instead of binary partitions, Σ and V could be partitioned into $\{T_0, T_1, ..., T_n\}$ and $\{V_0, V_1, ..., V_n\}$, respectively. In this case, Axiom 3 would specify a *single* tier in which all undirected edges must have one end. Axiom 5 would then hold for all tiers besides this tier. This results in 'paddle-wheel' APRs, like those defined by Pulleyblank (1986). Theories of feature geometry (Archangeli and Pulleyblank, 1994; Clements and Hume, 1995; Sagey, 1986) could also be accommodated for by positing additional structure on T. This, however, shall be left for future work.

5 Concatenation

This section defines a concatenation operation ($\circ$) based on that of Engelfriet and Vereijken (1997). Engelfriet and Vereijken (1997)'s operation merges nodes of graphs with specified beginning and end points; here, we use the tier structure to determine how the graphs are concatenated. We thus define $G_1 \circ G_2$ for two graphs G_1, G_2 in $GR(\Sigma)$ given a tier partition $T = \{T_m, T_t\}$ over Σ. The basic idea is to connect, if they exist, the last node of the first graph and the first node of the second graph for each tier. Such 'end nodes' with identical labels in the T_m tier alphabet are merged, whereas end nodes with labels in the timing tier alphabet and nodes with nonidentical labels in the melody tier alphabet are connected via a directed edge. As shown in §6.2 and §7, it is this 'merging' that derives both the OCP and spreading for APGs constructed this way.

As the concatenation operation is defined over graphs in $GR(\Sigma)$, it is at first very general and not of any phonological interest. However, we show in §6 that concatenation can be used to define a set of APGs that follow the axioms in §4, as shown in §6.2.

5.1 Definition

We assume that G_1 and G_2 are disjoint (i.e., that V_1 and V_2 are disjoint sets)—if G_2 is not disjoint with G_1, then we replace it with a graph isomorphic to G_2 that is disjoint with G_1.

We use two partial functions `first` : $GR(\Sigma) \times$

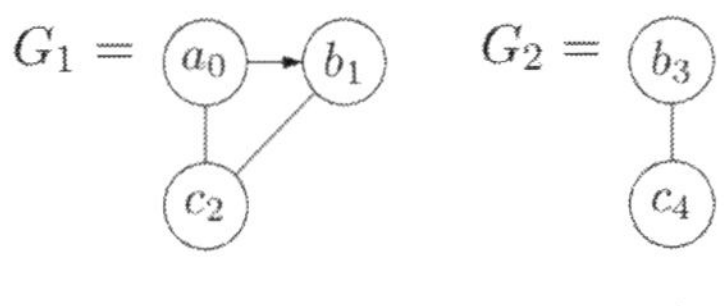

Figure 2: Two graphs in $GR(\Sigma)$

$T \to \mathbb{N}$ and $\texttt{last} : GR(\Sigma) \times T \to \mathbb{N}$ which pick out the first and last nodes on a particular tier in a graph.[3] Recall that V_i is the subset of V s.t. for all $v \in V_i$, $\ell(v) \in T_i$. Formally, $\texttt{first}(G, T_i) \stackrel{\text{def}}{=} v \in V_i$ s.t. $\forall v' \in V_i, v \preccurlyeq v'$ if such a v exists; otherwise it is undefined. Similarly, $\texttt{last}(G, T_i) \stackrel{\text{def}}{=} v \in V_i$ s.t. $\forall v' \in V_i, v' \preccurlyeq v$ if such a v exists; otherwise it is undefined. We shall sometimes refer to $\texttt{first}(G, T_i)$ (resp. $\texttt{last}(G, T_i)$) as the *first (last) node of G for tier alphabet* T_i.

Example 1 *Consider the alphabet* $\Sigma = \{a, b, c\}$ *and the tier partition* $T = \{T_m = \{a, b\}, T_t = \{c\}\}$. *Take graphs* G_1 *and* G_2 *where* $V_1 = \{0, 1, 2\}$ *and* $V_2 = \{3, 4\}$ *with edges and labeling as in Figure 2 Node indices are given as subscripts on the node labels.* $\texttt{last}(G_1, T_m) = 1$, *and* $\texttt{first}(G_2, T_t) = \texttt{last}(G_2, T_t) = 4$.

The concatenation operation combines the graphs, either merging or drawing arcs between the first and last nodes on each tier, depending on their labels. The operation can be broken down into multiple steps as follows. First, we define the graph $G_{1,2}$ as the pairwise union of G_1 and G_2. We denote $V_1 \cup V_2$ with $V_{1,2}$ and so on.

$$G_{1,2} = \langle \underbrace{V_1 \cup V_2}_{V_{1,2}}, \underbrace{E_1 \cup E_2}_{E_{1,2}}, \underbrace{A_1 \cup A_2}_{A_{1,2}}, \underbrace{\ell_1 \cup \ell_2}_{\ell_{1,2}} \rangle \quad (3)$$

Next, two binary relations over the nodes of $G_{1,2}$ are defined. R pairs the last element in G_1 and the first element in G_2 of each tier. R_{ID} is a restriction on R to pairs who share identical labels, excluding nodes whose labels are in T_t.

$$R \stackrel{\text{def}}{=} \{ \; (v, v') \in V_{1,2} \times V_{1,2} \mid v = \texttt{last}(G_1, T_i), v' = \texttt{first}(G_2, T_i), \text{for some } T_i \in T \; \} \quad (4)$$

[3]That these are partial functions will be most useful for dealing with graphs with no nodes on a particular tier; for example the empty graph, which is discussed below.

$$R_{ID} \stackrel{\text{def}}{=} \{ \; (v, v') \in R \mid \ell(v) = \ell(v'), \ell(v) \notin T_t \; \} \quad (5)$$

We also often refer to the complement of R_{ID} with respect to R; $R_{\overline{ID}} \stackrel{\text{def}}{=} R - R_{ID}$.

We can then use Engelfriet and Vereijken (1997)'s merging operation which reduces a graph G with any relation $R \subseteq V \times V$ over its nodes. Informally, nodes which stand in the relation are merged; everything else stays the same. Given any such relation R, we consider $\sim_R$, the finest equivalence relation on V containing R. In the usual way, let $[v]_{\sim_R} = \{v' | v \sim_R v'\}$. Here, we use $\sim_{R_{ID}}$, which assigns each node its own equivalence class, except for pairs $(v, v') \in R_{ID}$ of last and first nodes with identical labels, which are lumped together.

Example 2 *Continuing with* G_1 *and* G_2 *from Example 1,* $G_{1,2}$ *is given in Figure 3a. For* $G_{1,2}$, $R = \{(1, 3), (2, 4)\}$, $R_{ID} = \{(1, 3)\}$, *and so* $R_{\overline{ID}} = \{(2, 4)\}$. *The equivalence classes for* $\sim_{R_{ID}}$, $\{\{0\}, \{1, 3\}, \{2\}, \{4\}\}$, *are shown in Figure 3b.*

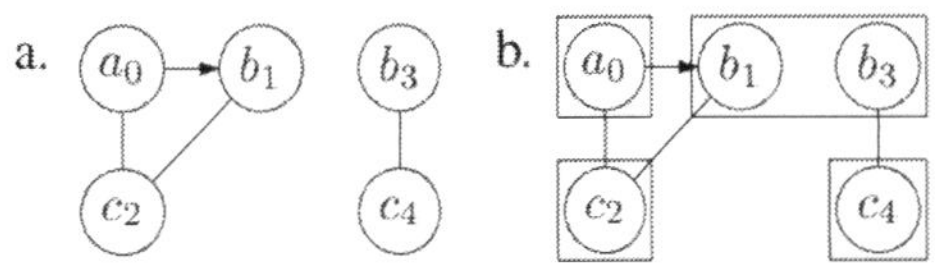

Figure 3: (a) $G_{1,2}$ and (b) $\sim_{R_{ID}}$ in $G_{1,2}$

Given a graph G and a relation $R \subseteq V \times V$, Engelfriet and Vereijken (1997) define $V/R = \{[v]_R | v \in V\}$. This simply 'merges' the nodes of V based on the equivalence relation $\sim_R$. G/R can then be defined as the graph reduced by this merged set of nodes; $\langle V/R, E, A, \ell \rangle$.

The final step is to add precedence arcs to connect $R_{\overline{ID}}$, the unmerged last and first nodes in $G_{1,2}/R_{ID}$. Again, $R_{\overline{ID}}$ is the pairs of last/first nodes on the melody tier that are not identical and the last/first pair on the timing tier, which are never merged.

Definition 2 ***(Concatenation of APGs).*** *The* concatenation $G_1 \circ G_2$ *of graphs* G_1 *and* G_2 *in* $GR(\Sigma)$ *is:*

$$G_1 \circ G_2 = \langle V_{1,2}/R_{ID}, E_{1,2}, A_{1,2} \cup R_{\overline{ID}}, \ell_{1,2} \rangle$$

Example 3 *The concatenation of* G_1 *and* G_2 *is given in Figure 4. The node numbered* $1, 3$ *represents the nodes from Fig. 3 which have been* *merged. Node also the added directed edge* $(2,4)$ *from* $R_{\overline{ID}}$ *in Example 2.*

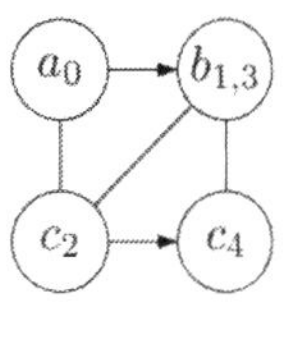

Figure 4: $G_1 \circ G_2$

Technically, the resulting set $V_{1,2}/R_{ID}$ is a set of *sets* of nodes representing the equivalence classes of $\sim_{R_{ID}}$; for example, $V_{1,2}/R_{ID}$ in Example 3 is $\{\{0\},\{1,3\},\{2\},\{4\}\}$. Represented strictly in this way, successive concatenations will yield sets of sets of sets of nodes, ad infinitum. For example, concatenating a third graph, such as G_3 in Figure 5 below, to $G_1 \circ G_2$ would further merge node $\{1,3\}$ with node 5 in G_3. Strictly speaking, the resulting node is $\{\{1,3\},\{5\}\}$. For clarity, we instead represent each node in this case as the union of the elements of each member of its equivalence class, e.g. $\{1,3,5\}$ for the concatenation $(G_1 \circ G_2) \circ G_3$ in Figure 5. This convenient renaming 'flattens out' the nested sets. It does not result in any loss of generality because union is associative. Also, it will be useful later when showing concatenation is associative for the particular class of graphs described in §6.

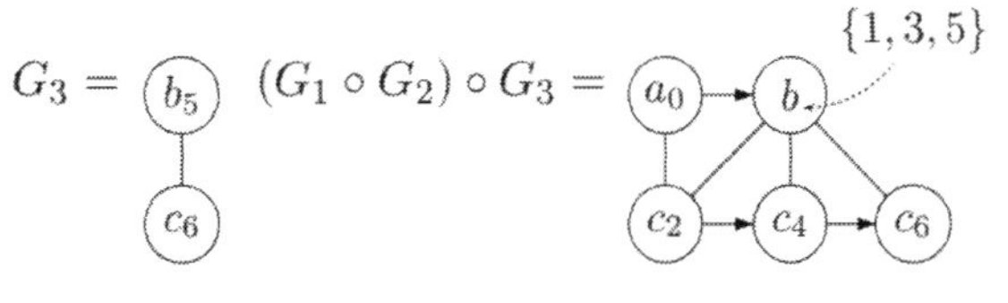

Figure 5: Concatenating a third graph G_3

Importantly, the relations R and R_{ID} do not depend on a binary partition over Σ; they only require that one partition T_t for the timing tier be specified. Thus, while the examples given here focus on two tiers, this operation is defined for graphs representing APRs with multiple melody tiers.

5.2 Properties

This section proves two important properties of concatentation, that G_λ is the identity for $\circ$, and that for any tier in both G_1 and G_2, $G_1 \circ G_2$ contains a string graph corresponding to those tiers.

Theorem 1 *G_λ is the identity element for the $\circ$ operation. That is, for any $G \in GR(\Sigma)$, $G \circ G_\lambda = G_\lambda \circ G = G$.*

Proof: Let $G = \langle V, E, A, \ell \rangle$. We first consider $G_\lambda \circ G$. Recall that the concatenation of two graphs is a modification of their disjoint union. From the properties of the union operation, we know that the disjoint union of G_λ and G is G. Note that $\texttt{first}(G_\lambda, T_i)$ and $\texttt{last}(G_\lambda, T_i)$ are undefined for all $T_i \in T$, because the set of nodes is empty in G_λ. Thus, $R = \emptyset$, and so $R_{ID} = R_{\overline{ID}} = \emptyset$. Because $R_{ID} = \emptyset$, $V/R = V$, because the smallest equivalence relation containing $\emptyset$ is $=$. Thus,

$$G_\lambda \circ G = \langle \quad (V/R) = V, (E \cup \emptyset) = E, \\ (A \cup \emptyset) = A, (\ell \cup \emptyset) = \ell \quad \rangle = G$$

$G \circ G_\lambda = G$ is similarly derived. □

The next lemma shows that concatenation preserves the string graph properties of any tiers in G_1 and G_2. This is important for showing the associativity of concatenation under certain graph classes, as will be discussed in §6.

Lemma 1 *Let U_i and V_i denote the set of all nodes in G_1 and G_2, respectively, with labels in some member T_i of T. If $G_1[U_i]$ and $G_2[V_i]$ are string graphs, or if U_i is empty and $G_2[V_i]$ is a string graph, or if $G_1[U_i]$ is a string graph and V_i is empty, then $(G_1 \circ G_2)[W_i]$ is a string graph, where W_i is the set of all nodes in $G_1 \circ G_2$ whose labels are in T_i. Furthermore, for any T_i, if $v = \texttt{first}(G_1, T_i)$, then $\texttt{first}(G_1 \circ G_2, T_i)$ is the unique node in $G_1 \circ G_2$ which contains v, and likewise for $\texttt{last}(G_2, T_i)$.*

Proof: This follows immediately from the definition of concatenation if $G_1[U_i]$ is a string graph and V_i is empty, because then $\texttt{first}(G_2, T_i)$ will be undefined and no member of U_i will appear in R, and thus all will appear in $G_1 \circ G_2$ unmodified and with no new arcs associated with them. Thus, $G_1[U_i] = (G_1 \circ G_2)[W_i]$ and so both are string graphs. The proof for the case in which U_i is empty and $G_2[V_i]$ is a string graph is very similar.

For the final case, recall that a graph G is a string graph iff its set of arcs A forms a total order $\preceq$ on its nodes V. For the case $G_1[U_i]$ and $G_2[V_i]$ are string graphs and $v_1 = \texttt{last}(G_1, T_i)$ and

$v_2 = \texttt{first}(G_2, T_i)$, then (v_1, v_2) appears in either R_{ID} or $R_{\overline{ID}}$. If the pair is in R_{ID} , v_1 and v_2 are merged into a node $v_{1,2}$ and no new arcs will be introduced to the set A_i of the arcs in $(G_1 \circ G_2)[W_i]$. So for the arc (v'_1, v_1) from $G_1[U_i]$ and (v_2, v'_2) from $G_2[V_i]$, the corresponding arcs in A_i are $(v'_1, v_{1,2})$ and $(v_{1,2}, v'_2)$, respectively, which maintains the total orders of both U_i and V_i. If $(v_1, v_2) \in R_{\overline{ID}}$, then (v'_1, v_1), (v_1, v_2), and (v_2, v'_2) are all in A_i, which also mantains the total order.

That for $v = \texttt{first}(G_1, T_i)$, $\texttt{first}(G_1 \circ G_2, T_i)$ is the unique node which contains v follows directly from the fact that the total order on U_i is maintained. Likewise for $v = \texttt{last}(G_2, T_i)$ and V_i. □

These properties allow us to treat sets of graphs parallel to sets of strings, as the next section shows.

6 APGs derived from concatenation

6.1 Alphabets of graph primitives

As Engelfriet and Vereijken (1997) observe, given a concatenation operation a class of graphs can be seen as an interpretation of a set of strings, where each symbol in the string corresponds to a graph primitive. We now define an APG graph primitive.

Definition 3 *Over an alphabet Σ and tier partition $T = \{T_t, T_m\}$, an* APG graph primitive *is a graph $G \in GR(\Sigma)$ which has the following properties:*

a. V_t is a singleton set $\{v_t\}$

b. $G[V_m]$ is a string graph or is empty

c. All $e \in E$ are of the form $\{v_m, v_t\}$, $v_m \in V_m$

We can then treat a finite set of primitives like an alphabet of symbols:

Definition 4 *An* alphabet of graph primitives *over $GR(\Sigma)$ is a finite set Γ of symbols and a naming function $g : \Gamma \rightarrow GR(\Sigma)$.*

An alphabet of APG graph primitives is thus Γ for which for all $\gamma \in \Gamma$, $g(\gamma)$ satisfies Definition 3.

Example 4 *As in the previous examples, consider the alphabet $\Sigma = \{a, b, c\}$ and the tier partition $T = \{T_m = \{a, b\}, T_t = \{c\}\}$. The alphabet of graph primitives $\Gamma = \{a, b, d\}$, where its naming function g is defined as in Figure 6, is an alphabet of APG graph primitives.*

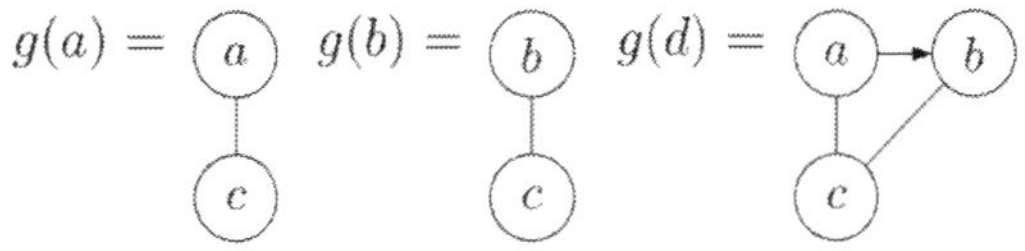

Figure 6: An example Γ and g

The strings in Γ^* thus represent a class of graphs, which we will call $APG(\Gamma)$. We define $APG(\Gamma)$by extending g to strings in Γ^*.

Definition 5 *For an alphabet of graph primitives Γ with naming function g, extend g to strings in Γ^* as follows. For $w \in \Gamma^*$, $g(w) \stackrel{def}{=}$*

- *G_λ if $w = \lambda$*
- *$g(u) \circ g(\gamma)$ if $w = u\gamma$, $u \in \Gamma^*, \gamma \in \Gamma$*

$APG(\Gamma)$ is thus $\{g(w) | w \in \Gamma^*\}$.

6.2 Derived properties

We now show that if Γ is an alphabet of APG graph primitives, then $APG(\Gamma)$ has a number of desirable properties. The following assumes Γ is an alphabet of APG graph primitives. First, we prove the following theorem stating that all graphs in $APG(\Gamma)$ follow Axioms 1 through 3 from §4 regarding the general structure of APGs.

Theorem 2 *For any $G \in APG(\Gamma)$, G satisfies Axiom 1 (that $\sim_A$ partitions V into at most two sets V_0 and V_1 such that $G[V_0]$ and $G[V_1]$ are string graphs), Axiom 2 (that the tiers of G correspond to the partition T), and Axiom 3 (that the ends of all undirected edges are between different tiers).*

Proof: That G satisfies Axioms 1 and 2 follows directly from parts (a) and (b) of Definition 3 and the fact that concatenation only adds arcs between nodes whose labels are in the same $T_i \in T$. That $G[V_0]$ and $G[V_1]$ are string graphs follows from parts (a) and (b) of Definition 3 and Lemma 1.

That G follows Axiom 3 follows directly from Part (c) of Definition 3 and the fact that concatenation adds no new undirected edges to E. □

Next, concatenation is associative over $APG(\Gamma)$. The following lemma allows one to prove Theorem 3 (associativity) below.

Lemma 2 *For any $u, v \in \Gamma^*$ denote $g(u), g(v) \in APG(\Gamma)$ with G_u and G_v respectively. Then for any $\gamma \in \Gamma$, $G_u \circ (G_v \circ G_\gamma) = (G_u \circ G_v) \circ G_\gamma$.*

Proof: Let $G = \langle V, E, A, \ell \rangle$ denote $(G_u \circ G_v) \circ G_\gamma$ and $G' = \langle V', E', A', \ell' \rangle$ denote $G_u \circ (G_v \circ G_\gamma)$. That $E = E'$ and $\ell = \ell'$ follow from Definition 2 of concatenation and associativity of union.

To show $V = V'$, there are seven relevent cases to consider. Let V_u, V_v, and V_γ denote the sets of nodes for G_u, G_v, and G_γ, respectively, and let v_u denote a node in V_u, etc. As merging is accomplished through grouping nodes into equivalence classes,all nodes in V or V' thus correspond to either **Cases 1–3** $\{v_u\}$, $\{v_v\}$, $\{v_\gamma\}$, **Cases 4–6** $\{v_u, v_v\}$, $\{v_v, v_\gamma\}$, $\{v_u, v_\gamma\}$, or **Case 7** $\{v_u, v_v, v_\gamma\}$ (recall from §5 we do not distinguish between nodes representing sets and nodes representing sets of sets).

As per the definition of concatenation, $V = ((V_u \cup V_v)/R_{ID\text{-}u,v} \cup V_\gamma)/R_{ID}$, where $R_{ID\text{-}u,v}$ is defined over $V_u \cup V_v$ and R_{ID} over $(V_u \cup V_v)/R_{ID\text{-}u,v}$. Likewise, $V' = (V_u \cup (V_v \cup V_\gamma)/R_{ID\text{-}v,\gamma})/R'_{ID}$.

Cases 1–3. We first establish that when $v \in V$ corresponds to a singleton set that $v \in V'$. Consider the case when $v \in V$ corresponds to $\{v_v\}$, when v_v has not been merged. For V, this is exactly the case in which there is no $(v_u, v_v) \in R_{ID\text{-}u,v}$ nor a $(\{v_v\}, v_\gamma) \in R_{ID}$. We show that this entails that there is neither a $(v_u, \{v_v\}) \in R'_{ID}$ nor a $(v_v, v_\gamma) \in R_{ID\text{-}v,\gamma}$, and so $\{v_v\} \in V'$. There is no $(\{v_v\}, v_\gamma) \in R_{ID}$ either when $\{v_v\}$ is not the last node in $G_u \circ G_v$ for any T_i or there is no v_γ with which it can merge. If $\{v_v\}$ is not the last node in $G_u \circ G_v$ for any T_i, then v_v is not the last node in G_v for any T_i, as by Theorem 2 and Lemma 1 the last node for T_i in $G_u \circ G_v$ must be the unique set which includes the last node for T_i in G_v. If there is no v_γ to merge with $\{v_v\}$, then there is no v_γ to merge with v_v either. Thus, there cannot be a $(v_v, v_\gamma) \in R_{ID\text{-}v,\gamma}$. Similarly, it follows that there is no $(v_u, \{v_v\}) \in R'_{ID}$. If there is no $(v_u, v_v) \in R_{ID\text{-}u,v}$, then either v_v is not the first node in G_v or there is no v_u with which it can merge. Thus, either $\{v_v\}$ is not the first node in $G_v \circ G_\gamma$ (again by Lemma 1) or there is no node v_u to merge with $\{v_v\}$, and so there is no $(v_u, \{v_v\}) \in R'_{ID}$. As there is neither a $(v_u, \{v_v\}) \in R'_{ID}$ nor a $(v_v, v_\gamma) \in R_{ID\text{-}v,\gamma}$, then $\{v_v\} \in V'$. The proofs that $v \in V$ implies $v \in V'$ when v corresponds to $\{v_u\}$ and $\{v_\gamma\}$ are very similar. The proofs that $v' \in V'$ implies $v' \in V$ for all three cases are identical.

The remaining cases deal with merged nodes. **Cases 4–6.** Consider the case in which $v \in V$ is $\{v_u, v_v\}$ corresponding to merged nodes from V_u and V_v. This is the case in which $(v_u, v_v) \in R_{ID\text{-}u,v}$ but there is no $(\{v_u, v_v\}, v_\gamma) \in R_{ID}$ for any v_γ. As before, if $\{v_u, v_v\}$ cannot be merged with some v_γ, then there is no v_γ to merge with v_v. Furthermore, if $(v_u, v_v) \in R_{ID\text{-}u,v}$, then v_u is the last node in G_u and v_v is the first node in G_v for some T_i. By Lemma 1, then $\{v_v\}$ is the first node in $G_v \circ G_\gamma$ for T_i, and so $(v_u, \{v_v\}) \in R'_{ID}$. Thus, $\{v_u, v_v\} \in V'$. The proof that $v \in V$ implies $v \in V'$ when $v = \{v_v, v_\gamma\}$ is very similar, as it is for $v = \{v_u, v_\gamma\}$. The latter is a special case in which V_v has no nodes for some T_i, but v_u and v_γ are compatible to merge. **Case 7.** For $v = \{v_u, v_v, v_\gamma\}$, both $(v_u, v_v) \in R_{ID\text{-}u,v}$ and $(\{v_u, v_v\}, v_\gamma) \in R_{ID}$. If $(\{v_u, v_v\}, v_\gamma) \in R_{ID}$, then through Lemma 1 $(v_v, v_\gamma) \in R_{ID\text{-}v,\gamma}$ and then $(v_u, \{v_v, v_\gamma\}) \in R'_{ID}$, so $\{v_u, v_v, v_\gamma\} \in V'$. The proofs that $v' \in V'$ implies $v' \in V$ for these cases are identical.

That $A = A'$ is very similar to the proof for $V = V'$. Let A_i denotes the set of arcs in $g(\gamma_i)$. Then $A = (A_u \cup A_v \cup R_{\overline{ID-u,v}}) \cup A_\gamma \cup R_{\overline{ID}}$, where $R_{\overline{ID-u,v}}$ is defined over $(V_u \cup V_v)/R_{ID\text{-}u,v}$ and $R_{\overline{ID}}$ is defined over $((V_u \cup V_v)/R_{ID\text{-}u,v} \cup V_\gamma)/R_{ID}$, and $V_u \cup V_v$ $A' = (A_u \cup (A_v \cup A_\gamma \cup R_{\overline{ID-v,\gamma}}) \cup R'_{\overline{ID}})$, where $R_{\overline{ID-v,\gamma}}$ and $R'_{\overline{ID}}$ are defined parallel to $R_{\overline{ID-u,v}}$ and $R_{\overline{ID}}$. As union is associative, it is sufficient to show that every pair $(v_u, v_v) \in R_{\overline{ID-u,v}}$ has a corresponding pair $(v_u, \{v_v\})$ or $(v_u, \{v_v, v_\gamma\})$ in $R'_{\overline{ID}}$ and vice versa, and that every pair $(\{v_v\}, v_\gamma)$ or $(\{v_u, v_v\}, v_\gamma)$ in $R_{\overline{ID}}$ has a corresponding pair $(v_v, v_\gamma) \in R_{\overline{ID-v,\gamma}}$, and vice versa. Both of these follow from the fact that $V = V'$ and Lemma 1 in the same way as merging nodes above. □

Next it is shown that graph concatenation is associative over arbitrary graphs in $APG(\Gamma)$ with the same kind of inductive argument which establishes concatenation is associative over strings.

Theorem 3 *The $\circ$ operation is associative over graphs in $APG(\Gamma)$. For any $u, v, w \in \Gamma^*$ denote $g(u), g(v), g(w) \in APG(\Gamma)$ with G_u, G_v, G_w re-*

spectively. Then $G_u \circ (G_v \circ G_w) = (G_u \circ G_v) \circ G_w$.

Proof: The proof is by induction on the size of w. For the base case, when $w = \lambda$, $G_w = G_\lambda$. Then $G_u \circ (G_v \circ G_w) = G_u \circ (G_v \circ G_\lambda)$, which equals $G_u \circ G_v$ by Theorem 1. It follows, again by Theorem 1, that $(G_u \circ G_v) \circ G_\lambda$. Hence the base case is proved.

Next we assume the inductive hypothesis that associativity holds for strings of length n and we consider any $w \in \Gamma^*$ of length $n+1$. Clearly there exists $x \in \Gamma^*$ of length n and $\gamma \in \Gamma$ so that $w = x\gamma$. Then $G_u \circ (G_v \circ G_w) = G_u \circ (G_v \circ (G_x \circ G_\gamma))$. By Lemma 2, this equals $G_u \circ ((G_v \circ G_x) \circ G_\gamma)$, which again by Lemma 2, equals $G_u \circ (G_v \circ G_x)) \circ G_\gamma$. Then, by the induction hypothesis, we have $((G_u \circ G_v) \circ G_x) \circ G_\gamma$, which again by the induction hypothesis, yields $(G_u \circ G_v) \circ (G_x \circ G_\gamma)$. This is of course is $(G_u \circ G_v) \circ G_w$. □

The next theorem states that any $G \in APG(\Gamma)$ follows the NCC.

Theorem 4 *For any $G \in APG(\Gamma)$, G satisfies the NCC (Axiom 4).*

Proof: The proof is by recursion on the length of $w \in \Gamma^*$. G_λ trivially satisfies the NCC because it has no nodes. For $g(\gamma)$ for any $\gamma \in \Gamma$, Definition 4 states that there is only one node v_t in V_t and this node must be one of the endpoints for each edge in E. Thus for any two edges $\{x, y\}$ and $\{x', y'\}$ in $g(\gamma)$ where $x \preccurlyeq x'$, it must be the case that $y = y' = v_t$, because directed edges only occur between nodes in tier V_m. Thus, any $g(\gamma)$ satisfies the NCC.

Next we assume it holds for $w \in \Gamma^*$ of length n and consider any $w \in \Gamma^*, \gamma \in \Gamma$. Then $g(w\gamma)$ satisfies the NCC because the graph concatenation operation does not add any undirected edges and because, by Lemma 1 concatenation preserves the order of each tier in $g(w)$ and $g(\gamma)$. □

The final theorem states that any $G \in APG(\Gamma)$ follows the OCP if the graph primitives do.

Theorem 5 *If $g(\gamma)$ for all $\gamma \in \Gamma$ satisfy the OCP (Axiom 5), then for any $G \in APG(\Gamma)$, G satisfies Axiom 5.*

Proof: The proof is again by recursion on the length of $w \in \Gamma^*$. The OCP is trivially satisfied for G_λ since it contains no nodes or arcs. The case when $|w| = 1$ is given as the condition of the theorem.

Assume that every $w \in \Gamma^*$ of length n satisfies the OCP. Now consider $G = g(w\gamma)$ with w of length n and $\gamma \in \Gamma$. To see that $G_u \circ G_\gamma$ satisfies the OCP, recall from Definition 2 of graph concatenation that the set of arcs for $G_1 \circ G_2$ is equal to $A_{1,2} \cup R_{\overline{ID}}$; i.e., the union of A_1 and A_2 and $R_{\overline{ID}}$. By definition $R_{\overline{ID}}$ only includes pairs of nodes (x, y) s.t. $\ell(x) \neq \ell(y)$, so if G_1 satisfies the OCP and G_2 satisfies the OCP $R_{\overline{ID}}$ will not add any arcs on V_m which violate the OCP (recall that the OCP only holds for tier V_m), and so $G_1 \circ G_2$ will also satisfy the OCP. □

Thus, the merging part of the concatenation preserves the OCP. One may wonder why the OCP is built in to the concatenation operation this way, instead of using string-like concatenation and then invoking a constraint that merges adjacent, like nodes in the resulting graph. Such a method, though, cannot capture violations of the OCP—all would be merged. The next section shows that the concatenation operation defined here can capture violations by concatenating OCP-violating graph primitives.

This section has thus proved the important properties of $APG(\Gamma)$. We now show how such an $APG(\Gamma)$ can be used to model autosegmental phenomena in natural language phonology.

7 Analysis of natural language phenomena

In this section we examine the extent to which the analysis presented here accounts for common and uncommon phenomena in phonological theory. The first two subsections examine spreading and contour tones, respectively, and demonstrate how both phenomena can be effectively represented with a $APG(\Gamma)$ for some Γ. It is also shown that the empirical generalization that there are only finitely many contour tones present in any given language is an automatic consequence of the finite alphabet Γ and the concatenation operation.

The third subsection addresses the few cases where OCP violations may be necessary to properly describe the language. It is sketched out how these cases could be accounted for by using special graph primitives or a second concatenation operation. Similarly, the fourth subsection addresses underpecification and floating tones. We conclude that these concepts can be represented in this approach. The caveat is that it is also observed as a consequence that gapped structures are also permitted. Again, we note that such gapped structures are also permitted with axioms given in §4 approaches above, and we discuss how a different concatenation operation may address this.

7.1 Spreading

The 'merging' of nodes on the melody tier models autosegmental spreading, in which one melody unit is associated to more than one timing tier unit. A classic example is Mende (Leben, 1973). Mende nouns separate into tone categories, three of which are shown in Table 1. The first rows show words whose syllables are all high-toned, the second rows show words whose syllables are all low-toned, and the third rows show words whose syllables start high and end low. In the following [á] transcribes a high tone, [à] a low tone, [â] a falling tone.

Monosyllables		Disyllables	
kɔ́	'war'	pɛ́lɛ́	'house'
kpà	'debt'	bɛ̀lɛ̀	'pants'
mbû	'owl'	ngílà	'dog'
Trisyllables			
háwámá	'waist'		
kpàkàlì	'three-legged chair'		
félàmà	'junction'		

Table 1: Mende word tone

An autosegmental analysis for this pattern is that a set number of melodies spread left-to-right over the tone-bearing units (TBUs; we assume that for Mende the TBU is the syllable, σ) of a word, as in Table 2. For example, the falling tone words [mbû] 'owl' and [félàmà] 'junction' have an HL melody. In this case, the H associates to the first syllable of the word, and the L associates to all remaining syllables.

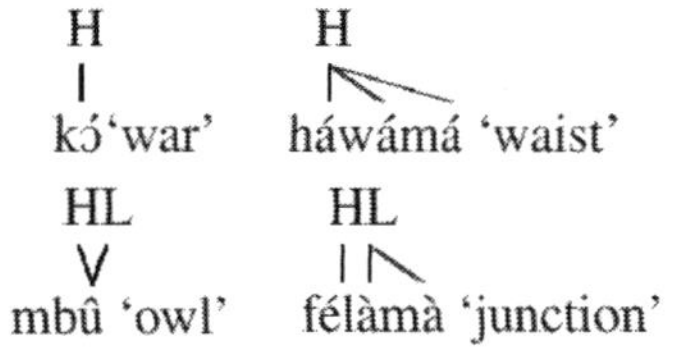

Table 2: APRs for four Mende words

The APRs in Table 2 can be generated with the alphabet of APG graph primitives Γ given in Figure 7. The alphabet is $\Sigma = \{\text{H}, \text{L}, \sigma\}$ and the tier partition $T = \{T_t, T_m\}$ where $T_t = \{\sigma\}$ and $T_m = \{\text{H}, \text{L}\}$. Note that for these APGs, we abstract away from consonants and vowels and focus on the TBU, σ.

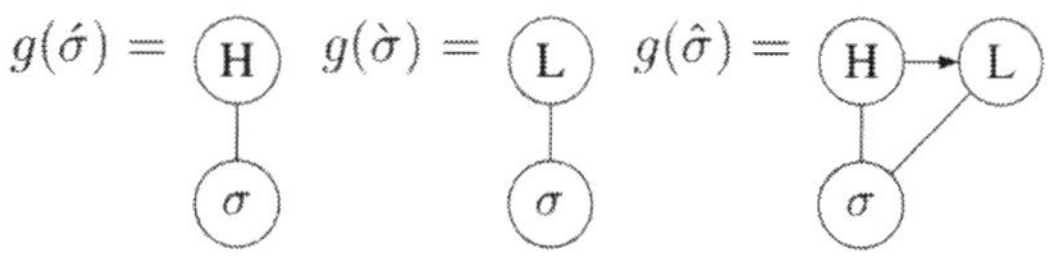

Figure 7: Γ and g for Mende

The APGs corresponding to the trisyllabic forms are thus $g(\acute{\sigma}\acute{\sigma}\acute{\sigma})$ and $g(\acute{\sigma}\grave{\sigma}\grave{\sigma})$, as in Figure 8.

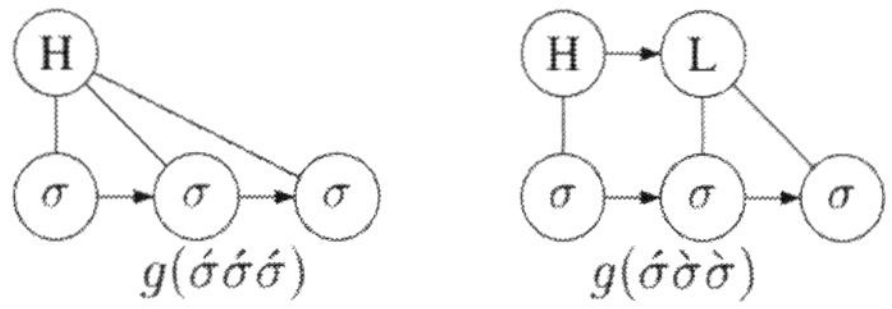

Figure 8: APGs for Mende APRs in Table 2

These spreading effects are achieved by, for example in $g(\acute{\sigma}\acute{\sigma}\acute{\sigma})$, the like H nodes from each $g(\acute{\sigma})$ merging during concatenation, resulting in a single H associated to multiple σ nodes (which are not merged, because $\sigma \in T_t$). Note that given Σ, T, Γ and g, we are able to generate APGs directly from the linear string of toned syllables.

7.2 Contours

Concatenation allows for unbounded spreading, as a single node on the melody tier may 'merge' any number of times. In contrast, concatenation does *not* allow for unbounded contours, as timing tier nodes do not merge. Figure 9 shows how concatenation obtains APGs corresponding to the APRs for the Mende words [mbû] 'owl' and [nyàhâ] 'woman'.

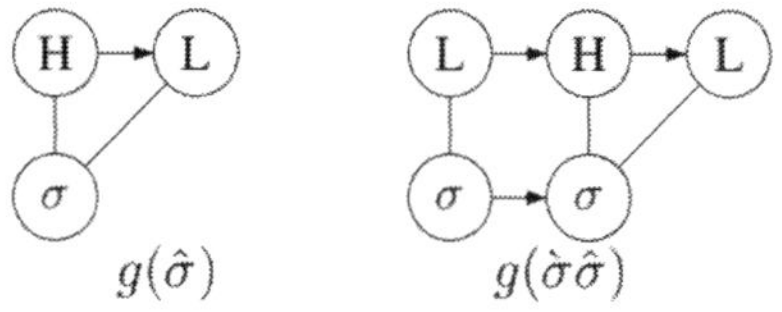

Figure 9: APGs for Mende contours

Importantly, any set of graphs is going to have a bound on the number of melody units a contour can have, which follows directly from the fact that Γ is finite, that each element of Γ has exactly one node on V_t, and so concatenation never creates new contours. Thus, for the example Γ we have been using for Mende, the graph in Figure 10 is *not* in $APG(\Gamma)$.

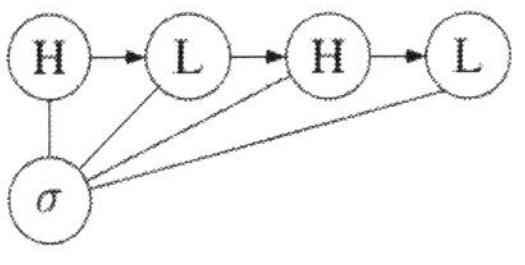

Figure 10: A graph not in $APG(\Gamma)$

While this is a natural property of graphs in $APG(\Gamma)$, the axiomatic approach to defining APRs requires a further axiom stating that for any language, the number of contours must be bound by some n. To our knowledge, the only explicit formalizations of such a constraint are by Jardine (2014) and Yli-Jyrä (2013) (the latter requiring that $n = 2$).

7.3 Violations of the OCP

As discussed in Odden (1986) and Meyers (1997), the OCP may not be an absolute universal. For example, Odden lists the contrasting APRs in Figure 11 for two nouns in Kishambaa (Odden, 1986, Fig. 13):

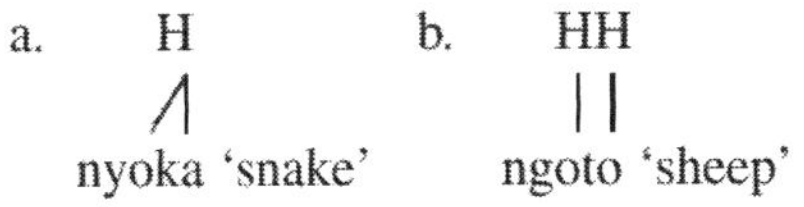

Figure 11: OCP violating forms in Kishambaa

This is partially motivated by the different surface pronunciation of the two forms: the first, Figure 11 (a) 'snake' is pronounced with two level H tones, nyóká, and 11 (b) 'sheep' is pronounced with a H followed by a downstepped H; ngó$^!$tó.

The corresponding graphs for these APRs, assuming the mora as the TBU, are given in Figure 12. Figure 12 (a) corresponds to Figure 11 (a), and Figure 12 (b) to Figure 11 (b).

Given an alphabet of graph primitives obeying the OCP, as the Γ for Mende in Figure 7, Figure 12 (a) is in $APG(\Gamma)$, but Figure 12 (b) is not, because it does not obey the OCP. Thus, Kishambaa is not describable with such a graph set $APG(\Gamma)$.

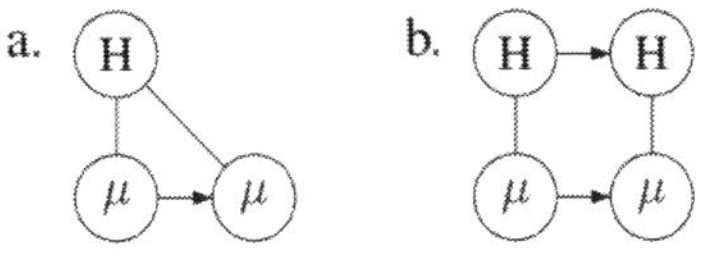

Figure 12: APGs for Kishambaa forms

There are at least two solutions to admitting graphs like in Kishambaa. One is to introduce OCP-violating graph primitives, as in Figure 13.

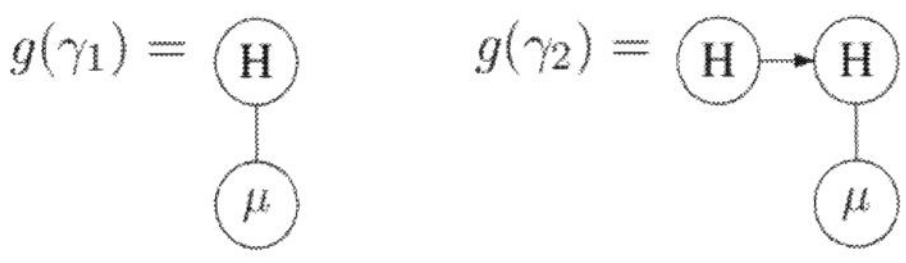

Figure 13: A Γ for Kishambaa

Given this alphabet of graph primitives, the spreading Kishambaa graph in Figure 12 (a) is $g(\gamma_1\gamma_1)$, and the OCP-violating (b) is $g(\gamma_1\gamma_2)$. The graph primitives follow the linear pronunciation of the morae; $g(\gamma_1\gamma_1)$ represents a sequence HH of two high-toned morae, and $g(\gamma_1\gamma_2)$ a sequence H$^!$H of a high followed by a downstepped high.

Another option is to define a second concatenation operation, in which there is no merging and directed edges are drawn between all last/first pairs. Spreading Kishambaa graph in Figure 12 (a) would be concatenated by the operation defined in this paper, and the OCP-violating Figure 12 (b) would be concatenated by this second no-merging operation. We shall leave it up to future work to compare the theoretical and empirical benefits of these approaches to OCP violations.

7.4 Underspecification and floating tones

Some graph primitives in Γ may not have any nodes in V_m; these represent *underspecified* timing units.

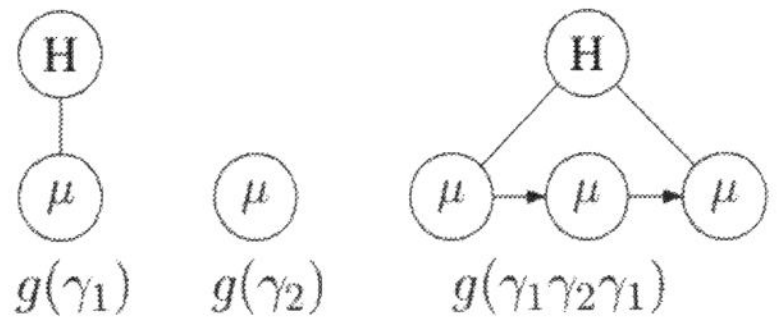

Figure 14: APGs with underspecification

However, such underspecified graph primitives

can give rise to 'gapped structures' via concatenation, as in $g(\gamma_1\gamma_2\gamma_1)$ in Figure 14. This can be seen as an unwelcome consequence as some researchers have argued against gapped structures (Archangeli and Pulleyblank, 1994). One solution could be to use a second concatenation operation which does not merge nodes, instead only drawing directed edges between the end nodes on each tier. This appears identical to the operation proposed in §7.3 for dealing with OCP violations. Again, studying additional concatenation operations will be left for future work.

Finally, graph primitives with more melody tier nodes than timing tier nodes can be used to generate floating tones, as in Figure 15.

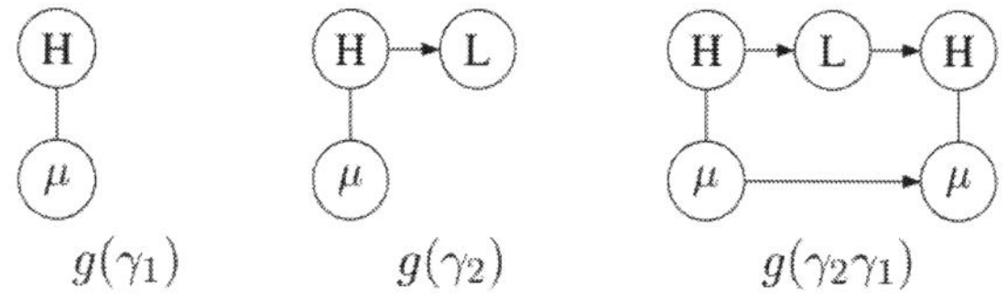

Figure 15: Generating APGs with floating tones

8 Dicussion

The examples in the previous section show several advantages to considering APRs through concatenation. One, as seen in Mende, simple cases allow direct translation of strings into APRs. Second, concatenation allows for unbounded spreading, as a single node on the melody tier may 'merge' any number of times. However, concatenation does *not* allow for unbounded contours, as timing tier nodes do not merge in this way. Thus, the number of contours is bounded by the number of graph primitives. This reflects the fact that languages exhibit unbounded spreading, but no language (to our knowledge) has an unbounded number of contour segments.

There are several avenues for future work. It was already mentioned that the set of valid autosegmental representations may be expanded by allowing a second concatenation operation. Also, while we have shown that every element of $APG(\Gamma)$ obeys the axioms in §4, it remains to be shown that for every graph which obeys those axioms, there is a finite alphabet which generates it.

Future work can also study the nature of transformations from underlying APGs with one alphabet to surface APGs with another (for instance it is known surface APGs can admit more contours than underlying APGs through association rules).

Another line of development concerns extending the analysis to feature geometry (Clements and Hume, 1995; Sagey, 1986), in which association lines also link featural autosegments and 'organizational' nodes, such as PLACE. Deriving a set of such operations would require more complex primitives and additional marking on the tier partition T, to denote timing tier nodes, organizational nodes, and melody nodes. The concatenation operation would then need to be revised to be sensitive to this marking. A more serious challenge would be adopting a concatenation-based framework for autosegmental morphology, which as mentioned in §2, disposes of the requirement that autosegments of a particular type must appear on a particular tier.

9 Conclusion

In this paper we addressed the question of what is the set of valid autosegmental representations looks like. In contrast to previous research, which explored this question axiomatically, we showed that the autosegmental representations can be generated recursively and constructively from a finite set of graph primitives, a concatenation operation, and an identity element for concatenation, much in the same way that strings can be so generated. Hence, the theory of free monoids may be fruitfully applied to APRs.

The advantages we wish to highlight are as follows. First, we proved that provided the finite set of primitives obey the NCC and the OCP, the autosegmental representations will as well. Second, we showed it also follows naturally from the nature of the alphabet and concatenation that new contour tones cannot be generated ad infinitum. Finally, this method makes clear the stringlike nature of autosegmental representations, and that their properties can be viewed as a consequence of this nature.

Acknowledgments

The authors would like to thank three reviewers for their insightful comments and suggestions. The first author acknowledges support from a University of Delaware Graduate Research Fellowship.

References

Diana Archangeli and Douglas Pulleyblank. 1994. *Grounded Phonology*. Cambridge: MIT Press.

Steven Bird and E. Klein. 1990. Phonological events. *Journal of Linguistics*, 26:33–56.

G. N. Clements and Elizabeth V. Hume. 1995. The internal organization of speech sounds. In John Goldsmith, editor, *The handbook of phonological theory*, pages 245–306. Oxford: Blackwell.

G. N. Clements. 1976. *Vowel Harmony in Nonlinear Generative Phonology: An Autosegmental Model*. Bloomington: Indiana University Linguistics Club Publications.

John Coleman and John Local. 1991. The "No Crossing Constraint" in autosegmental phonology. *Linguistics and Philosophy*, 14:295–338.

Joost Engelfriet and Hendrik Jan Hoogeboom. 2001. MSO definable string transductions and two-way finite-state transducers. *ACM Transations on Computational Logic*, 2:216–254, April.

Joost Engelfriet and Jan Joris Vereijken. 1997. Context-free graph grammars and concatenation of graphs. *Acta Informatica*, 34:773–803.

John Goldsmith. 1976. *Autosegmental Phonology*. Ph.D. thesis, Massachussets Institute of Technology.

John Goldsmith. 1979. *Autosegmental Phonology*. Gardland Press.

Michael Hammond. 1988. On deriving the Well-Formedness Condition. *Linguistic Inquiry*, 19(2):319–325.

Adam Jardine. 2014. Logic and the generative power of Autosegmental Phonology. In John Kingston, Claire Moore-Cantwell, Joe Pater, and Robert Staubs, editors, *Supplemental proceedings of the 2013 Meeting on Phonology (UMass Amherst)*, Proceedings of the Annual Meetings on Phonology. LSA.

András Kornai. 1995. *Formal Phonology*. Garland Publication.

W. R. Leben. 1973. *Suprasegmental phonology*. Ph.D. thesis, Massachussets Institute of Technology.

John J. McCarthy. 1979. *Formal Problems in Semitic Phonology and Morphology*. Ph.D. thesis, Massachusetts Institute of Technology.

John J. McCarthy. 1985. *Formal Problems in Semitic Phonology and Morphology*. New York: Garland.

John J. McCarthy. 1986. OCP effects: gemination and antigemination. *Linguistic Inquiry*, 17:207–263.

Scott Meyers. 1997. OCP effects in Optimality Theory. *Natural Language & Linguistic Theory*, 15(4):847–892.

David Odden. 1986. On the role of the Obligatory Contour Principle in phonological theory. *Language*, 62(2):353–383.

Douglas Pulleyblank. 1986. *Tone in Lexical Phonology*. Dordrecht: D. Reidel.

Elizabeth Sagey. 1986. *The Representation of Features and Relations in Non-Linear Phonology*. Ph.D. thesis, Massachusetts Institute of Technology.

Moira Yip. 2002. *Tone*. Cambridge University Press.

Anssi Yli-Jyrä. 2013. On finite-state tonology with autosegmental representations. In *Proceedings of the 11th International Conference on Finite State Methods and Natural Language Processing*, pages 90–98. Association for Computational Linguistics.

Syntactic Polygraphs
A Formalism Extending Both Constituency and Dependency

Sylvain Kahane
Université de Paris Ouest
Laboratoire Modyco (CNRS UMR 7114)
`sylvain@kahane.fr`

Nicolas Mazziotta
Universität Stuttgart
Institut für Linguistik/Romanistik
`nicolas.mazziotta@ulg.ac.be`

Abstract

Syntactic analyses describe *grouping operations* that explain how words are combined to form utterances. The nature of these operations depends on the approach. In a constituency-based approach, grouping operations are ordered, or *stratified*, part-whole relations. In a dependency-based approach, grouping operations identify a *governor* (or *head*), i.e. they are directed hierarchical relations between words. It is possible to convert a constituency tree into a dependency tree by dereifying the nodes, by identifying the governor and by removing the stratification of the part-whole relations. Polygraphs combine the two types of information into a single structure and are therefore a more powerful formalism. By relaxing constraints, polygraphs also allow to *underspecify* both kinds of information.

Keywords: syntactic structure, immediate constituents analysis, dependency tree, headedness, phrase structure tree, polygraph, reification, stratification, underspecification.

1 Introduction

Despite their differences, syntacticians agree on the fact that they study the way words combine to form utterances, either from a hierarchical point of view (abstract model) or focussing on word order.[1] Word order and distributional properties are beyond the scope of this study. The focus is on the means of modelling the hierarchy of words. Moreover, the aim of this paper is to explore the formal aspect of syntactic description rather than to propose grammatical rules or make predictions: the examples in this paper do not necessarily reflect the point of view of the authors as linguists, but are provided to illustrate that various formal points of view on syntactic combination can be encoded by the mathematical structures proposed here.

Dependency and constituency are often seen as alternative frameworks that describe how words combine to form utterances. For many linguists, choosing between these two alternatives is an important point that needs to be resolved early on[2], since dependency-based descriptions cannot be totally translated into a constituency-based counterpart and vice versa. By investigating the operations that allow this kind of conversion, an understanding of the differences between them as well as their similarities is promoted. The exploration of dependency and constituency provided here defines a new kind of linguistic structure: the *syntactic polygraph*. These structures encompass the expressive power of the other two formalisms.

In short, the aim of this paper is to compare immediate constituent analysis with dependency analysis and to provide a joint formalism that takes advantage of both approaches, making it possible to express more than either of the two formalisms would

[1] In this paper, *words* are considered to be minimal linguistic units. However, the argument is easily extendable to morphemes as in X-bar syntax since the introduction of IP instead of S; see also Groß (2011) a.o. for a syntactic analysis of morphemic combinations in dependency.

[2] However, phrase structure grammars have been the dominant paradigm in syntax since (Chomsky, 1957) and people working in PSGs generally do not discuss whether dependency could be an option. Most of the discussions on that subject have been conducted by syntacticians working in dependency syntax (Hudson, 1980; Mel'čuk, 1988; Osborne et al., 2011; Kahane, 2012).

Proceedings of the 14th Meeting on the Mathematics of Language (MoL 14), pages 152–164,
Chicago, USA, July 25–26, 2015. ©2015 Association for Computational Linguistics

allow. While a large part of the paper is devoted to presenting a new process of conversion between constituency and dependency, the conversion itself is not the main point. An extensive literature already exists on the conversion of syntactic treebanks from constituency-based formats to dependency-based formats and vice versa. The question in such studies is to see how to add missing information (for instance headedness) during conversion. Our concern here is to see how to encode information when you have it, how to make it explicit, structurally visible, and also how to avoid adding irrelevant information with formalisms that force you to express things you do not want to express.

Sec. 2 introduces the specific grouping operations of constituency and dependency from a linguistic perspective, with respect to the way they acknowledge combinations of words and gives a formal definition of syntactic trees. Sec. 3 evaluates the transformations that models undergo when converted from constituency to dependency and highlights the better expressiveness of polygraphs. Sec. 4 shows how to extend both formalisms, taking dependency trees as the starting point. The conclusion summarizes what is achieved in this paper (sec. 5).

2 How to group words: constituency and dependency

The minimal consensus between syntacticians is that words combine to form sets of words that follow some organizational rules allowing them to function together. Operations that describe these associations can be called *grouping operations*, for the sake of neutrality. Unlike words, which are observable units (see note 1), the *groups* that result from grouping operations are constructs that can also combine with other words or groups to build a hierarchy. Both constituency and dependency models acknowledge this, but the nature of the grouping operation, i.e. the constructed relation uniting the grouped elements, is somewhat different. This section gives a general linguistic definition of grouping operations in a constituency-based approach as well as in a dependency-based approach (sec. 2.1) and a general formal definition of the tree-like objects they use to represent groups (sec. 2.2).

2.1 Syntactic tree archetypes: linguistic definitions

This section describes how grouping operations differ in a constituency-based approach and in a dependency-based approach. To do this, it focuses on *strict* constituency and *strict* dependency, although many linguists actually blend these two ways of describing grouping operations.

Strict constituency: stratification of part-whole relations. This approach is grounded in Bloomfield's description of groups (Bloomfield uses the term *constructions*) in terms of *part-whole* relations (Bloomfield, 1933, § 10.2). Grouping operations result in complex units that *contain* lesser units called *immediate constituents*. The analysis thus consists of strata of constructions, each stratum being recursively divisible into smaller constructions until the lowest stratum is reached. This *stratification*[3] can be represented in various ways: bracketing, embedded boxes or trees. The latter representation (henceforth *constituency tree*) has become the most popular.[4]

(1) Mary loves red cars

The constituency tree of (1) appears as fig. 1(a): internal nodes represent various groups that can be labelled and terminal nodes represent words (POS are marked as superscripts to words and will not be discussed here), but edges express only one kind of link between linguistic elements: the part-whole relation. This approach generates nodes in the tree, each of them corresponding to a stratum.

Strict dependency: directed relations between words. In a dependency-based approach, grouping operations are essentially directed connections between words (Tesnière, 1959; Tesnière, 2015, ch. 1). Such an approach implies that some words are more important than others with respect to their

[3]The term *stratification* has the same meaning as in our previous work (Kahane, 1997), where stratification is applied in a dependency-based representation.

[4]There were no tree in Bloomfield (1933), nor in the famous paper by Wells (1947). According to Coseriu (1980, 48), the first tree-like representation of constituency appears in Nida (1949, 87) (the "tree" may rather be a polygraph, cf. sec. 3.1) and current trees with labels on nodes became popular after Gleason (1955) and Chomsky (1957); embedded boxes are used in Hockett (1958).

(a) Constituency tree *(b)* Dependency tree

Figure 1: Constituency and dependency trees

syntactic position. Therefore, the proposed hierarchy is not based on constructed part-whole relations, but on constructed *dependency* relations that associates pairs of *governor* and *dependent* words – see Mel'čuk (1988) for this terminology. Unlike part-whole relations, dependencies are associated with specific labels indicating the kind of grouping operation. The most common way to represent a set of dependencies is a tree-like structure (henceforth *dependency tree*).[5] The dependency tree of (1) appears as fig. 1(b), with traditional labels on top-down lines between governors (top) and dependents (bottom).

Hybrid models. Linguists often combine part-whole relations and governor-dependent dependencies. An extended review of all hybrid models is not needed: one example will suffice. X-bar trees analyse all phrases as a combination of a governor X (the *head*) and two dependents that aggregate with the head at their distinct strata (the *complement* at the $\bar{X}$ level and the *specifier* at the SX level). Although many scholars refer to this kind of modelling by using the word *constituency*, it should be clear that does not consist of *strict constituency*: it combines the two kinds of grouping operations that have just been introduced. This hybrid approach is based on Bloomfield's description of endocentric constructions (Bloomfield, 1933, § 12.10). The hybrid model introduced in this paper is compared with such headed constituency trees in sec. 3.2. Note that even if the headedness can be added in a constituency tree, it can only be introduced in the labelling and it will not be part of the topology of the structure as it would be in a classical dependency tree.

[5]Our dependency tree corresponds to the so-called surface-syntax structure in the Meaning-Text Theory (Mel'čuk, 1988). Other levels of representation in the same theory and other theories – e.g.: Word Grammar (Hudson, 2010) – may use graphs where a node can have several governors.

2.2 Syntactic trees as graphs: formal definitions

Intuitively it is obvious that dependency and constituency trees can be described as tree objects, i.e. constrained forms of graphs. The following analysis is grounded on the basic hypergraph structure (Bergé, 1973). A *hypergraph* on a set X is a set E of subparts of X. The elements of X are called the *nodes* and the elements of E are the *edges* of the hypergraph. An edge e is a subset of X whose elements are called its *vertices* (e.g. $e = \{x, y, z\}$, where $x, y, z \in X$). A *graph* is a hypergraph whose edges are pairs, i.e. edges have two vertices (e.g. $e = \{x, y\}$).

Since the distinction becomes important further on, it should be noted that nodes and vertices are different concepts, the latter being slots that the former occupy. In other words, a node can be the vertex of several edges. Vertices of a hypergraph can also be assigned a *type* that is bound to their association with an edge. An edge of a *typed hypergraph* is a subset of $X \times T$, where T is the set of types. A *directed graph* is a particular case of typed graph where each vertex is associated to a type carrying appropriate semantics: $T = \{source, target\}$; e.g. $e = \{(x, source), (y, target)\}$ with x and y in X, usually abbreviated in (x, y) and represented by an arrow from x to y ($x \rightarrow y$) or, as in a tree, by a vertical line with the source on top.

A *tree* is a connected directed graph such that every node but one, called the *root*, is the target of one and only one edge. In the graphical representation of such trees, the directions of the edges are expressed by a convention: the source is higher on the figure than its targets. Moreover, constituency and dependency trees are labelled trees, i.e. nodes and edges can be labelled.

3 Constituency and dependency dialectics

This section describes what formal operations must be applied to strict constituency trees (fig. 1(a)) in order to obtain a strict dependency tree (fig. 1(b)) and vice versa. Although it is possible to apply these operations to any strict constituency tree, the focus is on binary-branching trees and abstracts away from other constituency structures.[6] It is possible to convert a binary-branching strict constituency tree into a strict dependency tree in five steps:[7] converting the branching nodes into edges (sec. 3.1), specifying the direction of the edges (sec. 3.2), delinearizing the graph (sec. 3.3), selecting more specific vertices (sec. 3.4) and relabelling the edges (sec. 3.5). Some of these formal transformations alter the amount of information, either by addition or by substraction. Additionally, the graphical means used to express the syntactic structure achieves a variable level of semiotic coherence and readability.

There exists another way to convert a constituency tree into a dependency tree presented by Lecerf (1961), consisting of collapsing all the constituents with their lexical head, i.e. , in an X-bar approach, turning any specifier, complement or adjunct into a dependent of this head. This classical conversion gives the same result as the one presented here for headed binary-branching constituents, but it can only be applied to headed constituency trees. The conversion presented here can be applied to headed constituents as well as non-headed ones, preserving the non-headedness (sec. 4.2). Additionally, it shows that constituents and dependencies encode the same groups of words and that the dependency itself can be viewed as a grouping operation. The two conversion processes also differ for non-binary branching trees, giving interesting insight into the different uses of ternary branchings in constituency trees (sec. 4.3).

3.1 Node/edge conversions

From a purely formal point of view, graphs can undergo transformations that convert edges into nodes or nodes into edges, namely *reification* and *dereification*. As pointed out in sec. 2.1, constituency trees contain nodes that result from the description process. These nodes can be translated into edges by *dereification* without changing the structure of the analysis. To understand dereification, one has to be familiar with reification.

Reification. For the sake of clarity, reification will be illustrated by conversions on the semantic representation of (2). The focus will come back to syntactic trees when dereification is applied to them.

(2) Mary loves Peter

Modelling verbs as predicates is very common. In the semantic structure of (2), 'love' (the meaning of the lexeme LOVE) is a binary predicate taking 'Mary' and 'Peter' as arguments: 'love'('Mary', 'Peter'). The first argument, 'Mary', can be called the agent, and the second argument, 'Peter', the patient. This semantic representation can be encoded by different graphs according to different conventions that are provided here for illustration only (fig. 2): in (a), 'loves' is associated to an edge directed from the agent to the patient (represented as nodes); in (b), the meaning 'love' is a node, bound to two other nodes by edges that make semantic roles explicit through their label (*agent* and *patient*); in (c), *agent* and *patient* are represented as nodes and their relations to their sources and targets have been made explicit.[8]

The *reification* of an edge e is the conversion e into a node and creating a directed edge (e,x) for each vertex x of e. If the graph is typed, the edge representing a vertex relation typed a will be labeled by a. For instance the reification of a directed edge e gives a node e and a *source* and a *target* edge. The edge of graph (a) is reified in graph (b). The edges of graph (b) are reified in graph (c), where the numbers are simple indices that distinguish the relations.

Dereification. Dereification is the reverse operation. It can be applied to any oriented graph. Formally, a node n_0 that is the source of two edges

[6] A tree is *binary-branching* if each node is the source of 0 or 2 edges.

[7] Provided that dereification is the first step, the conversion can perform the other steps in any order.

[8] Representations like (b) are structurally similar to the semantic graphs of Meaning-Text Theory – however, Mel'čuk (1988) merely numbers the edges; representation (c) is similar to Sowa's conceptual graphs (Sowa, 2000) – the labels with numbers conventionally identify arguments of the predicate structure.

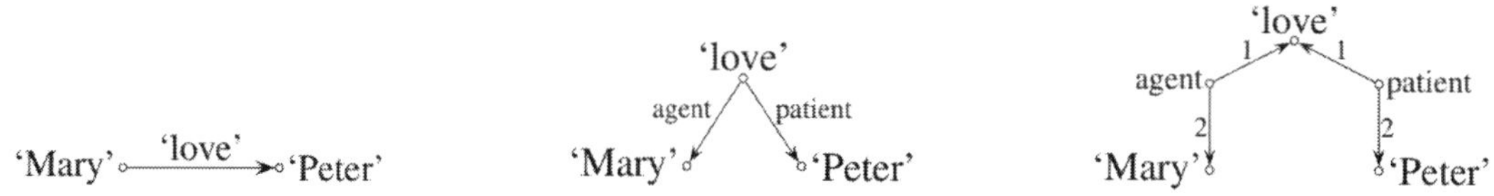

(a) *'love'* labels an edge *(b)* *'love'* labels a node *(c)* *agent* and *patient* label nodes

Figure 2: Various structures corresponding to 'love'('Mary', 'Peter')

$e_1 = (n_0, n_1)$ and $e_2 = (n_0, n_2)$ will become an edge $e_0 = \{n_1, n_2\}$. If e_1 and e_2 have the labels a and b, e_0 becomes a typed edge $\{(n_1, a), (n_2, b)\}$.

One can dereify all non-terminal nodes of the constituency tree. The terminal nodes (expressing words) remain nodes, whereas the nodes corresponding to groups become edges. The constituency tree in fig. 1(a) can be dereified since a tree is a directed graph (see fig. 3(a), where the orientation of the part-whole relation is explicit). By dereifying the NP representation (*red cars*), the NP node and the two part-whole edges of which it is the source become a single edge (fig. 3(b)) that represent the grouping operation rather than its result. This NP edge remains undirected because the immediate constituents have the same hierarchical status in a strict constituency approach (both are the target of part-whole directed edges sharing the same source). Applying dereification to all nodes expressing constructions results in the structure in fig. 3(c). The starting constituency tree is binary-branching, hence the edges obtained are binary. However, the result is not exactly a tree: vertices can be edges as well as nodes. Henceforth, this type of structure will be called a *polygraph*.[9] In a polygraph, an edge e_1 can have another edge e_2 as vertex. In other words, if a polygraph has a set X of nodes and a set E of edges, each edge e in E is associated with a set of vertices in $X \cup E$; e.g.: $e_1 = \{x, e_2\}$, where $e_2 = \{y, z\}$ with $x, y, z \in X$.

From a formal point of view, the resulting polygraph expresses exactly the same stratification as the constituency tree, and will be called a *constituency polygraph*. However, it achieves two straightforward semiotic correspondences:

- Nodes always correspond to words (i.e. "observable" units, see note 1).
- Edges always correspond to grouping operations/constructions (i.e. linguistic analysis).

Since it shares these characteristics with dependency trees, the constituency polygraph can be considered as a better starting point for the conversion that will be performed.

3.2 Specification of the governors: directing the edges

The constituency polygraph does not represent the governor-dependent relations of the dependency approach. To be able to do so, the polygraph must be directed, by typing the vertices of each edge with either the type *source* or the type *target* (see sec. 2.2).[10] Graphically, lines are turned into arrows to express the direction of the edges according to the target dependency tree (fig. 1(a)). This operation results in fig. 4(a) – which will be compared to fig. 4(b) below. In fig. 4(a), the nodes of the polygraph are also linearly ordered.

Contrary to the dereification process, direction specification actually adds information to the structure:

- Edges still express grouping operations and the stratification of the constituency tree.
- In addition, edges now express governor-dependent relations.

The directed polygraph expresses the same contents as a *headed constituency tree* (sec. 2.1). Such a tree

[9] The term *polygraph* is used in category theory for a family of objects that are a particular case of the structures called polygraphs here (Burroni, 1993; Bonfante and Guiraud, 2008).

[10] Or *governor/dependent*, *head/non-head*, *H/NH*, etc. The types need to be interpreted to express the direction of the edges. It should be noted that even in a directed graph defined only by tuples (and not unordered sets), associating the direction of the edges with linguistic dependency is also a reading convention.

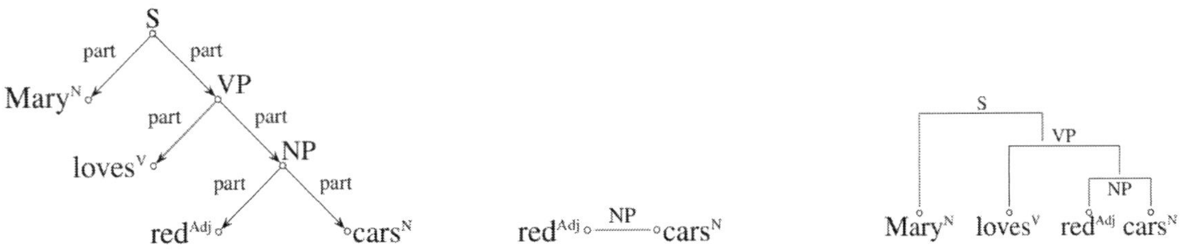

(a) Direction in the constituency tree (b) Dereified NP (c) Constituency polygraph

Figure 3: Dereification of constituency trees

(a) Ordered directed polygraph (b) Headed constituency tree

Figure 4: Directed polygraph and headed constituency tree

combines constituency with dependency by specifying the governor (or *head*) at each level of aggregation. In fig. 4(b), each governor is identified by the label H on the part-whole relation linking it to the source construction.[11] When the dereification is applied to a constituent C with two immediate constituents A and B, A being the head, C becomes the edge $\{(A,H),(B,NH)\}$. If H and NH are interpreted as, namely, *source* and *target* typing, the structure becomes a directed polygraph (fig. 4(a)).

Again, the comparison of the two structures with respect to the information they carry must be supplemented with the evaluation of their relative semiotic efficiency. The directed polygraph is more coherent (because of the aforementioned correspondences *nodes/observed units* and *edges/constructed analyses*). It is also more readable:

- All directed edges have exactly the same interpretation (direction and label).
- All directed edges express relations of the same kind, whereas the edges in the headed constituent tree express both part-whole relations and headedness.

3.3 Graphical delinearization

The structures in fig. 4(a) actually combine two structures: the proper polygraph, which expresses the hierarchical order – i.e. the structural order of Tesnière (1959) – and the linear order, which is represented by the position of the nodes on the same horizontal line. Since the focus is on the comparison between strict constituency and dependency, the disposition of the nodes is not relevant: changing it will not modify the hierarchical structure. The directed polygraph in fig. 4 can be drawn as in fig. 5(a) by abstracting away from word order and placing the source of directed edge above the target.[12]

3.4 Underspecification of the grouping orders: node selection

The vertices of an edge of a polygraph can be nodes or edges. To convert this structure into a tree that has the same shape as the target dependency tree in fig. 1(b), one simply has to select nodes as vertices for each edge. Nodes will be selected using the direction. A node, called the *top node*, is associated

[11] The conceptual means to define the governor and its graphical expression may vary, e.g., $\bar{X}$ and X labels in phrase structure trees (Jackendoff, 1977) or labelled edges such as H and NH in HPSG (Pollard and Sag, 1994).

[12] Such representations were first proposed by Tesnière (1959, ch. 65 and ch. 108) for encoding scope relations in its dependency-based representation.

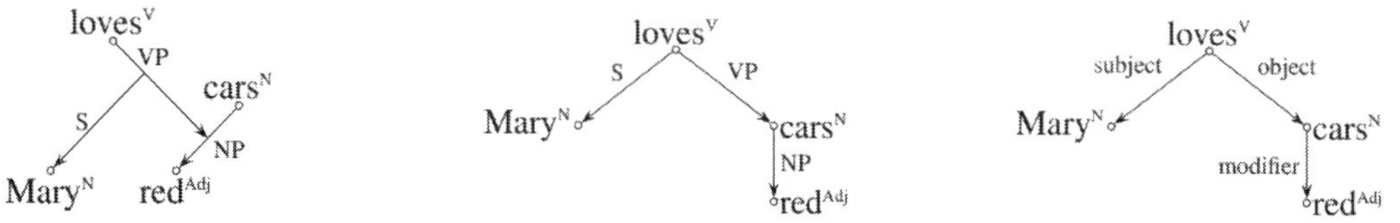

(a) Directed constituency polygraph *(b)* Destratified constituency polygraph *(c)* Dependency tree

Figure 5: From directed polygraph to dependency tree

to each edge recursively: if $e = (v_1, v_2)$, $\text{top}(e) = \text{top}(e_1)$ and if n is a node, $\text{top}(n) = n$.[13] To each directed polygraph, a graph is associated by replacing any edge (v_1, v_2) by the edge $(\text{top}(v_1), \text{top}(v_2))$.

For instance, in the polygraph in fig. 5(a), $\text{top}(S) = \text{top}(VP) = loves$ and $\text{top}(NP) = cars$. The corresponding graph after selection is then the tree in fig. 5(b). From the geometrical point of view, the selection can be viewed as a down-shifting of edges along other edges.

Selection corresponds to deleting the order of grouping operations in the constituency tree. The operation substracts information from the original constituency tree: namely, its stratification. The polygraph expresses the stratification of groups and each instance of selection underspecifies this order for a couple of grouping operations.

Geometrically, the edge *VP* can be slid along the edge *S* (and *loves* becomes the source of *S*) and the edge *NP* can be slid along *VP* (and *cars* becomes the target of *VP*), as seen in the tree in fig. 5(b).

3.5 Relabelling

By relabelling the edges of the polygraph obtained after selection with names of grammatical functions, one generates fig. 5(c), which is the dependency tree of fig. 1(b) plus arrows that express the directions of the governor/dependent relations. We can see that the grouping operation named *S* in phrase structure grammar (Chomsky, 1957) is the grouping operation that dependency grammar names the *subject* relation (Mel'čuk, 1988). In works preceding Chomsky's formalization (Chomsky, 1957), groupings are more clearly interpreted as constructions. Bloomfield (1933, §12, 2) calls this grouping *actor-action construction* – Bloomfield's *action* does not correspond exactly to a VP constituent, since the *direct object* relation (like *watch me*) is called the *action-goal construction* in §12.8. Even Chomsky (1957, ch. 4) introduces the constituency tree as a derivation tree. Following Vijay-Shanker (1992), one can extend the notion of derivation tree to TAG, where each node must be interpreted as a rule. In other words, the S node of the derivation tree can be interpreted as $\alpha_S = [S \rightarrow NP + VP]$ and the relation between S and VP as the substitution of a rule α_{VP} into a rule α_S.

Constituency and dependency are two ways to encode grouping operations on the same words. Constituency focuses on the stratification, that is the respective order of the grouping operations (a verb groups with its object before grouping with its subject), while dependency focuses on headedness, that is the governor-dependent relation inside each group (a verb governs its subject and its object). Consequently, when a word is defined as a governor in the latter approach, it is aggregated with each of its dependents in several groups at the same level. Constituency generally acknowledges the same associations, but it has the means to stratify it through the definition of smaller nested groups.

The conversion is complete and reverting the process would allow one to go from the dependency tree to the constituency tree.

3.6 Intermediate conclusion

The five steps can be summarized with respect to the information carried by each strict formalism (the most important steps[14] appear in fig. 6). Polygraphs, which have been used as a temporary step in the process can be regarded as an independent modelling

[13]The top node of an edge is nothing else than the lexical head of the corresponding constituent. This step, the replacement of a constituent by its lexical head, is equivalent to the step in Lecerf's procedure collapsing all the nodes with their common lexical head.

[14]Delinearization and relabelling are not represented.

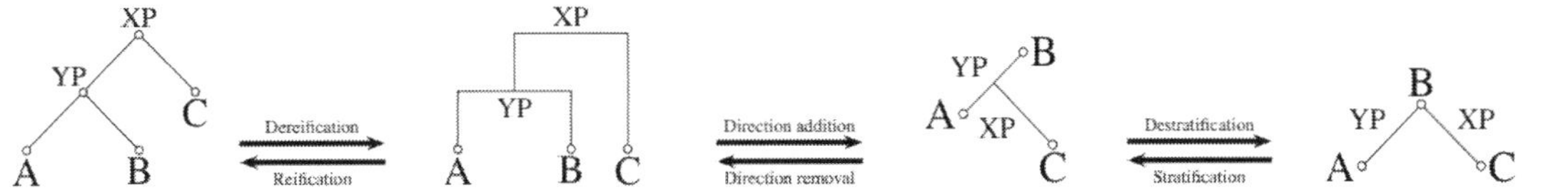

(a) Constituency tree *(b)* Undirected polygraph *(c)* Directed polygraph *(d)* Dependency tree

Figure 6: Synthesis of constituency tree/dependency tree conversion steps

tool that achieves better expressiveness.

Descriptive powers compared *Governor specification* adds dependency information (sec. 3.2), that is *headedness*, to the polygraph and *selection* deletes the order of grouping operations (sec. 3.4), that is *stratification*, from the polygraph. What this means is that neither strict constituency trees nor strict dependency trees can really be considered to be a more powerful formalism, because neither of them can express everything that can be expressed by the other. A formalism A is *more powerful* than B if there exists a one-to-one map from A to B preserving the topology of the structures in A. In other words, every structure *a* written with the formalism A can be converted into a structure *b* written in the formalism B without losing any information, that is, *a* can be recovered from *b*.

Constituency is able to acknowledge stratification; e.g.: in fig. 1(a), *red cars* aggregates with *loves* before *Mary* aggregates with the rest of the sentence. Choosing between one order or the other is an important methodological issue that should not be overlooked (Gleason, 1955; Wells, 1947). The drawback of constituency is that part-whole relations do not express any hierarchy between immediate constituents of a single construction (*Mary* and *loves red cars* share the same level). On the other hand, dependency provides an explicit way to describe the relative importance of individual words in comparison with the other words that surround them. Therefore, identifying the head of a group is a major concern (Mel'čuk, 2009, 27-34) and is even *compulsory* in a strict dependency approach.

Polygraph as a synthesis Polygraphs can be considered a gateway for the conversion of one type of structure into the other. They can contain both the stratificational information of constituency trees and the dependency structure. They can potentially express everything that both strict approaches can. Directed polygraphs can be drawn as in fig. 5(a), a representation close to a dependency tree, in order to assert the hierarchical order. But an ordered polygraph can also be represented with the nodes linearly ordered (fig. 4(a)) and looks more like a constituency tree.

Seen as an intermediate stage in a conversion process, polygraphs offer no free option: from strict constituency to strict dependency, the direction of all edges *must* be specified and information about the order of grouping operations *is necessarily* lost. Nevertheless, polygraphs can be considered as an *independent modelling tool* that allows one to choose whether or not specific governors and the order of grouping operations must be specified. They therefore allow *underspecification* in addition to combining the descriptive power of concurrent formalisms.

- A dependency tree is a directed polygraph that is completely destratified.
- A constituency tree is a stratified polygraph that specifies no governor.

As an independent modelling tool, polygraphs achieve better expressiveness than strict constituency and strict dependency alone, for they satisfy the following requirements:

Descriptive power Polygraphs can structurally express everything that constituency and dependency can express (grouping, headedness, stratification).

Underspecification Polygraphs can underspecify what is considered as non necessary (headedness, stratification)

The latter property is as important as the former: a formalism that forces one to specify irrelevant information can be as problematic as a formalism that does not allow one to specify relevant information.[15]

[15]Derivation trees provide an example of underspecification

(a) Original polygraph *(b)* Reified version

Figure 7: Stratification of dependency trees

4 Extending dependency trees

As demonstrated in sec. 3, it is possible with polygraphs to encode both dependency and constituency, i.e. *grouping*, *headedness* and *stratification*. The aim of this section is to show what can be achieved by using polygraphs as an independent formalism rather than as a gateway for converting constituency trees into dependency trees. In a dependency-oriented perspective, it shows in which ways and for what purpose polygraphs can extend the formalism of dependency trees.

4.1 Stratification of dependency trees

Stratification can be acknowledged by polygraphs. It is not clear what advantage can be obtained by systematically forcing the verb to combine with its subject after its object as is done in PSGs.[16] DGs are attached to the verb centrality, defended at least since Tesnière (1959, ch. 49), who fought strongly against the bi-partition of the sentence into subject and predicate, i.e. NP and VP. Tesnière's point of view could be reformulated by saying that stratification of the verbal subcategorization in current constituency-based approaches is an artefact of the encoding of the syntactic structure by a binary-branching constituency tree (Kahane and Osborne, 2015, xxxix-xlii).

In other words, the fact that dependency trees make it possible to formalize binary groupings without needing stratification is viewed as a strong advantage by DGs.

This preliminary remark does not mean that stratification cannot be useful sometimes. Consider the following example:

(3) red cars that you saw yesterday

As already noticed by Coseriu (1980, 55), Tesnière (1959, ch. 65) proposed to represent the syntactic structure of (3) with a polygraph-like stemma (redrawn here in fig. 7(a)) (Tesnière, 1959, stemma 149) and provided the following justification: "By this process, the phrase *red cars that you saw yesterday* can be analyzed structurally in such a way that the connecting line extending upward from the subordinate clause reaches the connection line connecting *red* to *cars*. This means that *that you saw yesterday* is connected not to *cars* but to *red cars*, since what you saw yesterday was not cars, but red cars." Indeed, this representation says that *cars* combines first with *red* and after with the relative clause, as shown also by the constituency tree in fig. 7(b) which is obtained by reification of the relations in polygraph in fig. 7(a).

4.2 Headless grouping

The lively debate about the DP vs. NP analysis of the determiner-noun construction – the still open discussion started 30 years ago (Hudson, 1984; Abney, 1987) – could be resolved by admitting that both the determiner and the noun have head features. Neither of them should be favored. This leads to propose a syntactic structure such as fig. 8(d) for sentence (4):

(4) Mary drives a red car.

This polygraph can be obtained starting from the constituency tree in fig. 8(a). Dereification outputs the ordered polygraph of fig. 8(b) where only a

that every formal linguist knows about. Context-free grammars were defined by Chomsky (1957) as rewriting systems. A *derivation* in a CFG is a string of rewriting steps. The *derivation tree* is a better formalism of representation of the derivation process, because it masks the order in which irrelevant steps occur in a derivation and "keeps only what is essential to understand the phrase structure" (ibid.: ch. 4). That is why the derivation tree is much more adequate than the proper derivation for a syntactic representation of the structure of a sentence.

[16] As far as incremental parsing is concerned, which is cognitively very motivated, it seems clear that the verb will combine with its subject before its object in SVO languages.

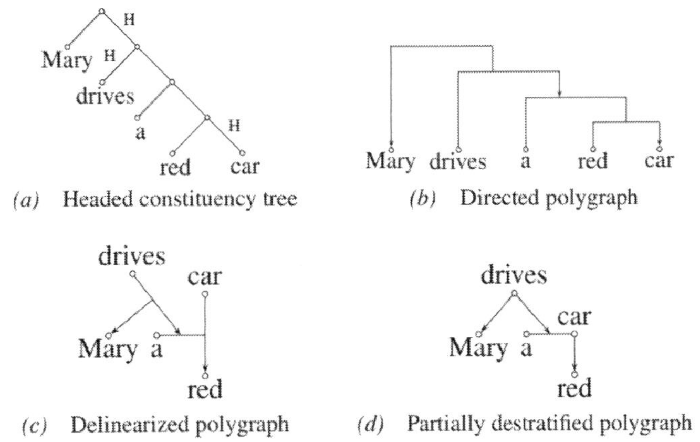

(a) Headed constituency tree *(b)* Directed polygraph
(c) Delinearized polygraph *(d)* Partially destratified polygraph

Figure 8: Underspecified determiner-noun headedness

headed grouping outputs a directed dependency. Delinearization outputs the representation of fig. 8(c), where directed edges are represented by vertical lines with the head on top, while the non-directed edges are represented by horizontal lines. Selection results in the structure depicted in fig. 8(d). Non-directed edges cannot undergo selection and remain governed as a whole.

From the geometrical point of view, it means that non-directed edges cannot undergo any selection, which justifies representing them horizontally. Reed and Kellogg (1877) already use horizontal convention in their famous diagram, where the subject-verb and the object-verb combinations are non hierarchically organized. A linguist wishing to express a "headless" (or exocentric) structure, such as the former Chomskyan scheme $S \rightarrow NP + VP$, simply cannot do so with a dependency tree. The transformation exposed in sec. 3 applied to such a constituency tree (fig. 9(a)) produces a polygraph very similar to the Reed-Kellogg diagram (fig. 9(b)).[17]

(5) I think Mary loves cars.

The horizontal convention was also used by Tesnière (1959, part II) for coordination (see sec. 4.3).

4.3 Ternary grouping

In a dependency tree, every grouping is a binary grouping between a head word and a non-head word, encoded by a dependency. Some linguists consider that function words are just markers of constructions. For instance, *to* in (6) could be analyzed not as the head of a PP but rather as the marker of the combination between *talked* and its indirect object.[18]

(6) Mary talked to Peter.

This results in the constituency tree in fig. 10(a), with a ternary branching where *talked* is the head (*H*), *to* the marker (*M*), and *Peter* a non-head word. Dereification and selection result in the structure in fig. 10(b), i.e. a *hyperpolygraph*. A *hyperpolygraph* is to a polygraph what a hypergraph is to a graph. In this case, hyperpolygraph contains a ternary "dependency" with three vertices: a governor, a dependent, and a third vertex put on the edge, occupied by the marker *to*, which must not be interpreted as a label. Such a convention already appears in dependency-based representations, even in some very early attempts (Kern, 1883, 17) – see also Débili (1982) for a more recent example. The representation in fig. 10(c) is also valid: the marker *to* depends on the relation it marks (Kahane and Mazziotta, submitted Depling 2015), indicating that the marker cannot occur without the relation. Such a polygraph cannot be reified into a (constituency) tree.

Another example combines ternary grouping with non-directed grouping. In symmetrical analyses of

[17]Note that Nida (1966, 17-47) consists of dozens of similar diagrams, that contrast a.o. endocentric structures and exocentric ones.

[18]A ternary grouping can be justified because the three elements can be grouped pairwise following different criteria: the verb and the object are linked by a semantic relation, the marker can be associated with the verb (*the guy Mary talked to*) as well as the object (*To Peter, Mary should not talk*).

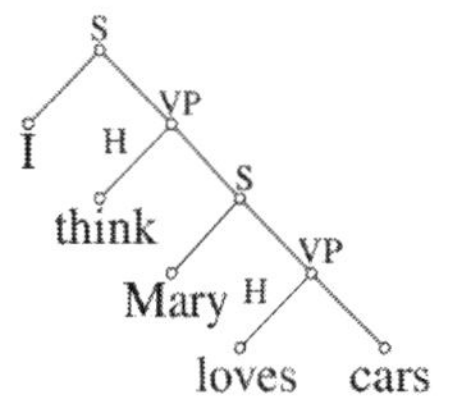

(a) Partially headed constituency tree

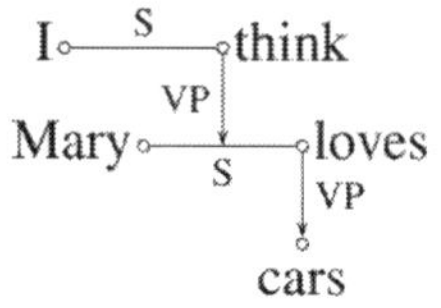

(b) Partially directed polygraph

Figure 9: Underspecified subject-verb headedness

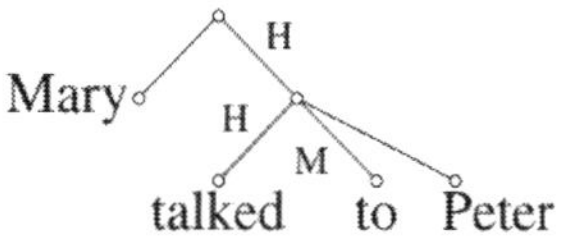

(a) Constituency tree

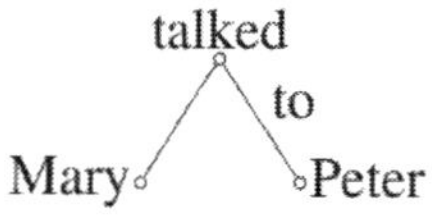

(b) Hyperpolygraph with marker node

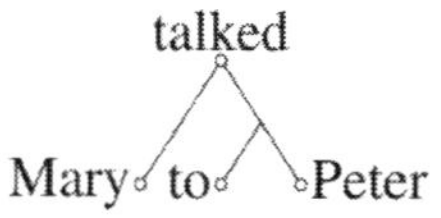

(c) Polygraph with marking relation

Figure 10: Ternary grouping

coordination, conjuncts are considered as co-heads, while the coordinating conjunction is a marker, as proposed by Jackendoff (1977).

(7) Mary and Peter got married.

Fig. 11 shows a constituency tree for (7) and the corresponding polygraph after dereification and sliding. A horizontal line is again used to encode a grouping between two co-heads. The resulting representation is similar to what Tesnière proposes from his early works (Tesnière, 1934) to his main book (Tesnière, 1959, ch. 38).

5 Conclusion

The goal of this paper was to present a new formalism that subsumes both constituency and dependency, and which can be then used to extend these two formalisms in order to encode syntactic analyses.

Two goals have been achieved. First, a new way to associate a dependency tree to a headed binary-branching constituency tree has been proposed. Contrary to the previous one from Lecerf (1961), this one proves that every dependency corresponds to a grouping.

Second, the formalism of polygraphs extends both formalisms, constituency trees as well as dependency trees, making stratification as well as headedness explicit. Interestingly these structures are similar to representations that have been previously proposed by other linguists (Reed and Kellogg, 1877; Kern, 1883; Nida, 1949; Nida, 1966; Tesnière, 1959; Débili, 1982) and polygraphs give them a mathematical foundation – see Mazziotta (2014) for a more comprehensive application of polygraphs to Tesnière's model. Beyond that, the formalism of polygraphs can be the basis for the development of new formal grammars, especially lexicalized formal grammars such as TAG, that derive the structure of a sentence by combination of elementary structures. In particular, it seems more accurate for some problematic constructions (such as unbounded dependencies or complex determiners) to be formalized with a derivation structure which is a polygraph rather than a tree or even a graph.[19]

Acknowledgement

The authors would like to thank the following people for their suggestions and corrections (in no particular order): Kim Gerdes, François Lareau, Timothy Osborne and Pierre Swiggers.

[19] As stated earlier (sec. 3.5), CFG derivations have been extended to TAGs (Vijay-Shanker, 1992). However, even for TAGS, it appears that derivations are too complex to be fully formalized by a tree; e.g. predicative adjunction demonstrates the limits of the tree structure in this context (Schabes and Shieber, 1994). This question cannot be developed here; Kahane (2013) suggests polygraphic derivation structures in a dependency-based formalism.

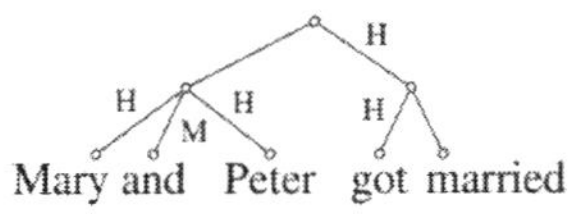

(a) Headed constituency tree

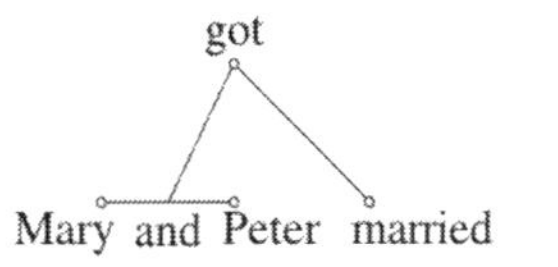

(b) Polygraph with marker node

Figure 11: Ternary grouping with co-heads

References

Steven P. Abney. 1987. *The English Noun Phrase in its Sentential Aspect.* Ph.D. thesis, Massachusetts Institute of Technology.

Claude Bergé. 1973. *Graphs and hypergraphs.* North-Holland, Amsterdam.

Leonard Bloomfield. 1933. *Language.* The University of Chicago Press.

Guillaume Bonfante and Yves Guiraud. 2008. Intensional properties of polygraphs. *Electronic Notes in Theoretical Computer Science*, 203(1):65–77.

Albert Burroni. 1993. Higher-dimensional word problems with applications to equational logic. *Theoretical computer science*, 115(1):43–62.

Noam Chomsky. 1957. *Syntactic structures.* Mouton, The Hague.

Eugenio Coseriu. 1980. Un précurseur méconnu de la syntaxe structurale: H. Tiktin. In *Recherches de linguistique: hommage à Maurice Leroy*, pages 48–62. Éditions de l'Université de Bruxelles, Bruxelles.

Fathi Débili. 1982. *Analyse syntaxico-sémantique fondée sur une acquisition automatique de relations lexicales-sémantiques.* Ph.D. thesis, Université Paris Sud.

Henry A. Gleason. 1955. *An Introduction to Descriptive linguistics.* Holt, Rinehart and Winston.

Thomas Groß. 2011. Catenae in morphology. In Kim Gerdes, Elena Hajičová, and Leo Wanner, editors, *Proceedings of Depling 2011, International Conference on Dependency Linguistics, Barcelona*, pages 47–57. Barcelona.

Charles F. Hockett. 1958. *A course in modern linguistics.* The MacMillan Company.

Richard A. Hudson. 1980. Constituency and dependency. *Linguistics*, 18(3-4):179–198.

Richard Hudson. 1984. *Word Grammar.* Blackwell, Oxford.

Richard Hudson. 2010. *An introduction to word grammar.* Cambridge Textbooks in Linguistics. Cambridge University Press, Cambridge.

Ray Jackendoff. 1977. *X-bar syntax: A study of phrase structure.* MIT Press, Cambridge, MA.

Sylvain Kahane and Nicolas Mazziotta. submitted (Depling 2015). Dependency-based analyses for function words - introducing the polygraphic approach.

Sylvain Kahane and Timothy Osborne. 2015. Translators' introduction. In *Elements of structural syntax* (Tesnière, 2015), pages xxix–lxxiv.

Sylvain Kahane. 1997. Bubble trees and syntactic representations. In *Proceedings of Mathematics of Language (MOL5) Meeting*, pages 70–76. Citeseer.

Sylvain Kahane. 2012. Why to choose dependency rather than constituency for syntax: a formal point of view. In J. Apresjan, M.-C. L'Homme, M.-C. Iomdin, J. Milićević, A. Polguère, and L. Wanner, editors, *Meanings, Texts, and other exciting things: A Festschrift to Commemorate the 80th Anniversary of Professor Igor A. Mel'čuk*, pages 257–272. Languages of Slavic Culture, Moscow.

Sylvain Kahane. 2013. Predicative adjunction in a modular dependency grammar. In *2nd international conference on Dependency Linguistics (DepLing)*, pages 137–146.

Franz Kern. 1883. *Zur Methodik des deutschen Unterrichts.* Nicolai, Berlin.

Yves Lecerf. 1961. Une représentation algébrique de la structure des phrases dans diverses langues naturelles. *Comptes rendus de l'Académie des Sciences*, 252(2):232–235.

Nicolas Mazziotta. 2014. Nature et structure des relations syntaxiques dans le modèle de Lucien Tesnière. *Modèles linguistiques*, 69:123–152.

Igor Mel'čuk. 1988. *Dependency syntax: theory and practice.* State University of New York, Albany.

Igor Mel'čuk. 2009. Dependency in natural language. In Alain Polguère and Igor Mel'čuk, editors, *Dependency in linguistic description*, pages 1–110. John Benjamins, Amsterdam and Philadelphia.

Eugene Nida. 1949. *Morphology: the descriptive analysis of words.* University of Michigan press, Ann Arbor, 2 edition.

Eugene Nida. 1966. *A synopsys of English Syntax.* Mouton and Co., London, The Hague, Paris, 2 edition.

Timothy Osborne, Michael Putnam, and Thomas M. Gross. 2011. Bare phrase structure, label-less trees, and specifier-less syntax. is minimalism becoming a dependency grammar? *The Linguistic Review*, 28(3):315–364.

Carl Pollard and Ivan A. Sag. 1994. *Head-driven phrase structure grammar*. University of Chicago Press.

Alonso Reed and Brainerd Kellogg. 1877. *Higher Lessons in English: A Work on English Grammar and Composition*. Clark and Maynard, New-York.

Yves Schabes and Stuart M. Shieber. 1994. An alternative conception of tree-adjoining derivation. *Computational Linguistics*, 20(1):91–124.

John F. Sowa. 2000. *Knowledge Representation: Logical, Philosophical, and Computational Foundations*. Brooks Cole Publishing Co., Pacific Grove, CA.

Lucien Tesnière. 1934. Comment construire une syntaxe. *Bulletin de la Faculté des Lettres de Strasbourg*, 7:219–229.

Lucien Tesnière. 1959. *Éléments de syntaxe structurale*. Klincksieck, Paris.

Lucien Tesnière. 2015. *Elements of structural syntax, translated by Timothy Osborne and Sylvain Kahane*. Benjamins, Amsterdam/Philadelphia.

K. Vijay-Shanker. 1992. Using descriptions of trees in a tree adjoining grammar. *Computational Linguistics*, 18(4):481–517.

Rulon S. Wells. 1947. Immediate constituents. *Language*, 23(2):81–117.

Author Index